Nanofluidic
Microfl

Nanofluidics and Microfluidics

Systems and Applications

Shaurya Prakash and Junghoon Yeom

AMSTERDAM • BOSTON • HEIDELBERG • LONDON
NEW YORK • OXFORD • PARIS • SAN DIEGO
SAN FRANCISCO • SINGAPORE • SYDNEY • TOKYO

ELSEVIER

William Andrews is an Imprint of Elsevier

William Andrew is an imprint of Elsevier
225 Wyman Street, Waltham, 02451, USA
The Boulevard, Langford Lane, Kidlington, Oxford OX5 1GB, UK

First edition 2014

Library of Congress Cataloging-in-Publication Data
A catalog record for this book is available from the Library of Congress

British Library Cataloguing in Publication Data
A catalogue record for this book is available from the British Library

ISBN: 978-1-437-74469-9

Contents

Preface

Microfluidics and nanofluidics span a broad array of disciplines including mechanical, materials, and electrical engineering, surface science, chemistry, physics, and biology. The interplay of all these disciplines has led to several topics being covered by this multidisciplinary area and has consequently led to several different types of devices and systems for myriad applications. Our intent in writing this book is to present a comprehensive resource that integrates essentials of microfluidic and nanofluidic systems in a single text. Therefore, the goal is to present the underlying theory, discuss the main (and a large variety of) fabrication methods that are commonly used and finally present a snapshot of the variety of applications. Throughout the book several case studies are presented to highlight the development of systems and subsequent use in applications. Therefore, the rationale behind the book is to demonstrate the use of multiple disciplines and topics for constructing a broad array of systems for solving science, engineering, and technology challenges.

The microfluidics and nanofluidics community comprises a growing group of active, creative, and pioneering researchers. Therefore, this area has seen tremendous growth in the past decade evidenced by the large number of scientific papers and books that have been published. Our book presents a unique look at this vast body of scientific literature from a systems perspective with the underlying assumption that fundamental theories and fabrication go hand in hand for construction and operation of many practical devices. Therefore, it would be useful to view the advances through the prism of "global" advances in science and technology using our case studies as an example rather than advances in individual disciplines. The direct consequence of our approach is that this book covers a large range of topics. It is likely that we have omitted some material either for clarity or lack of space but we assist the interested reader by pointing them to a select bibliography at the end of each chapter. Finally, we expect the book to be adopted in introductory courses on microfluidics and nanofluidics as it covers all the essentials. Therefore, some chapters present exercises in the text with additional problems available on the companion website. In addition, the companion website (which includes all chapters in digital form) presents a unique advancement in sharing knowledge through books. We present a purely digital chapter that discusses several experimental methods relevant to microfluidics and nanofluidics. The goal of this e-chapter is to demonstrate a variety of laboratory scale techniques that are usually lost in the print format. Therefore, our book presents a first comprehensive effort at viewing the discipline of microfluidics and nanofluidics from an application standpoint driven by a systems approach by drawing upon underlying theories and advances in fabrication.

Shaurya Prakash and Junghoon Yeom

About the Authors

Shaurya Prakash received his B.S. in mechanical engineering from the University of Arkansas, Fayetteville, in 2001, and went on to receive his M.S. and Ph.D. degrees also in mechanical engineering from the University of Illinois, Urbana-Champaign, in 2003 and 2007, respectively. He has been on the faculty at the Ohio State University since 2009 in the Department of Mechanical and Aerospace Engineering. He is the director of the Microsystems and Nanosystems Laboratory at OSU, and has published over 20 archival journal articles, 50 conference papers, and 4 book-chapters on various aspects of microsystems and nanosystems. His research focuses on developing microsystems and nanosystems for applications in water purification, renewable and alternate energy, and health-care systems. His group addresses fundamental scientific questions and strives to develop new technologies for problems important to modern societal needs. The research work is multidisciplinary and often requires bridging several fields including mechanical engineering, surface chemistry, and materials science.

Junghoon Yeom received his B.S. in mechanical design and production engineering from Yonsei University, Seoul, Korea, in 2000, and his M.S. and Ph.D. in mechanical engineering from the University of Illinois, Urbana-Champaign, in 2003 and 2007, respectively. He is an assistant professor in the Department of Mechanical Engineering at Michigan State University and the director of Nanomaterials, Nanomanufacturing, Nanodevice Laboratory. Prior to joining MSU, Dr. Yeom was an NRC research associate at Naval Research Laboratory in Washington DC and a research scientist at Cbana Labs Inc. in Illinois. He was also a visiting scholar in the Department of Mechanical Science and Engineering at University of Illinois, Urbana-Champaign, and affiliated with the nano-CEMMS center. His research group focuses on developing unconventional nanomanufacturing platforms, photocatalysis, and gas sensing and separation. He is a coauthor of a book chapter, over 20 journal publications, and more than 50 conference proceedings in the field of chemical sensors, NEMS/MEMS, micro/nanofluidics, fuel cells, and gas sorbents technology.

Acknowledgments

Any author of a book will disclose that writing a book, even a single volume such as this one, is an arduous task. We acknowledge the support of many key individuals in completing this task.

We owe many thanks to several special people. We would like to thank our families for their enormous patience and constant encouragement during the entire process, from the initial concept to writing and going through all the details of publishing. Shaurya's wife, Ruchika, and Jung's wife, Hyokyoung, have been a pillar of strength and have provided wonderful feedback on several aspects of writing, while managing many responsibilities and allowing us the time to undertake this academic exercise. No words can express our heartfelt thanks to both of them. We also appreciate the support of our parents who nurtured the curiosity in us to grow as a scientist and an engineer, which was instrumental in initiating this effort and taking it to completion.

Both of us thank the entire microfluidics and nanofluidics community of researchers for providing us with ample creative and wonderful scientific material to discuss. Among this group, a few select individuals have influenced us both professionally and personally that made the task of writing this book an enjoyable journey. In particular, we thank our Ph.D. advisor, Prof. Mark Shannon at the University of Illinois, for teaching us the value of communicating scientific results and sharing our knowledge with the broad scientific community. He also was instrumental in training us with all the fundamentals on building successful systems. We are both deeply indebted to Mark. We also would like to take this opportunity to thank our graduate students, co-workers, collaborators, research mentors, and friends who made numerous contributions to the book and whose work appears in several places including the Select bibliography.

Without all these wonderful people, this book would not be possible.

Shaurya Prakash and Junghoon Yeom

Nomenclature List

Symbols

γ_{LF}	interfacial energy the liquid–fluid interface
γ_{SF}	interfacial energy for the solid–fluid interface
γ_{SL}	interfacial energy for the solid–liquid interface
ΔW_{SLF}	adhesion energy per unit area of contact
ε_0	permittivity of free space
ε_m	complex dielectric permittivity of the medium
ε_p	complex dielectric permittivity of the particle
ε_r	dielectric permittivity
ζ	zeta potential
η	liquid viscosity
θ	fraction of surface sites occupied
θ_c	contact angle
$\theta_{c,R}$	contact angle for a rough surface
κ	inverse of the Debye length
λ_D	Debye length
μ_{EOF}	electroosmotic mobility
μ_i	ionic mobility
μ-Na*fl*	microfluidic and nanofluidic
ρ	fluid density
ρ_l	liquid density
ρ_e	charge density
ρ_F	gas (other fluid in the capillary) density
σ	surface tension
σ_s	surface charge
τ_{shear}	shear stress
φ_s	surface potential at the wall
A_{Fi}	fractional area of phase *i*
Bo	Bond number
b	a constant related to the adsorbent
c	concentration of the adsorbing species near the surface
C_d	differential capacitance
D_{AB}	binary diffusion coefficient of the species of interest
De	Dean number

D_i	diffusion coefficient
E	electric field
F	force
F	Faraday's constant
f_B	applied body forces
f_{CM}	Clausius–Mossotti factor
F_D	drag force
F_{DEP}	dielectrophoretic force
g	acceleration due to gravity
I	ionic strength of electrolyte
J_A	total current density
J_i	molar flux of the ith species
k_a	adsorption rate constants
k_B	Boltzmann constant
k_d	desorption rate constants
L	length
l_c	characteristic length
n_i	local concentration
p	applied pressure
q	charge
q_e	elementary charge
R	gas constant
R_1 and R_2	radii of curvature of the surface along any two orthogonal tangents differential capacitance, C_d
Re	Reynolds number
SA	surface area
S_R	surface roughness factor
St	Strouhal number
S_t	Stokes number
T	absolute temperature
t_d	diffusion time scale
u	bulk fluid velocity
u, v, w	being the x, y, and z components of the velocity vector
V	volume
V	applied potential
z_i	valence

Common Abbreviations

AFM	atomic force microscope
Al	aluminum
Au	gold
Cr	chromium
CVD	chemical vapor deposition
DRIE	deep reactive ion-etching
EDL	electric double layer
FTIR	Fourier transform infrared
IPA	isopropanol
NCAM	nanocapillary array membrane
PDMS	poly dimethyl siloxane
PI	polyimide
PMMA	poly (methyl methacrylate)
PR	photoresist
RIE	reactive ion-etching
SEM	scanning electron microscope
Si	silicon
SSN	solid-state nanopore
TEM	transmission electron microscope
XPS	X-ray photoelectron spectroscopy

CHAPTER

Introduction

1

CHAPTER OUTLINE

1.1 Length scales

Microfluidic and nanofluidic (μ-Na*fl*) systems are defined as systems with functional components with operational or critical dimensions in the 1−100 μm range for microfluidics and 1−100 nm for nanofluidics, respectively. Therefore, we now have the ability to study and systematically manipulate exceedingly small volumes (approaching the order of zeptoliters or 10^{-21} l has been discussed in literature and listed in several bibliographic references throughout this book) of fluids and other species. Consequently, the ability to engineer processes and phenomena that operate at fundamental molecular lengths driving a host of applications in chemical, biological, and particle separations, sensors, energy generation and harvesting, environmental remediation, water purification, and at the interface of several science and engineering disciplines is being pursued. Figure 1.1 shows a conceptual plot that depicts how the interplay between critical length scales and subsequent device volumes can drive several applications for μ-Na*fl* systems.

An identifying feature of all μ-Na*fl* systems is the surface-area-to-volume (*SA/V*) ratio. Consider two examples: (1) a simple circular cross-section nanopipe with a diameter of 10 nm and a length of 1 μm will have a *SA/V* ratio on the order of $10^7\ m^{-1}$ and (2) a microchannel with a rectangular cross-section with a width of 100 μm, depth of 20 μm, and a length of 1 mm will have an *SA/V* ratio on the order of $10^5\ m^{-1}$. The discussion for *SA/V* ratios is pertinent because several forces and related phenomena important to fluid transport at these length scales change as SA/*V* ratios increase, as the governing principles dominating these phenomena assume different relative magnitudes. For example, in the

S. Prakash & J. Yeom: Nanofluidics and Microfluidics. DOI: http://dx.doi.org/10.1016/B978-1-4377-4469-9.00001-9

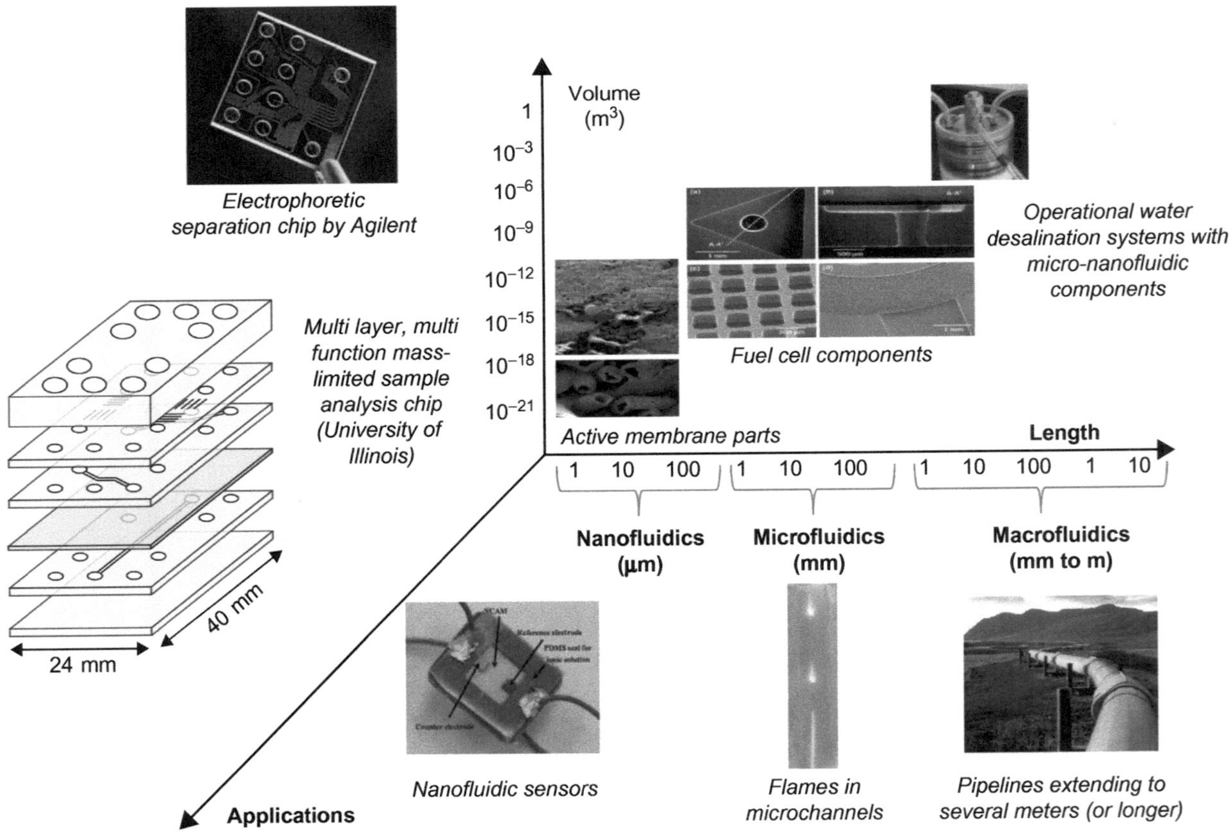

FIGURE 1.1

A conceptual figure which shows the common length scales spanned by microfluidics and nanofluidics along with a few examples of devices and systems. The macroscale pipeline is used as an example to provide a reference.

nanopipe example above, if the walls of the nanopipe have an electric charge, the surface of the channel will exert an electrostatic force. Since the charge is distributed over the channel or pipe area along the walls, the areas charge density becomes an important consideration. Therefore, to set up a simple scaling law comparing any surface-area term to a volume term we see that,

$$\begin{aligned} &\text{SA term} \propto l_c^2 \\ &\text{and} \\ &V \text{ term} \propto l_c^3 \end{aligned} \tag{1.1}$$

where l_c denotes characteristic length.

Consequently,

$$\frac{\text{SA}}{V} \propto \frac{l_c^2}{l_c^3} = \frac{1}{l_c} \tag{1.2}$$

Equation (1.2) implies that surface-driven terms (see Chapter 3 for more details) will dominate as volumes decrease with reducing characteristic lengths. In Figure 1.2, we illustrate the concept of various phenomena that are influenced by the SA/*V* ratio and can be important in designing and constructing μ-Na*fl* systems. In turn, this gives insight to how the equations and fundamental principles can be used for implementing the ideas to building successful devices and systems.

Following our discussion, scaling analysis can therefore provide insight to how fluid phenomena in μ-Na*fl* may occur in contrast to the macroscale counterparts. It should be noted that scaling analyses typically provide broad ideas and trends but more detailed experimentation and analysis may be needed for specific details.

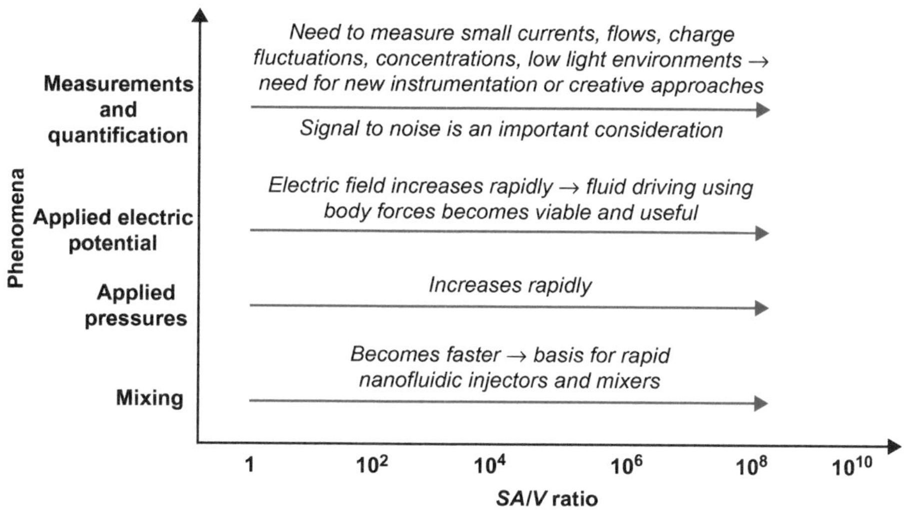

FIGURE 1.2

A conceptual schematic showing some of the important phenomena that influence μ-Na*fl* systems as function of *SA/V* ratios.

Figure 1.2 discusses several critical aspects of the scaling analysis in a brief, pictorial representation. Let us begin by considering mixing. Often in several μ-Na*fl* systems (see examples in Chapters 5 and 6), there is a need to bring in distinct fluid streams and allow these to mix. Due to confinement of the fluid in a device with large SA/V ratios, often the viscous and surface effects will dominate the flow phenomena (see Chapter 3 for further details), and consequently, mixing is largely driven by molecular diffusion. The time scales for molecular diffusion can be expressed as

$$t_{\mathrm{d}} = \frac{l_{\mathrm{c}}^2}{D_{AB}} \tag{1.3}$$

where t_{d} is the diffusion time scale, l_{c} is the characteristic length, and D_{AB} is the diffusion coefficient of the species of interest. Equation (1.3) shows the diffusion time scales as the square of the characteristic length. Consequently, as we approach the nanoscale with decreasing l_{c}, time scales needed for diffusion decrease rapidly. For example, if we compare two devices and, one device has half that of the other, the smallest l_{c} device needs four times less time to mix the same species by diffusion as compared to the device with larger l_{c}. This scaling has permitted development of rapid mixers and injectors for nanoscale mixing and schemes for generating vortex-like structures at the interface of micro- and nanochannels to enhance mixing.

As we discuss in detail in Chapter 2, μ-Na*fl* systems with aqueous electrolyte solutions usually operate using principles of electrokinetics or using an applied electric potential to generate a body force that drives the fluid flow as opposed to pressure commonly used at the macroscale. Therefore, let us consider the scaling of a force due to the electric field. From electrostatics, the force F on a charge (let us say an ion or particle in the flow field; see Figure 1.3.) q in an electric field E is given by,

$$\vec{F} = q\vec{E} \tag{1.4}$$

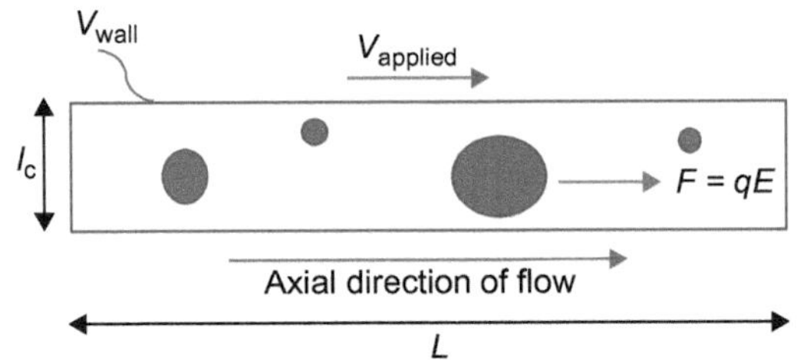

FIGURE 1.3

A schematic depicting a channel with flow and the forces due to an electric field shown in the axial direction. Note that the potential at the wall due to a finite wall surface density, σ_{s}, will also cause a force on any particles in the fluid. A more detailed discussion of the forces and the consequences in terms of flow phenomena and formation of electric double layers are discussed in Chapters 2 and 3.

For a simple scaling analysis here, we will consider a 1-D field and look at the magnitudes, so we will not consider the directionality or the vector nature of the forces in this discussion. The electric field is usually scaled as a function of the length, L across the potential drop; therefore, for an applied potential V the electric field is given by,

$$E = \frac{V}{L} \tag{1.5}$$

It can be noted from Eq. (1.5) that the electric field scales inversely with the length over which the potential acts or for the forces,

$$F = q\frac{V}{l_c} \tag{1.6}$$

where we have now replaced L with l_c as the forces act over the characteristic length of interest. Therefore, it is clear that the electric field force will scale inversely with the length, that is, the smaller the characteristic length, the higher the force due to a potential V. Contrast this with the gravitational forces that may act in a μ-Na*fl* system. One useful non-dimensional parameter to consider is the Bond number, *Bo* which compares gravitational forces to surface tension forces. Equation (1.7) describes *Bo* as:

$$Bo = \frac{(\rho_l - \rho_F)gl_c^2}{\sigma} \tag{1.7}$$

where σ is the surface tension, g is the acceleration due to gravity, and ρ_l is the density of the liquid or gas droplet within the surrounding fluid of density, ρ_F.

Note that the numerator of Eq. (1.7) describes the gravitational force term with respect to the surface tension force in the denominator. Since the numerator scales as square of the characteristic length, the importance of gravitational forces decreases rapidly with shrinking device length scales.

In Chapter 6, we discuss separations as one of the major application areas for micro- and nanofluidics. It is also useful to look at some of the forces that influence separation phenomena such as electrophoresis. Equation (1.4) describes the net electrical force on a charged particle. Now, if the particle is a sphere of radius a and moves with a velocity U in a fluid, the drag force, F_D, opposing the motion is given by,

$$F_D = 6\pi\eta aU \tag{1.8}$$

where η is the fluid viscosity, and when the electrostatic force balances the drag, the velocity is given by,

$$U = \frac{qE}{6\pi\eta a} \tag{1.9}$$

Therefore, from Eqs. (1.8) and (1.9) we once again note the scaling for forces (and a measurable velocity) on the characteristic length, a, for the spherical particle. The discussion here applies to charged particles. In many cases, polarizable or dielectric particles have also been studied. In addition, there could be time-varying (e.g., AC potentials) electric potentials and consequently, electric fields that are also time-varying have been used. In such cases, phenomena like dielectrophoresis can become important. The dielectrophoretic force arises on a particle due to dielectric polarization in the presence of a time-varying (usually non-uniform) electric field. The dielectrophoretic force, F_{DEP}, on a spherical particle is given by,

$$F_{\mathrm{DEP}} = 2\pi a^3 \varepsilon_{\mathrm{m}}\, Re[f_{\mathrm{CM}}]\nabla(\vec{E}\cdot\vec{E})$$
$$f_{\mathrm{CM}} = \frac{\varepsilon_{\mathrm{p}} - \varepsilon_{\mathrm{m}}}{\varepsilon_{\mathrm{p}} + 2\varepsilon_{\mathrm{m}}} \tag{1.10}$$

where f_{CM} is the Clausius–Mossotti factor and ε_{p} denotes the complex dielectric permittivity of the particle, and ε_{m} denotes the complex dielectric permittivity of the medium. Once again, we note the scaling depends on the size of the particle, and in contrast to the electrophoretic force, the scaling varies as the cube of the size of the particle. Physically, f_{CM} can be considered to provide a measure of the frequency dependence of the polarization effects as the complex permittivities vary with the frequency of the electric potentials causing the field and are expressed as,

$$\varepsilon = \varepsilon_{\mathrm{R}} - \mathrm{i}\frac{\sigma}{\omega} \tag{1.11}$$

where ε_{R} is the real part of the permittivity, σ denotes the electrical conductivity, and ω is the frequency.

The above discussion shows that length scales can play a critical role in how different phenomena occur at the micro- and nanoscale. Scaling analysis can help provide a quick comparative analysis of the dominance of one aspect or force over another. For example, scaling analysis has been shown to be useful in studying effects of velocity, as the velocity field in electroosmotic flow (see Chapter 2 for definitions) of a solution with ionic strength of 1 M through a 45 nm channel is similar to that of a 0.2 mM solution through a 1 μm channel. Much more detail is available on dimensional and scaling analysis in Conlisk's book as listed in the bibiliography. One caveat we would like to point out is that the scaling analysis or use of non-dimensional parameters is inherently based on the continuum assumption, and so care must be taken in predicting or extrapolating trends when such an assumption might not be valid at nanoscale.

1.2 Scope and layout of the book

The main theme for this book is structured around the idea of "systems." Therefore, a "system" is defined as the main functional unit of interest. Consequently, the system can be a small particle or a region in space where fluidic phenomena occur, or a device component, or a complete operational unit that can perform multiple functions. The context for each type of system is discussed throughout the book.

In Chapter 2, we present a brief theoretical background needed for μ-Na*fl* systems. All essentials are covered here; however, for an in-depth discussion the interested reader is directed to the vast bibliography throughout the book. Chapter 3 discusses the role of interfaces and presents an overview of how internal interfaces (e.g., surfaces) and external interfaces (e.g., connections to the outside world) can influence fluidic phenomena. At the same time, the challenges and opportunities that arise due to interfacial phenomena and the theory behind these phenomena are also discussed. Chapter 4 presents a detailed discussion on the micro- and nanofabrication aspects needed to design and construct these devices. Chapters 5 and 6 present an overview for fluid manipulation, lab-on-chip, and energy- and environment-related applications using a large variety of examples from the scientific literature to demonstrate the versatility of μ-Na*fl* systems. In Chapter 5, we present two case studies that discuss details about an integrated gas-sensing and chromatography system and a microscale nanofluidic flow regime sensor. The case studies capture the multi-component fabrication. In addition, the case studies also highlight the complexity of μ-Na*fl* systems that can perform several unit operations on a single platform.

1.3 Future outlook

Nascent scientific fields like micro- and nanofluidics (really started as a field in 1990s) can evolve rapidly and grow in directions unforeseen by the best scientists and engineers. Therefore, our attempts to gaze into the future are fraught with grave risks. However, a few trends are clearly emerging. With continued and rapid advances in micro- and nanofabrication, scaling of devices or device arrays to meet requirements for high-throughput applications are now a reality. μ-Na*fl* devices continue to grow in areas of sensors, medical and bio-related applications, and energy and environment applications. This growth is built on several years of work on micropumps, microvalves, and thermal-fluid phenomena at the micro- and nanoscale.

On the theoretical and computational side, techniques like molecular dynamics elucidate the fundamentals at length scales where the continuum assumptions do not work. Continuum methods provide insight to experimentally observed phenomena, and large-scale system modeling along with multiscale models can be one way to bridge all the various theories, length scales, and experimental efforts.

Finally, while several microscale systems have been built, nanoscale systems still operate for the most part, at the individual component level. Therefore, truly integrated nanosystems or hybrid micro-nanosystems present an opportunity for further growth. We believe that the potential is vast and is challenged only by the ingenuity and imagination of fellow scientists and engineers.

References

[1] Conlisk AT. Essentials of micro- and nanofluidics: with applications to the biological and chemical sciences. Cambridge, UK: Cambridge University Press; 2012.

[2] Prakash S, Pinti M, Bellman K. Variable cross-section nanopores fabricated in silicon nitride membranes using a transmission electron microscope. J Micromech Microeng 2012; 22.

[3] Prakash S, Yeom J, Shannon M.A, editors. A microfabricated impedance sensor for ionic transport in nanopores. 11th solid-state sensors, actuators, and microsystems workshop. Hilton Head Island, SC;2006.

Select Bibiliography

Abgrall P, Nguyen N-T. Nanofluidics. Artech House; 2012.

Israelachvili J. Intermolecular and surface forces. London: Academic Press; 1991.

Madou M. Fundamentals of microfabrication. Boca Raton, Florida: CRC Press; 1997.

Madou MJ. Fundamentals of microfabrication and nanotechnology. Boca Raton, FL: CRC Press; 2012.

Nguyen N-T, Wereley ST. Fundamentals and applications of Microfluidics. Artech House; 2010.

Piruska A, et al. Nanofluidics in chemical analysis. Chem Soc Rev 2010;39:1060–72.

Prakash S, Piruska A, Gatimu EN, Bohn PW, Sweedler JW, Shannon MA. Nanofluidics: systems and applications. IEEE Sensors J 2008;8(5):441–50.

CHAPTER

Fundamentals for Microscale and Nanoscale Flows

2

CHAPTER OUTLINE

2.1 Introduction

μ-Na*fl* systems are governed by physical laws that describe the following basic problem: in a confined space with critical dimensions on the micrometer or nanometer length scales, how is fluid behavior quantified for relative motion between fluids, and a fluid and solid? Consequently, questions pertaining to forces between solids and fluids, within the fluids, and any conditions affecting the microfluidic or nanofluidic system must be addressed. Obviously, with such a broad definition, μ-Na*fl* systems present a complex, vast, and rapidly evolving research area. Therefore, microfluidics and nanofluidics relies on subject areas such as fluid mechanics, heat and mass transfer, surface chemistry, electrochemistry, and electrostatics for a complete understanding and description. Furthermore, the multidisciplinary nature and breadth of activity for microfluidics and nanofluidics is expressed in the numerous studies being published every year in attempts to answer different aspects of the fundamental question posed above. In this chapter, we present a brief summary of the underlying physical laws that can be used to describe most μ-Na*fl* systems.

This chapter will begin with a short review of essential concepts from fluid mechanics, followed by a review of main ideas from electrostatics, electric double layer (EDL) theory, electrochemistry, and surface chemistry relevant to μ-Na*fl* systems.

S. Prakash & J. Yeom: Nanofluidics and Microfluidics. DOI: http://dx.doi.org/10.1016/B978-1-4377-4469-9.00002-0

2.2 Definition of a fluid

A fluid is defined as a substance that deforms continuously under shear stress (see Figure 2.1) [1, 2]. Both gases and liquids constitute categories of fluids. Fluids can be further classified as either Newtonian or non-Newtonian. Newtonian fluids follow a linear relationship between the shear stress applied to the fluid and the velocity profile that develops along the *y*-direction, which is the coordinate perpendicular to the wall of a channel.

$$\tau_{\text{shear}} = \eta \frac{du}{dy} \tag{2.1}$$

where τ_{shear} is the shear stress, η is the viscosity of the fluid and represents the friction or resistance of fluid to the applied shear stress, and u is the velocity of the fluid.

2.3 Pressure-driven flows

It was discussed in Chapter 1 that μ-Na*fl* systems comprise small conduits that carry the working fluid. Therefore, fluid phenomena in μ-Na*fl* systems can be driven by a variety of methods such as applied pressure, electric, or magnetic fields.

Let us begin by setting up some ground rules for the governing equations. We will consider μ-Na*fl* systems with isothermal, laminar, incompressible flows for isotropic, Newtonian fluids with constant viscosity. Furthermore, we will limit ourselves to Cartesian coordinates for illustrative purposes. Under such limits, the Navier–Stokes equations (momentum equations) in the vector form describing fluid flow can be expressed as:

$$\rho \frac{D\vec{u}}{Dt} = -\nabla p + \eta \nabla^2 \vec{u} + \vec{f}_B \tag{2.2}$$

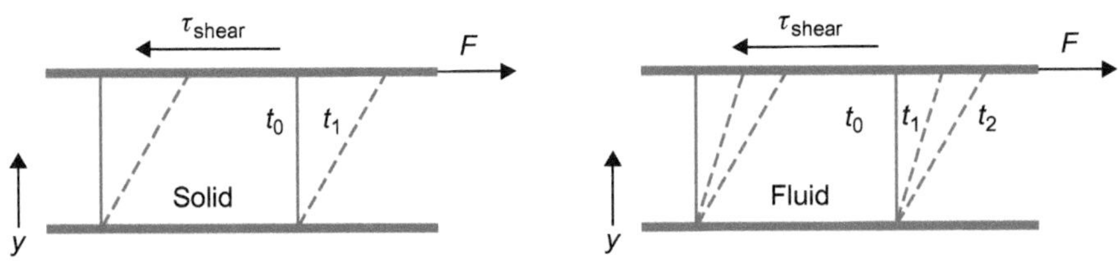

FIGURE 2.1

Schematic representation of the difference between a solid and a fluid as a constant force, *F*, is applied causing a shear stress between the solid or fluid confined between two plates. In this conceptual diagram the bottom plate is considered to be rigidly fixed while the top plate can move under the influence of the applied force. Equation (2.1) discusses the mathematical relation between the shear stress in the fluid and the fluid property called viscosity.

Source: Figures adapted from Fox, McDonald, & Pritchard Introduction to Fluid Mechanics, *6th edition, and Munson et al.* Fundamentals of Fluid Mechanics.

Or, can be written out explicitly as

$$\rho \frac{\partial \vec{u}}{\partial t} + \rho \vec{u} \cdot \nabla \vec{u} = -\nabla p + \eta \nabla^2 \vec{u} + \vec{f}_{\mathrm{B}} \tag{2.3}$$

where ρ is the fluid density, D/Dt denotes the material derivative for velocity vector $\vec{u}$, p is the applied pressure, and $\vec{f}_{\mathrm{B}}$ denotes any applied body forces. Equation (2.2) can be expanded out in the component form in Cartesian coordinates and is expressed as

$$\begin{aligned}
\rho\left[\frac{\partial u}{\partial t} + u\frac{\partial u}{\partial x} + v\frac{\partial u}{\partial y} + w\frac{\partial u}{\partial z}\right] &= -\frac{\partial p}{\partial x} + \eta\left[\frac{\partial^2 u}{\partial x^2} + \frac{\partial^2 u}{\partial y^2} + \frac{\partial^2 u}{\partial z^2}\right] + f_{\mathrm{B},x} \\
\rho\left[\frac{\partial v}{\partial t} + u\frac{\partial v}{\partial x} + v\frac{\partial v}{\partial y} + w\frac{\partial v}{\partial z}\right] &= -\frac{\partial p}{\partial y} + \eta\left[\frac{\partial^2 v}{\partial x^2} + \frac{\partial^2 v}{\partial y^2} + \frac{\partial^2 v}{\partial z^2}\right] + f_{\mathrm{B},y} \\
\rho\left[\frac{\partial w}{\partial t} + u\frac{\partial w}{\partial x} + v\frac{\partial w}{\partial y} + w\frac{\partial w}{\partial z}\right] &= -\frac{\partial p}{\partial z} + \eta\left[\frac{\partial^2 w}{\partial x^2} + \frac{\partial^2 w}{\partial y^2} + \frac{\partial^2 w}{\partial z^2}\right] + f_{\mathrm{B},z}
\end{aligned} \tag{2.4}$$

with u, v, and w being the x, y, and z components of the velocity vector respectively. Equation (2.4) is most commonly used to describe flows in many different configurations and is often coupled to other equations to account for conservation of mass through the continuity equation and applied body forces depending on the specific problem being solved. If the μ-Na*fl* system is at steady state, the only driving force is pressure, with no body force, and combining continuity (conservation of mass), then the above equation reduces to:

$$\begin{aligned}
-\frac{\partial p}{\partial x} + \eta\left[\frac{\partial^2 u}{\partial x^2} + \frac{\partial^2 u}{\partial y^2} + \frac{\partial^2 u}{\partial z^2}\right] &= 0 \\
-\frac{\partial p}{\partial y} + \eta\left[\frac{\partial^2 v}{\partial x^2} + \frac{\partial^2 v}{\partial y^2} + \frac{\partial^2 v}{\partial z^2}\right] &= 0 \\
-\frac{\partial p}{\partial z} + \eta\left[\frac{\partial^2 w}{\partial x^2} + \frac{\partial^2 w}{\partial y^2} + \frac{\partial^2 w}{\partial z^2}\right] &= 0
\end{aligned} \tag{2.5}$$

Now, for a fully developed flow in a slit-like channel that is long ($L \gg h$) and wide ($W \gg h$) compared to the height of the channel, h, with negligible velocity v and w components of velocity,

$$-\frac{\partial p}{\partial x} + \eta \frac{\partial^2 u}{\partial y^2} = 0 \tag{2.6}$$

and can be integrated twice in y to yield the following solution,

$$u(y) = \frac{y^2}{2\eta}\frac{\mathrm{d}p}{\mathrm{d}x} + \frac{Ay}{\eta} + B \tag{2.7}$$

To determine the constants of integration A and B, we will need to consider the boundary conditions for Eq. (2.6). For most flows including microchannel flows, the no-slip boundary condition has been demonstrated to be valid, i.e. the velocity of the fluid at the wall is same as the velocity of the wall. In most practical applications of microfluidics, the wall of the channel is stationary, then by applying the no-slip condition to $y = 0$ and h, we obtain the classic Poiseuille flow with the parabolic flow profile

$$u(y) = \frac{h^2}{2\eta}\frac{\mathrm{d}p}{\mathrm{d}x}\left\{\left(\frac{y}{h}\right)^2 - \left(\frac{y}{h}\right)\right\} \tag{2.8}$$

The velocity can be integrated along the height of the channel to provide the flow rate in the axial direction going through the width of the channel

$$\dot{Q} = W\int_0^h u(y)\mathrm{d}y = \frac{h^3 W}{12\eta}\frac{\mathrm{d}p}{\mathrm{d}x} \tag{2.9}$$

Equation (2.9) shows that the flow rate scales with the third power of channel height and linearly with channel width. Consequently, for nanochannels of a fixed width, pressure per unit length through progressively smaller channels grows as the cube of the height. For example, as shown in Figure 2.2 in order to achieve a flow rate of 1 ml/s for water in a channel of width 100 μm and depth 1 nm the applied pressure must be 16 orders of magnitude higher than atmospheric pressure. In fact, for flow rates of water greater than 1 pl/s in a channel with 100 nm depth pressures approaching a few atmospheres are needed. While pressure-driven flows are used in some microfluidic devices, for nanochannels pressure as a driving force can become impractical. So, how are fluids driven in nanofluidic and most microfluidic devices? The answer lies once again with the Navier–Stokes equations. The body force term, f_B, can be manipulated to provide adequate driving force in the absence or in conjunction with pressure. Most commonly, an electric field is used in μ-Na*fl* systems. This coupling with an electric field leads to several interesting aspects to the resulting fluid mechanics, which will be discussed later. However, pressure-driven flows continue to be important for developing new systems including streaming potential energy conversion systems with more details discussed in applications later in the book. Therefore, as a quick reference a summary of velocity profiles and flow rate equations for pressure-driven flow are summarized in Table 2.1.

2.4 Low Reynolds number flows

One non-dimensional parameter often used to characterize flow regimes as a ratio of the inertial forces transferred by the velocity (or momentum) of the fluid to the viscous (or frictional forces) is the Reynolds number, *Re*, and is given by

$$Re = \frac{\rho V l_c}{\eta}, \tag{2.10}$$

where ρ is the fluid density, η is the fluid viscosity, l_c is the characteristic length for the flow, and V is the fluid velocity. As seen from Eq. (2.10), *Re* is directly proportional to the characteristic length, l_c, of the flow. With l_c decreasing as the channel size becomes smaller, the inertial forces decrease, and as a result *Re* becomes smaller. The direct consequence is that for given flow parameters of fixed velocity and fluid type, a decreasing l_c implies increasing influence of viscous forces in contrast to the inertial forces. The physical significance for flow analysis presented by a low *Re*, which for microfluidics is typically $Re < 10$ (with

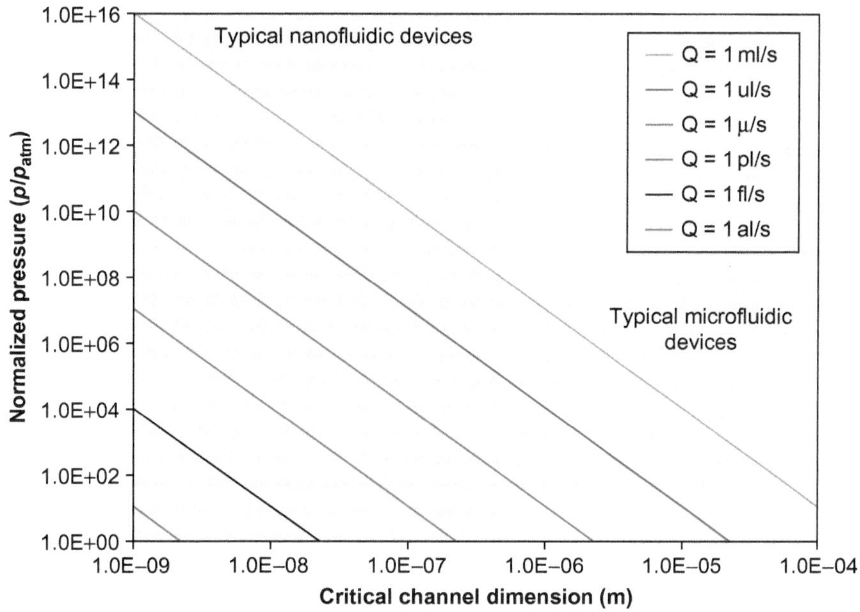

FIGURE 2.2

A log–log plot showing scaling of pressure for water flowing in a slit-like channel of width 100 μm and length 1 cm. The applied pressure is normalized to atmospheric pressure, $p_{atm} = 101$ kPa, with fluid properties taken for water at room temperature. The critical channel dimension refers to the height of the channel and broadly distinguishes between common length scales for microfluidics and nanofluidics.

many applications working at $Re < 1$ and $Re \to 0$), is that μ-Na*fl* phenomena are classified as Stokes' flows or creeping flows with analytical tools considering the fluid as a massless but viscous fluid.

In order to evaluate velocity profiles for Stokes' flows, we start with Eq. (2.3). However, we begin our analysis by discussing the physical implications of Stokes' flows. As discussed in Chapter 1, scaling analyses with non-dimensional parameters permit a relative comparison for the various terms in a governing equation. Therefore, there is a need to non-dimensionalize Eq. (2.3). By defining a characteristic length scale for the flow, l_c, a characteristic time scale, t_c, with a reference flow velocity, V, the following nondimensional parameters can be defined:

$$\begin{aligned} u^* &= \frac{u}{V}, \quad p^* = \frac{p}{(\eta V/l_c)} \\ t^* &= \frac{t}{t_c}, \quad \nabla^* = l_c \nabla \\ l^* &= \frac{l}{l_c}, \quad \nabla^{*2} = l_c^2 \nabla^2 \end{aligned} \tag{2.11}$$

Substituting back in Eq. (2.3), and dividing through by $\eta V/l_c^2$ gives,

$$\frac{\rho V l_c^2}{\eta V t_c}\frac{\partial u^*}{\partial t^*} + \frac{\rho V l_c}{\eta} u^* \cdot \nabla^* u^* = -\nabla^* p^* + \nabla^{*2} u^* + \frac{f_B}{(\eta V/l_c^2)} \tag{2.12}$$

Table 2.1 Summary of Velocity Profiles and Flow Rate Governing Equations for Pressure-Driven Flow for Circular, Rectangular, and Triangular Channel Cross-Sections

Channel Cross-Section	Flow Profile Descriptions
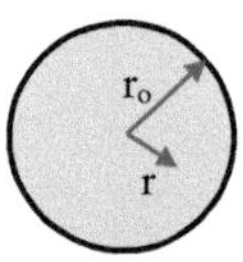	$u(r) = -\frac{1}{4\eta}\frac{dp}{dx}(r_0^2 - r^2)$ $\dot{Q} = \frac{\pi r_0^4}{8\eta}\left(-\frac{dp}{dx}\right)$
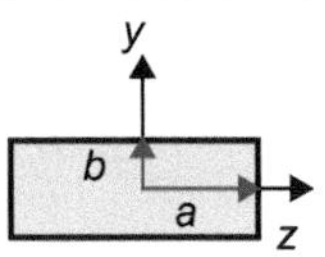	$u(y,z) = \frac{16a^2}{\eta\pi^3}\left(-\frac{dp}{dx}\right)\sum_{i=1,3,5,\ldots}^{\infty}(-1)^{(i-1/2)}\left[1 - \frac{\cosh(i\pi z/2a)}{\cosh(i\pi b/2a)}\right]\frac{\cos(i\pi y/2a)}{i^3}$ $\dot{Q} = \frac{4ba^3}{3\eta}\left(-\frac{dp}{dx}\right)\left[1 - \frac{192a}{\pi^5 b}\sum_{i=1,3,5,\ldots}^{\infty}\frac{\tanh(i\pi b/2a)}{i^5}\right]$
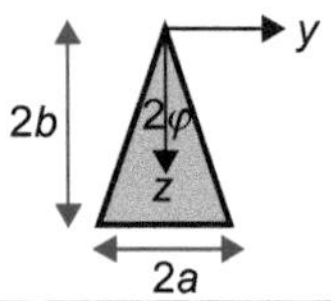	$u(y,z) = \frac{1}{\eta}\left(-\frac{dp}{dx}\right)\frac{y^2 - z^2\tan^2\varphi}{1 - \tan^2\varphi}\left[\left(\frac{z}{2b}\right)^{B-2} - 1\right]$ $\dot{Q} = \frac{4ab^3}{3\eta}\left(-\frac{dp}{dx}\right)\frac{(B-2)\tan^2\varphi}{(B+2)(1-\tan^2\varphi)}, \quad B = \sqrt{4 + \frac{5}{2}\left(\frac{1}{\tan^2\varphi} - 1\right)}$

Apply the definition of *Re* in Eq. (2.10), and use the definition of the Strouhal number as,

$$St = \frac{V}{l_c t_c}, \tag{2.13}$$

to obtain,

$$\frac{Re}{St}\frac{\partial u^*}{\partial t^*} + Re\, u^* \cdot \nabla^* u^* = -\nabla^* p^* + \nabla^{*2} u^* + \frac{f_B}{(\eta V / l_c^2)} \tag{2.14}$$

For Eq. (2.14) to be non-dimensional, f_B must have dimensions of Pa/m. Recalling our discussion that most microscale and nanoscale flows are driven by electric fields, f_B can be written in terms of the charge density, ρ_e, contained in the fluid and the applied electric field, *E*. Therefore, the governing equation becomes

$$\frac{Re}{St}\frac{\partial u^*}{\partial t^*} + Re\, u^* \cdot \nabla^* u^* = -\nabla^* p^* + \nabla^{*2} u^* + \frac{\rho_e \vec{E}}{(\eta V / l_c^2)} \tag{2.15}$$

Notation alert: We have been using ρ for fluid density and ρ_e for charge density.

As $Re \rightarrow 0$, the left-hand side of Eq. (2.15) vanishes, and the equation reduces to,

$$0 = -\nabla^* p^* + \nabla^{*2} u^* + \frac{\rho_e \vec{E}}{(\eta V / l_c^2)} \tag{2.16}$$

But, if the flow is purely driven by an electric field, then the pressure gradient can be neglected to get,

$$\nabla^{*2} u^* + \frac{\rho_e \vec{E}}{(\eta V / l_c^2)} = 0, \tag{2.17}$$

which can be re-written in the dimensional form as

$$\nabla^2 u = -\frac{\rho_e \vec{E}}{\eta} \tag{2.18}$$

Note that as $Re \rightarrow 0$, it eliminates the convection and the time-dependent (unsteady) terms from Eq. (2.14). Also, the velocity profile varies directly with the applied electric field and is a function of the charge density. Therefore, methods and techniques that alter ρ_e can be powerful tools to manipulate microscale and nanoscale flows. See chapter 3 for more details on changing controlling ρ_e in the solution phase and also the charge density on channel walls.

Macroscale discussions for Stokes' flow do not usually explicitly consider the body forces since most macroscale flows are pressure driven. Equation (2.15) in the absence of a body force and with $Re \rightarrow 0$ would reduce to

$$\nabla^* p^* = \nabla^{*2} u^* \tag{2.19}$$

with classic fluid mechanics textbooks discussing a variety of solutions for Stokes' flow under these conditions. See Table 2.2 for a summary of some of the common pressure-driven flows following the creeping flow conditions.

2.5 Electrokinetic phenomena

In the previous section, it was pointed out that using pressure to drive flows in μ-Na*fl* systems may be impractical. Electric potential or applied voltages that generate electric fields are more commonly used. The coupling between an electric and fluid field leads to the field of electrokinetic flows. There are four primary kinds of electrokinetic phenomena that are important in μ-Na*fl* systems. These are defined as:

- *Electroosmosis* is defined as the movement of liquid relative to a stationary charged surface. The direction of electroosmotic flow can be different from the direction of bulk flow.
- *Electrophoresis* is defined as the movement of a charged surface, typically charged particles, relative to stationary fluid. If the charged particles have different charge-to-mass (q/m) ratios it leads to separation of particles into distinct groups as a function of q/m ratio.
- *Streaming potential* is defined as the induced electric field when ionic solutions (e.g., aqueous electrolytes) are made to flow by external forces such as pressure along a stationary charged surface. This phenomenon is opposite of electroosmosis.
- *Sedimentation potential* is defined as the induced electric field when charged surfaces or particles move with respect to stationary fluid. This phenomenon is the opposite of electrophoresis.

Table 2.2 Summary of Velocity Profiles and Flow Rate Governing Equations for Purely Electrokinetically Driven Flow for Circular and Slit-Like Channel Cross-Sections

Channel Cross-Section	Flow Profile Descriptions
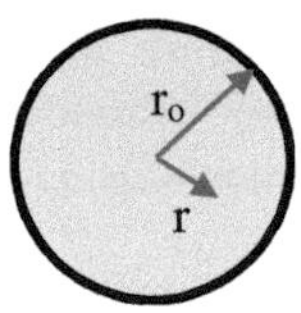	$u(r) = \frac{\varepsilon\zeta E_x}{\eta}\left(\frac{I_0(\kappa r)}{I_0(\kappa r_0)} - 1\right)$ $\dot{Q} = \frac{\pi\varepsilon\zeta r_0^2 E_x}{\eta}\left(\frac{2I_1(\kappa r_0)}{(\kappa r_0)I_0(\kappa r_0)} - 1\right)$
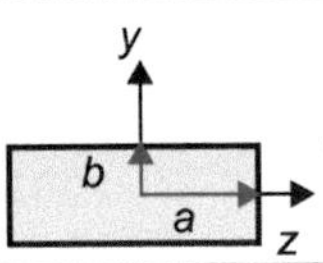 $a \gg b$ called "slit-like" channel	$u(y) = \frac{\varepsilon\zeta E_x}{\eta}\left(\frac{\cosh(\kappa y)}{\cosh(\kappa b)} - 1\right)$ $\dot{Q} = \frac{2\varepsilon\zeta b E_x}{\eta}\left(\frac{\tanh(\kappa b)}{\kappa b} - 1\right)$

2.6 The electric double layer

The definitions for electrokinetic phenomena follow from colloidal literature. Interaction of electric fields in aqueous solutions to move colloidal particles is considered to be first studied by Reuss in early 1800s. Understanding and applying the basic definitions for electrokinetic phenomena to μ-Na*fl* devices requires a few additional tools and concepts, among which the idea of the electric double layer (EDL) is an important starting point. The idea of the EDL also first arose with colloidal systems and electrode-solution interfaces. The earliest models are ascribed to Helmholtz in 1853 and Quincke in 1861 due to adsorption of ions to metallic electrodes in contact with electrolyte solutions governed by the Gibbs isotherms. The EDL as shown in Figure 2.3 shows an example substrate with a net negative charge with positive counterions electrostatically adhered to the surface [3]. This surface layer of counterions is considered to be an immobile layer and is referred to as the Stern layer. The loci of the first ionic layer that is mobile form the shear or slip plane and the potential at that plane is defined as the zeta, ζ, potential.

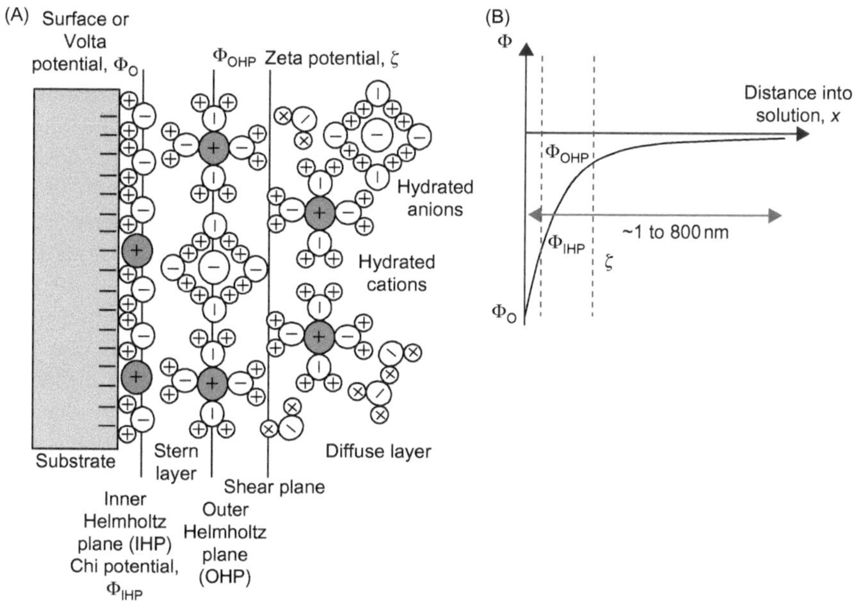

FIGURE 2.3

The figure shows a conceptual schematic for the EDL. The ions are presented as idealized rigid spheres. (A) Potentials, with locations as commonly defined, beginning at the solid surface and extending to the bulk fluid, are indicated on the schematic. (B) Another schematic representation emphasizes the potential distributions beginning with the surface potential. The typical thickness of an EDL for most applications in microfluidics and nanofluidics varies from nearly 1–800 nm for pure water.

Source: Figure from Prakash et al. Nanofluidics, *2008.*

While the concept of the EDL has been around for a long time, the first mathematical model is attributed to Gouy and Chapman in the pre-World War I (1910–1913) era [4, 5]. The Gouy–Chapman model prescribes a surface potential to the substrate of interest and allows for ions in contact to follow a Boltzmann distribution giving rise to an exponential potential decay. Analytical expressions for ion and potential distribution for a charged surface-electrolyte system have been obtained by using the mathematical constructs that have been developed and evaluated extensively; which will be discussed in the following paragraphs.

Before a discussion of the basic mathematical descriptions of the EDL, a few qualitative points about the EDL should be noted. First, for most μ-Na*fl* systems the wall potential is assumed to be the same as the ζ potential. The ζ potential in actuality is at the slip or shear plane which is at a finite distance from the physical channel wall. Typically, the ζ potential will have a magnitude lower than the actual wall potential, ϕ_0. However, ions between the physical wall and the slip plane are bound by strong electrostatic forces and are therefore considered immobile in a region of space called the Stern layer. Consequently, the first motion of ions and therefore related fluid (since ions in solution are hydrated) occurs at the slip plane. Second, while the bulk solution within a microchannel or a nanochannel may remain electrically neutral, i.e., equal concentrations of anions and cations, the region of space occupied by ions near the surface is not electrically neutral and contains an excess of counterions with the co-ion concentrations depleted due to electrostatic repulsion from the channel walls. Third, as discussed above, the Stern layer comprises ions with charge polarity opposite to the wall due to electrostatic attraction. The region of the fluid beyond the Stern layer is called the diffuse layer. The diffuse layer is separated from the Stern layer by the OHP. One conceptual idea to explore here would be the potential at the OHP, ϕ_{OHP}, in contrast to the slip plane. Since the ζ potential arises during fluid motion (hence the name slip or shear plane) and the potential at the OHP is due to the formation of a static EDL, it would be expected that the potentials are different. Given that the slip plane is located at most within a few molecular diameters from the OHP, it is likely that only a small potential drop exists between the OHP and the slip plane. Therefore, most practical systems consider the location of the OHP identical to the slip plane. In order to capture the essence of the discussion above, Figure 2.3 shows the slip plane as being distinct from the slip plane. For further reading the reader is referred to Lyklema's book (1995), which discusses the interpretation of electrokinetic potentials with respect to the wall potentials. It should be noted that the above planes and regions described are not rigid physical boundaries, but conceptual frameworks that have been developed to explain existing data and models. Next, a few basic mathematical tools used for describing a simple EDL are presented.

In most μ-Na*fl* systems an aqueous electrolyte solution will be confined in a channel or pore with microscale or nanoscale critical dimensions. In such cases, one needs to evaluate the ionic concentration and potentials that affect transport phenomena starting at a wall and transferring into the solution. As an example, a discussion for a planar wall is presented next.

The Poisson equation relates the potential ϕ, to the charge density of mobile ions, ρ_e in a medium with dielectric permittivity, ε_r, and ε_0 denotes permissivity of free space,

$$\nabla^2\phi = -\frac{\rho_e}{\varepsilon_0\varepsilon_r} \tag{2.20}$$

The ion organization within the diffuse part of the EDL is considered to follow a Boltzmann distribution. It should be noted that the Boltzmann distribution in thermodynamics is defined for indistinguishable particles in an ensemble at a given thermodynamic state and describes the most probable distribution of particles. For a species i with the bulk concentration $n_{i,\infty}$ far from the charged surface with equal cations and anions (i.e., condition of electroneutrality is met) the Boltzmann distribution gives

$$n_i = n_{i,\infty}\, e^{\left\{-\frac{z_i q_e \phi}{k_B T}\right\}} \tag{2.21}$$

where n_i is the local concentration, z_i is the valence, q_e the elementary charge, k_B is the Boltzmann constant, and T is the absolute temperature. Equation (2.21) shows that once the local concentration is known, the electric potential at that location can be determined from the Boltzmann distribution. The charge density ρ_e in Eq. (2.20) should account for the summation of all N ionic species present in the solution, i.e.,

$$\rho_e = \sum_{i=1}^{N} q_e z_i n_i \tag{2.22}$$

Therefore, by using Eqs. (2.21) and (2.22) in Eq. (2.20) we obtain the Poisson–Boltzmann (P–B) equation given by

$$\nabla^2\phi = -\frac{1}{\varepsilon_0\varepsilon_r}\sum_{i=1}^{N} q_e z_i n_{i,\infty}\, e^{\left\{-\frac{z_i q_e \phi}{k_B T}\right\}} \tag{2.23}$$

Due to the restrictions on the Boltzmann distribution driven by thermodynamics, Eq. (2.23) is valid typically in the diffuse part of the EDL. It should be noted that the P–B equation describes the electric potential distribution in the diffuse layer due to a charged wall. A closer look at the P–B equation shows that the left-hand side describes the divergence in the gradient of the electric potential, ϕ, since the Laplacian operator can be written as

$$\text{div}(\text{grad}\ \phi) = \nabla^2\phi \tag{2.24}$$

Conceptually, Eq. (2.24) can be visualized as the spreading of the electric field as the electric field is the gradient of the electric potential. In order to solve

Eq. (2.23), boundary conditions must be identified. Most commonly, boundary conditions arise at the wall or at OHP (which acts as a virtual wall for the start of the diffuse layer) with the potential being defined as the ζ potential. A second boundary condition often considers the bulk solution to be electrically neutral and so the boundary condition far from the surfaces takes $\phi = 0$. Also, since the P–B equation is a nonlinear equation, a full solution is numerical. Therefore, computing analytical solutions requires some simplifications. For example, for a 1-D configuration in Cartesian coordinates with a symmetric, i.e., $z{:}z$ electrolyte with the surface potential, ϕ_s, at $x = 0$ and with the potential decaying to 0 far from the surface (the boundary conditions described above), the solution to Eq. (2.23) is given by

$$\phi = 2\ln\left[\frac{1 + e^{-\kappa x}\tanh\{(\phi_s/4)\}}{1 - e^{-\kappa x}\tanh\{(\phi_s/4)\}}\right], \tag{2.25}$$

as reported by Masliyah and Bhattacharjee in their book [6]. In Eq. (2.25), κ is the inverse of the Debye length, λ_D.

2.7 Debye length

The concept of the Debye length is an essential concept for μ-Na*fl* systems. The Debye length is the distance from the charged surface where the potential has decayed to the $1/e$ point or 36.7% of the surface potential. For a symmetric, $z{:}z$ electrolyte the Debye length can be calculated as in Eq. (2.14).

$$\kappa = \frac{1}{\lambda_D} = \sqrt{\frac{2q_e^2 z^2 n_\infty}{\varepsilon_0 \varepsilon_r k_B T}} \tag{2.26}$$

If all the constants are evaluated and the concentration is expressed in terms of the solution molarity the expression can be rewritten as

$$\kappa^{-1} = \lambda_D = \left(\frac{0.304}{z\sqrt{M}}\right), \tag{2.27}$$

with the resulting Debye length being expressed in nanometers. Equation (2.27) shows that λ_D scales inversely with the square root of concentration (or molarity) and also scales inversely with the valence of the ions in solution for $z{:}z$ electrolytes. Therefore, even at the same concentration, the Debye length for a divalent ion is half that for the monovalent ion. To put λ_D into perspective for operational systems, at an electrolyte concentration of 1 μM for a 1:1 electrolyte $\lambda_D = 304$ nm but at 1 M, it is 0.3 nm.

Many μ-Na*fl* systems use electrolytes that are not simple $z{:}z$ electrolytes like NaCl or KCl. These comprise ions with non-unity valences and perhaps a mixture of cations and anions, most likely in applications related to biology. For example, phosphate buffers are common. These buffers contain a mixture of potassium monohydrogen phosphate (K_2HPO_4) and potassium dihydrogen phosphate (KH_2PO_4). In such cases, the simple formula presented in Eq. (2.27) is no longer valid for calculating λ_D. A different formulation should be used and is given by

$$\lambda_D = \sqrt{\frac{\varepsilon_0 \varepsilon_r k_B T}{q_e^2 \sum_{i=1}^{N} z_i^2 n_{i,\infty}}} \tag{2.28}$$

As Eq. (2.28) shows all ions of valence z_i need to be accounted for as they exist in solution. However, in many cases the concentration is expressed as a molarity, and therefore $n_{i\infty}$ may need to be explicitly related to the concentration in mol/l, which can be accomplished by using the relationship that includes NA as Avogadro's constant,

$$n_{i,\infty} = 1000\ M_i N_A \tag{2.29}$$

Another alternate form for calculating λ_D uses the ionic strength of an electrolyte solution, I

$$\lambda_D = \sqrt{\frac{\varepsilon_0 \varepsilon_r k_B T}{2 N_A q_e^2 I}}, \tag{2.30}$$

and is related to the concentration as

$$I = \frac{1}{2} \sum_{i=1}^{N} z_i^2 M_i \tag{2.31}$$

However, many references in literature carry notations relying on physical chemistry and use a different set of constants and so,

$$\lambda_D = \sqrt{\frac{\varepsilon_0 \varepsilon_r RT}{2F^2 I}}, \tag{2.32}$$

where R is the gas constant and F is Faraday's constant, and can be related to the constants in Eq. (2.30) by the following relations,

$$\begin{aligned} R &= k_B N_A \\ F &= q_e N_A \end{aligned} \tag{2.33}$$

Note that Eqs. (2.28), (2.30), and (2.32) calculate the same quantity but by using different notations and constants depending on how the specific details are presented. The interested reader is also directed to Conlisk's book (2012; see bibliography) that provides additional discussions on several related concepts. In literature, all forms of the equations are used with the chemistry-oriented literature relying more on molar concentrations or ionic strengths and the physics-oriented literature using more of the number density formulation. For practical devices and systems, the governing equation typically follows Eq. (2.32) since working solutions are often quantified in terms of concentrations and ionic strengths.

Physically, the Debye screening length, λ_D, is the $1/e$ length scale for the potential from solution of the linearized P–B equation, i.e., the distance from the wall where the potential decays to nearly 36.7% of the initial value, Φ_0. λ_D can vary from angstroms for aqueous electrolyte solutions at molar concentrations of $O(1\ \text{M})$ to microns for nanopure or high purity deionized water, often at a reported resistivity of 18 MΩ. One point to note for the calculation of λ_D is that the bulk solution concentration (or ionic strength) is used to calculate λ_D. In fact, in many practical systems altering bulk electrolyte concentration to control λ_D is a fairly common approach, and will be discussed in further detail in the chapter on experimental methods.

Therefore, even for nanofluidic systems where the effect of large λ_D perhaps be exploited to achieve permselectivity (see Chapter 5), it may be desirable to work with dilute solutions given the discussion of how the Debye length scales with electrolyte concentration. More details on applications related to varying EDL can be found in Chapter 5. As a rule of thumb, the EDL thickness is usually taken to be at four times λ_D, but many applications also consider EDL thickness to be the same as λ_D.

Despite the advances in nanofabrication, experimental methods have not quantified completely the structure and distributions of ions in aqueous solutions near surfaces. Therefore, quantitative descriptions of the EDL arise from either simulations or from models that have been developed and refined over the last century. One of the earliest models as mentioned above was developed by Helmholtz. The model was simplistic and assumed a linear drop in potential starting at the surface and reaching electroneutrality in the solution bulk within a fixed distance. The model was refined in the early 1900s by Gouy and Chapman [4, 5].

Let us consider a relatively straightforward analysis for a common nanofluidic system where we assume an aqueous electrolyte solution within the nanochannels with electric potential at the middle of the channel approaching zero. If we apply the Gouy–Chapman theory to the EDL at the nanochannel wall, the surface change density, θ_s, is given by,

$$\sigma_s = 2.57 \times 8.4\sqrt{M\varepsilon_r T}\,\sinh\left\{\frac{\phi_s}{2}\right\} \tag{2.34}$$

where M is the solution concentration in mol/l, T is the absolute temperature, ε_r is the relative permittivity of the solvent, and ϕ_s is the non-dimensional surface potential at the wall. The constant on the right hand side of Eq. (2.34) reflects the constants have already been evaluated accounting for the permittivity of free space, ε_0, Avogadro's number, N_A, and the Boltzmann constant, k_B (please refer to the nomenclature list for a complete listing of terms used in equations). From Figure 2.4 it can be seen that the Gouy–Chapman model predicts that the surface charge density keeps increasing with increasing surface potential and bulk electrolyte concentration. Consequently, at high surface potentials and high bulk electrolyte concentrations an unphysical concentration of counterions (e.g., calculated by Eq. (2.21)) would be needed to shield the surface charge. Even though the Gouy–Chapman model allows for more counterions to be adsorbed to the walls than co-ions, the monotonic increase in surface charge density and the failure of the Gouy–Chapman model is due to the fact that the model does not account for the finite size of ions, i.e., it assumes ions as point charges. The direct consequence of neglecting the finite size of ions is that the point charges approach the surface with no size or space limit. Therefore, as per the Gouy–Chapman model an infinite number of ions can be accommodated at the surface to neutralize the surface charge. However, the closest distance of approach of charge to a charged surface, especially within a medium, is determined by the mutual electrostatic forces. Hence, any model of ion distribution near a surface must address the finite size of ions. In 1923–1924, Stern modified the Gouy–Chapman model to account for a layer of immobile ions that form the first layer of adsorbed ions to a surface with charge density σ_s with typical thickness t_1. The solvated (or hydrated) ions (see Figure 2.5 for a conceptual schematic of hydrated ions) are then limited to a closest distance of approach t_2 and form a diffuse layer beyond

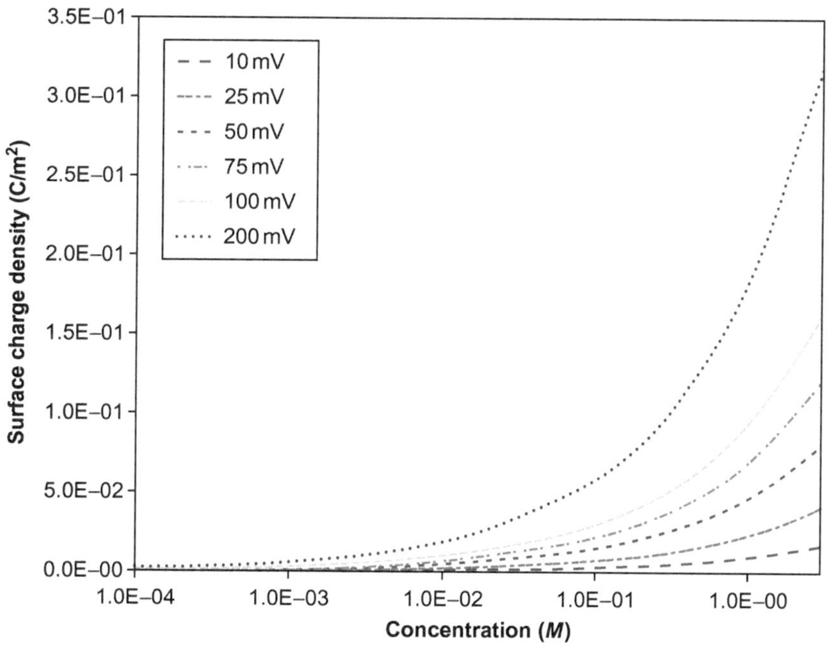

FIGURE 2.4

A plot of surface charge density as function of bulk electrolyte concentration for various surface potentials as predicted by the Gouy–Chapman model is shown here. It should be noted that at high surface potential with increasing electrolyte concentration increasingly unphysical charge densities are predicted. A fully ionized surface is estimated to have a charge density of 0.2 C/m^2.

Source: Israelachvili JN. Intermolecular and surface forces, *1985.*

the immobile Stern layer. The Gouy–Chapman formalism still applies to this diffuse layer.

The above discussion should elucidate and once again emphasize that the concept of the EDL is a *model* with related governing equations describing ion and potential distribution near surfaces and therefore depending on the complexity of the system under study it may require more than two layers to fully describe the ion and related potential distributions. So, a logical question that comes to mind is, why even study the Gouy–Chapman model if it does not capture the correct physics? The answer is simple: the Gouy–Chapman model provides several physical insights that have enabled and continue to enable rapid advances in microfluidics and nanofluidics. For example, the Gouy–Chapman model for the EDL predicts that with increasing ionic valence, z_i, or increasing bulk concentration, M, the diffuse layer will be closer to the surface and the potential decays faster beginning from the surface. Furthermore, concentration dependence follows a square-root dependence, i.e., if the molarity is increased four times the surface charge density as predicted by Eq. (2.34) would increase by two times for the same surface potential within the diffuse layer.

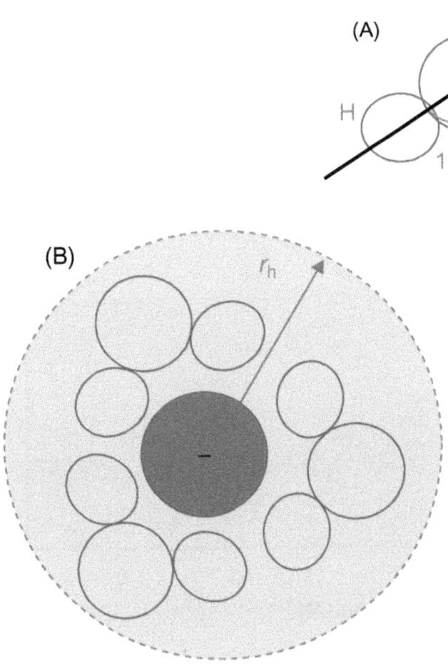

FIGURE 2.5

(A) Schematic showing a representation for a water molecule. (B) and (C) Schematic depicts formation of a hydration sphere around an anion and cation respectively leading to an effective radius of hydration, r_h. Within the hydration sphere, note orientation of the water molecules. The anions are expected to have a larger effective radius (or diameter) than cations with the same net charge.

We started this discussion by focusing on four types of electrokinetic phenomena. One more parameter needs to be discussed explicitly before we can delve deeper about the electrokinetic phenomena. One practical parameter of interest as defined in Figure 2.3 is the zeta, ζ, potential at the shear or slip plane. The ζ potential is an interfacial property usually considered to be the wall potential in microfluidics. Following the discussion above for the Gouy–Chapman EDL model, along with use of the P–B equation coupled to the fluid flow with the no-slip condition, leads to the Eq. (2.34), which relates Φ_s to σ_s. Taking the analysis one step further with fluid properties uniform and with no change in properties from the wall to bulk, it can be shown that $\zeta = \Phi_s$.

It should be noted that while the ζ potential is correlated to the wall potential, Φ_s, the two are distinct quantities. The approximation $\zeta \sim \Phi_s$ holds fairly well in microfluidic systems with thin EDLs. However, in nanofluidic systems the approximation may fail.

By contrast to microfluidic systems, nanofluidic systems operate at length scales with critical dimensions on the order of 100 nm or less with device length scales approaching λ_D for dilute solutions. As fabrication methods continue to target smaller critical dimensions, it is conceivable that in the near future moderate to high concentrations, i.e., λ_D in the sub-10 nm range, will yield operational fluidic devices with similar physical dimensions. With sufficiently small channels or dilute solutions, the shear or slip plane may not lie within the physical limits of the nanofluidic system thereby leading to an ill-defined concept for the zeta potential. The more relevant parameter in such cases, and perhaps for all nanofluidics, is σ_s. Therefore, if we use Gouy–Chapman model to predict the potential as a function of distance x from the surface then by eliminating ϕ_s between Eqs. (2.25) and (2.34) we obtain

$$\phi(x) = 2\ln\left[\frac{1 + e^{-\kappa x}\tanh\left\{\frac{1}{2}\sinh^{-1}(\sigma_s/\alpha\sqrt{M\varepsilon_r T})\right\}}{1 - e^{-\kappa x}\tanh\left\{\frac{1}{2}\sinh^{-1}(\sigma_s/\alpha\sqrt{M\varepsilon_r T})\right\}}\right] \tag{2.35}$$

with $\alpha = 257.84\left(\frac{J}{mol.k.m^3}\right)^{1/2}$ or the constant expressed in Eq. (2.34).

One of the major limitations of the Gouy–Chapman theory was the treatment of ions in solution as point objects. The correction to make is to consider the finite size of ions. So, the question becomes what is the size of an ion? A typical ion has a diameter on the order of 1 Å with ionic diameter or radii increasing with increasing atomic number in a periodic table within a group. Anions have a greater size than cations. The ionic radius also changes with the coordination of ions with the immediate neighbors. In most μ-Na*fl* systems the most common solvent is water, which is a polar molecule. Therefore, in the presence of an ion, water molecules tend to realign and change orientation so that the opposite polarity between the ion and the water molecules coordinate. In effect, this realignment of the water molecules causes the formation of a hydration sphere as depicted in Figure 2.5 around the ions and the effective ionic diameter grows to the order of 3 Å. Equation (2.21) can be rewritten after accounting for the finite size of ions to yield

$$n_i = \frac{n_{i,\infty}\, e^{\{-((z_i q_e \phi)/(k_B T))\}}}{1 + \sum_{i=1}^{N} (n_{i,\infty}/n_{i,\max})[e^{\{-((z_i q_e \phi)/(k_B T))\}} - 1]} \tag{2.36}$$

where $n_{i,\max}$ is the maximum ion concentration of hydration radius, r_h, that can be packed near a surface and can be expressed, as r_h can also be considered as the maximum packing length for each hard sphere representing the ion

$$n_{i,\max} = \frac{1}{N_A r_h^3} \tag{2.37}$$

The denominator in Eq. (2.36) is the volume taken up by the finite size ions and is considered a correction to the standard Boltzmann distribution. The interested reader is also directed to Kirby's book (2010; see bibliography) for additional discussions on related concepts.

2.8 Electrokinetic phenomena: revisited

Now that we have collected a few basic tools that can help us better analyze and understand electrokinetic phenomena, let us look at the four electrokinetic phenomena outlined above.

Electroosmosis couples the fluid field mathematically explained by the Navier–Stokes equations to the electric field and ion distributions mathematically evaluated by the P–B equation for a given EDL model. Typically, an external field is applied in a direction parallel to the flow.

The electric field generates a Coulombic or electrostatic force within the EDL causing ions to move and exert a force on the surrounding fluid and thereby inducing bulk motion. Therefore, the overall velocity profiles deviate from standard pipe-flow equations for pressure-driven flow and become dependent on the chemical features (ion distributions, charge density, etc.) at the wall or the fluid-wall interface. At the nanoscale effects of channel size to EDL thickness may also begin to dominate. For example, for a cylindrical nanochannel with radius a, in the thick EDL limit or diffuse double layer regime ($\kappa a \ll 1$), transport is dominated by electroosmotic flow, and in the limit of a compact double layer (or conditions similar to a microchannel with thin EDL with $\kappa a \gg 1$), electromigration is more important.

Furthermore, effects such as Joule heating, which is the heating of a fluid due to flowing current, become important. Joule heating can cause local gradients in temperature adding various complexities to analyzing flows. For example, for a cylindrical channel the temperature rise across the channel will scale as the square of the applied electric field magnitude, E, and will also scale inversely with the specific heat of the electrolyte solution in the channel. The longer the duration in time that the field acts on the channel, the more heat is generated.

As discussed in Section 2.2, there are four main types of electrokinetic phenomena. Once the governing theories for electrostatics as above have been employed and the potential distributions obtained, it is time to consider using the fluid mechanics theories. The solution of the coupled system of equations discussed in Sections 1.3, 2.1, 2.3, and 2.4 lead to velocity and flow rate expressions for a variety of electrokinetic flows. These were summarized in Table 2.2.

Table 2.2 presents a few interesting observations. It should be noted that for the two common geometries of microchannels and nanochannels typically used the velocity profiles scale linearly with the applied axial (along the x-direction) electric field. In addition, the ζ potential also plays an important role in mediating flow profiles. A few comments about the summary presented in Table 2.2 are in order.

It should be noted that the above equations are valid for a variety of assumptions. In particular, a $z{:}z$ symmetric electrolyte is assumed for potential calculations that are subject to the P–B equation with laminar flow and constant fluid properties. Furthermore, it is assumed that the potential at the physical wall is the same as the potential at the shear plane, i.e., the ζ potential. Therefore, the boundary conditions for the circular cross-section for solving potential profiles are

$$\phi|_{r=a} = \zeta \tag{2.38}$$

and,

$$\phi|_{y=\pm h} = \zeta \tag{2.39}$$

for the slit-like channel. In addition, boundary conditions must also be placed on the fluid flow equations as discussed above, and these are given by

$$u|_{r=a} = 0; \quad \left.\frac{du}{dr}\right|_{r=0} = 0 \tag{2.40}$$

for the circular cross-section, and

$$u|_{y=\pm h} = 0; \quad \left.\frac{du}{dy}\right|_{y=0} = 0 \tag{2.41}$$

for the slit-like channels. Therefore, inherently we assume there is no shear stress at the fluid mid-plane and the physical wall is not moving and thus the fluid is in a no-slip condition at the wall.

Electrophoresis was defined as the motion of a particle under the influence of an electric field, depending on the particle charge polarity. In μ-Na*fl* devices, the most common electrolytes are aqueous solutions containing salt ions, which can be treated as particles for analysis in dilute solutions. Figure 2.6 shows a schematic that depicts particle motion under the influence of an electric field. It should be noted that different particles move through a channel at different rates based on the electrophoretic velocity (see equations below) and these differences form the basis

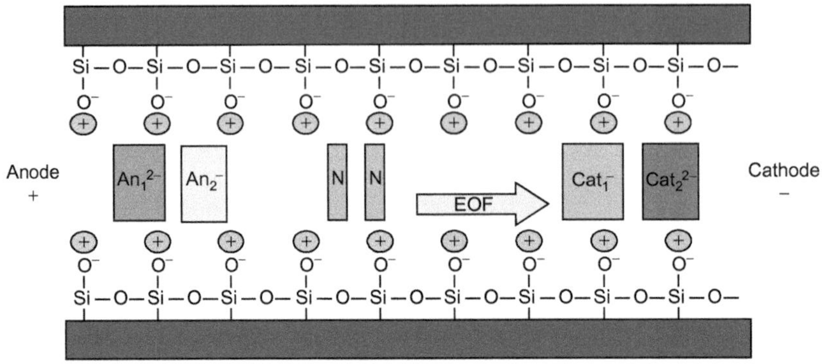

FIGURE 2.6

The conceptual schematic shows effect of an electric field to confined flow of an aqueous electrolyte in a silica channel with silanol groups ionized at the wall to reveal a negative charge density. Ions and particles migrate to electrodes based on their charge polarity and mobility. The motion of particles (and ions) causes electroosmosis through motion of the fluid being dragged by particles. The shielding of surface charge by the counterions from the bulk electrolyte is also shown.

of several separation techniques and applications discussed in more detail in Chapter 5. For a particle with zeta potential ζ_P, in a channel with thin EDL (the thickness of EDL is much smaller than channel critical dimensions), the electrophoretic particle velocity is given by the Smoluchowski equation

$$u_{EP} = \frac{\varepsilon \zeta_P E_x}{\eta} \tag{2.42}$$

A useful parameter to consider for all electrokinetic phenomena is the mobility of a particle, and is considered as the particle velocity per unit electric field, and is expressed as

$$\mu_{EP} = \frac{\varepsilon \zeta_P}{\eta} \tag{2.43}$$

with μ_{EP} is the electrophoretic mobility. Note, if ζ_P is replaced by the channel zeta potential, ζ, then Eq. (2.43) can be rewritten for the electroosmotic mobility, μ_{EOF}.

$$\mu_{EOF} = \frac{\varepsilon \zeta}{\eta} \tag{2.44}$$

Equation (2.44) is often called the Helmholtz–Smoluchowski equation. A more general expression for electroosmotic or electrophoretic velocities is given by Henry's equation, expressed in the form for electrophoretic velocity as

$$u_{EP} = \frac{2\varepsilon \zeta_P E_x}{3\eta} f(\kappa b) \tag{2.45}$$

For a slit-like channel as discussed in Table 2.2, Henry's equation is usually implemented in two common limiting cases,

$$\begin{aligned} \kappa b \ll 1;\ f(\kappa b) &= 1 \\ \kappa b \gg 1;\ f(\kappa b) &= 3/2 \end{aligned} \tag{2.46}$$

For other cases, solutions need to be derived and additional solutions have been previously reported in the literature.

It can easily be envisioned based on the discussion above that microfluidics and nanofluidics could lead to several applications for manipulating particles of all types, properties, and geometries. For example, in many μ-Na*fl* systems there is a need to understand the forces and resulting particle velocity between a flat wall and a spherical particle of radius r_0 interacting with the flat wall with separation d. Results for such geometries have been derived previously,

$$u(d) = \pi\varepsilon r_0(\zeta_P^2 + \zeta^2)\left\{\frac{2\zeta_P\zeta}{\zeta_P^2 + \zeta^2}\ln\left(\frac{1 + e^{-\kappa d}}{1 - e^{-\kappa d}}\right) + \ln\left(1 - e^{-2\kappa d}\right)\right\} \tag{2.47}$$

and the EDL force is expressed as,

$$F_{EDL} = 2\pi\varepsilon r_0\kappa(\zeta_P^2 + \zeta^2)\left\{\frac{2\zeta_P\zeta}{\zeta_P^2 + \zeta^2}\left(\frac{e^{-\kappa d}}{1 - e^{-2\kappa d}}\right) - \left(\frac{e^{-2\kappa d}}{1 - e^{-2\kappa d}}\right)\right\} \tag{2.48}$$

Therefore, it is no surprise that several exciting reports for such applications exist and are discussed in Chapters 5 and 6.

Above, we have discussed electroosmosis and electrophoresis in significant detail due to the direct applicability of these to several μ-Na*fl* systems. Another aspect of electric fields affecting particle transport is seen in dielectrophoresis. Typically, particles that are not electrically charged but can polarize under the influence of an electric field can also be manipulated. The dielectrophoretic velocity is defined similar to the electrophoretic velocity in terms of the dielectrophoretic mobility,

$$u_{DEP} = \mu_{DEP}\nabla E^2 \tag{2.49}$$

The force, F_{DEP}, on a spherical particle of radius a under a DC field is given by the expression,

$$\vec{F}_{DEP} = 2\pi\varepsilon_m a^3\, Re(f_{CM})\nabla(\vec{E}\cdot\vec{E}) \tag{2.50}$$

where f_{CM} is the Clausius–Mossotti factor and is the polarization coefficient expressed as,

$$f_{CM} = \frac{\varepsilon_P - \varepsilon_m}{\varepsilon_P + 2\varepsilon_m} \tag{2.51}$$

with ε_m being the permittivity of the medium and ε_P the permittivity of the particle. Note that the permittivities can be complex numbers and Eq. (2.50) considers the real part of f_{CM}.

2.9 Coupling species transport and fluid mechanics

In the previous sections, we have discussed the coupling between electrostatics and fluid mechanics in detail. We also alluded to the fact that most microchannels and even more so in nanochannels electrokinetic phenomena are important due to the inability of pressure to drive significant flows without very high pressure gradients being applied, which is usually impractical for most real devices. Since the electrolyte comprises some salt in water as the most common medium, it is also important to consider species transport in μ-Nafl systems.

As we now know, any charged surface in contact with a liquid forms an EDL. Ionic transport is classically described by the Nernst–Planck equation and the velocity profiles are often obtained by using the Navier–Stokes equations, which are the equations of motion. The potential distribution in the nanochannel is critical because the dominant transport mechanisms are electrokinetically driven. The ionic and potential distributions are obtained from the Poisson equation. Frequently, a Boltzmann distribution of charged species is assumed yielding the P–B equation, as discussed above.

The molar flux of the ith species, J_i, is given by a form of the 1-D Nernst–Planck equation, such that,

$$J_i = -D_i \frac{\partial c_i}{\partial x} + c_i u - z_i \mu_i c_i \frac{\partial \Phi}{\partial x} \tag{2.52}$$

where c_i is the molar concentration of the ith species, Φ is the electrical potential, u is a bulk fluid velocity to be generally obtained from Navier–Stokes equations, and z_i is the valence of the ith species. Einstein's relation connects the ionic mobility, μ_i, to the diffusion coefficient, D_i, such that,

$$\mu_i = \frac{FD_i}{RT} \tag{2.53}$$

where F is Faraday's constant, R is the universal gas constant, and T is on the absolute temperature scale. The electrical potential is generally considered the sum of two potentials,

$$\Phi = \phi + \psi \tag{2.54}$$

where ϕ is the applied and ψ is the local potential. Summing fluxes over all the species totaling in number to M, and accounting for the valence of each species, provides the total current density, J_A, within a fluidic channel,

$$J_A = F \sum_{i=1}^{M} z_i J_i \tag{2.55}$$

which combined with Eq. (2.52) yields,

$$J_A = F \sum_{i=1}^{M} \left\{ -D_i z_i \frac{\partial c_i}{\partial x} + z_i c_i u - \mu_i z_i^2 c_i \frac{\partial \Phi}{\partial x} \right\} \tag{2.56}$$

Note that the electric potential term follows a squared dependence on ionic valence while all other terms follow linearly. Therefore, in several limiting cases the valence of ions can strongly influence the flow characteristics. For example, it is now known that divalent cations adsorbed to negatively charged microchannel walls can change the wall zeta potential by nearly 20 mV, as divalent cation concentrations (replacing 1–8% of the monovalent cations), for two monovalent electrolyte solutions at different pH and I the NaCl and $Na_2B_4O_7$ solutions, were studied.

Now, let us look at some important limiting cases as these often arise in the μ-Na*fl* systems. In the case of a concentration gradient being the only driving force, an electrical current arises such that,

$$J_A = F \sum_{i=1}^{M} \left\{ -D_i z_i \frac{\partial c_i}{\partial x} \right\} \tag{2.57}$$

This equation can also be obtained from Fick's laws of diffusion. For purely electrokinetic flow, the diffusion and electroosmotic terms can be often neglected in comparison to the electrophoretic term, yielding

$$J_A = F \sum_{i=1}^{M} \left\{ -\mu_i z_i^2 c_i \frac{\partial \Phi}{\partial x} \right\} \tag{2.58}$$

Consider a single, cylindrical nanocapillary with the applied potential across it being the only driver for transport of ionic species. In such a case, the velocity of the species in the axial direction is given by the results summarized in Table 2.2. As the expression for $u(r)$ implies, the transport velocity is proportional to the ζ potential and the permittivity of the solution, varies inversely with the viscosity, and is exponentially affected by the shielding length. One caveat for transport in nanochannels is that the concept of the slip or shear plane (Figure 2.3) becomes ill defined when the EDLs from opposed walls within nanopores begin to overlap, and/or structural layering of molecules and charges are a significant portion of the molecules within the channel. In these circumstances of nanoscale transport with interacting or overlapping EDLs, it is more useful to consider the surface charge or the surface potential. We will discuss extensively the role for surface charge and surface properties in Chapter 3.

2.10 Numerical simulations and advanced modeling methods

The theoretical background presented above assumes the validity of the continuum assumption, i.e., that the ions are treated as hard spheres and point charges and water is considered as a continuous matter. Typically, for confined nanoscale transport if the size of the species of interest is about 10 times smaller than the characteristic system dimension the continuum approach has been found to be approximately valid. Furthermore, any interactions between the surface and the analytes are considered through the boundary conditions. These assumptions neglect ion–ion, ion–wall, ion–water interactions, and effects of hydrogen bonding between water molecules.

In most cases, when the continuum assumption holds governing equations can be solved numerically if analytical solutions are difficult or impossible to compute. In such cases, the solution is a description of the magnitude of the flow variables throughout the flow field. Several computational fluid dynamics packages are available based on both finite volume and finite element methods. OpenFOAM, STAR-CD, ANSYS-Fluent, ANSYS-CFX, and COMSOL multiphysics are some commonly used packages and present examples for the current availability of numerical packages.

However, with continued advances in fabrication technologies practical device length scales along with functional component length scales continue to approach molecular dimensions. Consequently, the molecular details of a fluid flow need to be explored. Furthermore, at these length scales it is important to consider the individual molecular nature of the species being studied. Since a fluid is composed of molecules, one method to calculate static or dynamic properties is to compute the motion of the fluid constituents from first principles by using the fundamental equations of motion.

Molecular dynamics or MD is the term employed to describe the solution of the classical equations of motion for a set particles representing every single atom (or particle as defined above) in a system of interest. The available energy for the particles is distributed between potential and kinetic energy, and the molecules or atoms are thus able to overcome barriers separating minima if the barrier height is less than the total energy minus the potential energy. In MD, the forces acting on the atoms are computed from functions called potentials or force fields which are usually parametrized from empirical studies or from quantum chemistry methods. Usually, the interactions between particles are divided in two types, bonding and nonbonding interactions. Bonding interactions are described using constraining algorithms which constrain the motion of the particles. Non-bonding interactions are short and long range with short interactions including the van der Waals forces and long-range interactions including the Coulomb or electrostatic forces.

Depending on the nature of molecular information desired other advanced numerical capabilities may need to be employed. For example, the Monte Carlo method is a stochastic strategy that relies on probabilities and is not a deterministic method like MD. The Monte Carlo strategy provides a relatively simple mechanism for determining the nondynamical properties of fluids through the use of calculated thermodynamic data. However, if the dimensions of the system of interest are below the typical molecular size (0.3 nm) then one must undertake fundamental calculations for the quantum mechanical behavior. Methods such as density functional theory allow the calculations of atomistic and molecular properties based on the electron density only.

In summary to this section, it should be noted that no one method exists to answer all questions about fluid and transport interactions in confined spaces. Additionally, one must evaluate the advanced numerical techniques in terms of the computation costs involved. Generally, advanced techniques such as MD are yet unable to simulate realistic systems with dimensions on the order of 100 nm with more than 100,000 atoms. Simulated timescales can reach a few nanoseconds but for most practical systems these are still very short. However, several researchers have reported a plethora of interesting information. For example, recent work by Zambrano et al. (2012) showed that for a surface with discontinuous changed patches, MD with approximations yields velocity profiles and chloride ion concentrations that are strongly influenced by the surface change as shown in Figures 2.7 and 2.8.

Therefore, most advanced simulations are limited in scope. For example, for simple hydrated ions such as Na^+:H_2O in a nanochannel with critical dimensions on the order of 10 nm or greater, the continuum assumptions hold for dilute solutions. However, contrasting criteria for the breakdown of the continuum methods

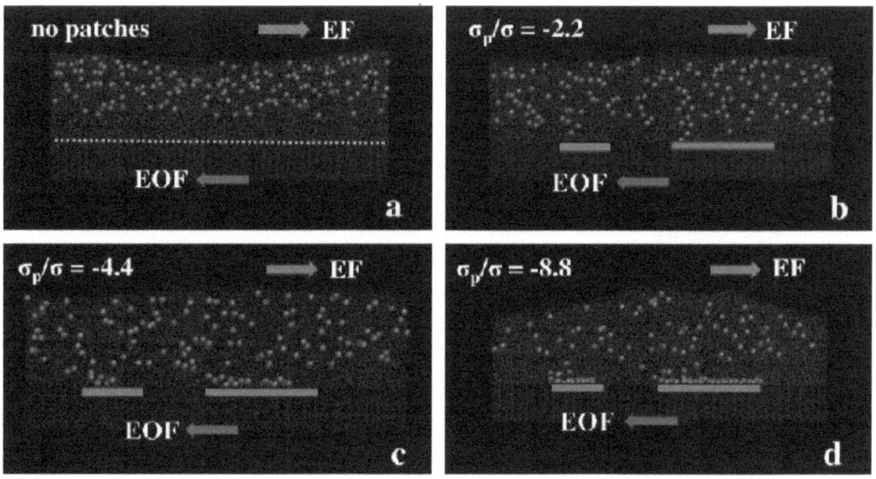

FIGURE 2.7

A snapshot of images from an MD simulation reported by Zambrano et al. (2012) shows the distribution of chloride ions as function of surface charge with respect to a native silica wall in figure a. The axial electric field denoted by the green arrow was 0.1V/nm and the system consisted of 14,000 water molecules (not shown) and 150 chloride ions. The direction of electroosmotic flow is shown by the blue arrow and the red patches denote the location of surface patches with the left patch being 3.5 nm long and the right patch being 8 nm long, and patch gap is at 3.5 nm.

Figure source: Zambrano et al. Microfluidics and Nanofluidics, *2012*

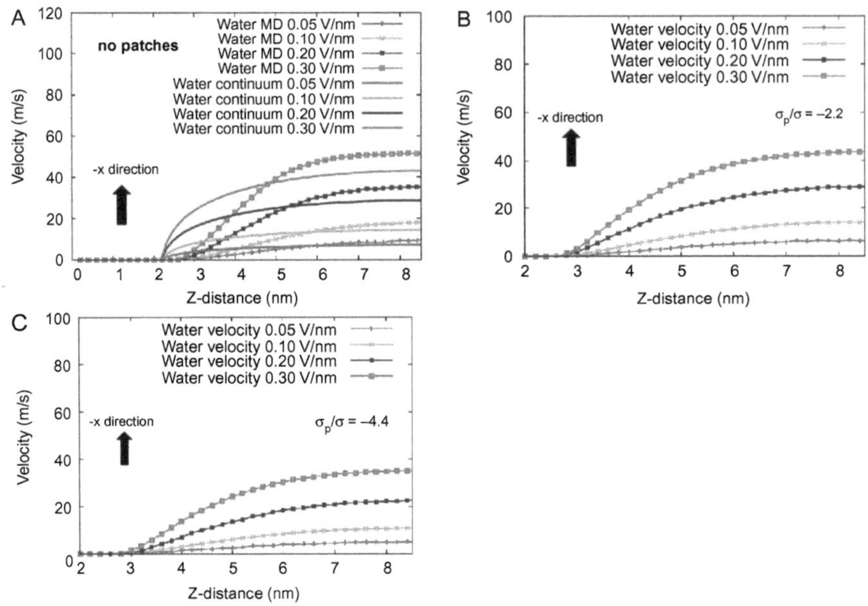

FIGURE 2.8

Velocity profiles calculated by using MD simulations and continuum calculations. Figure A shows a comparison between velocity profiles for water computed using the two methods, and it can be seen that qualitatively the velocity profiles have similar shape away from the wall. Figures B and C show MD calculations for the velocity profiles for water as function of surface charge density for a variety of axial fields. The images here correspond to the system configuration simulated as described in Figure 2.7.

Source: Zambrano et al. Microfluidics and Nanofluidics, *2012.*

have also been presented. Employing advanced numerical techniques to nanofluidics has demonstrated several interesting phenomena. Effects of finite ion size and hydration on ion distribution and velocity, charge inversion and flow reversal, density fluctuations, role of fluid density on particle transport, role of surface charge on transport of ions and water, layering and orientation of fluids and ions with changes in fluid density near surfaces due to surface energy interactions, and effect of surface roughness in nanochannels have been investigated.

EXERCISES

Problem 1

Consider a wide, but shallow channel such that the channel width, $w \gg 2a$, the channel height. Channel cross-section is as shown below with the coordinate system marked in the figure below. Using the derivation for Eq. (2.18), write down the governing differential equation for the velocity, V_z, under the influence of an applied electric field, E_z. Compare your answer to the summary in Table 2.2.

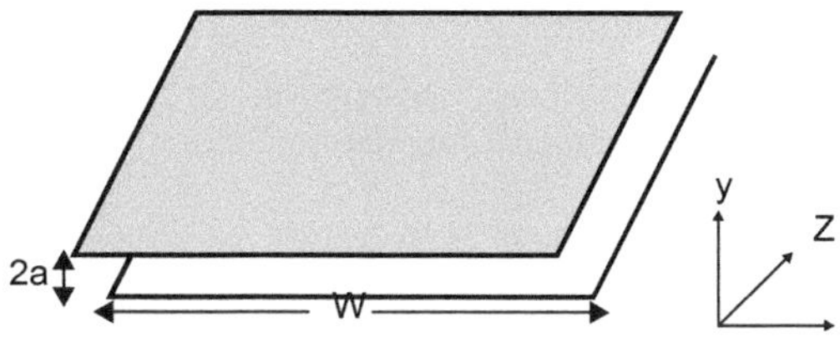

Problem 2

List the terms that can be neglected from Navier–Stokes equations to obtain the governing equations for low Reynolds number flows.

Problem 3

Use Eq. 2.32 to calculate λ_D for ionic strength of 0.01, 0.1, 0.2, and 0.5 M NaCl and $CaCl_2$ solutions.

References

[1] Fox RW, Pritchard PJ, McDonald AT. Introduction to fluid mechanics. 6th ed NY: Wiley; 2009.
[2] Munson BR, Rothmayer A, Okiishi T, Huebsch W. Fundamentals of fluid mechanics. 6th edition NY: Wiley; 2009.
[3] Prakash S, Piruska A, Gatimu EN, Bohn PW, Sweedler JW, Shannon MA. Nanofluidics: systems and applications. IEEE Sensors J 2008;8(5):441–50.
[4] Gouy M. Sur la constitution de la charge électrique a la surface d'un électrolyte. J de Physique et Appliquee 1910;9(4):457–68.
[5] Chapman DL. A contribution to the theory of electrocapillarity. Philos Mag 1913;V25 (148):475–81.
[6] Masliyah JH, Bhattacharjee S. Electrokinetic and colloid transport phenomena. Hoboken NJ: John Wiley & Sons; 2006.

Select bibliography

Abgrall P, Nguyen NT. Nanofluidics. Norwood, MA: Artech House; 2009.
Arrhenius S. Text-book of electrochemistry. London: Longman's Green and Co.; 1902.
Bard AJ, Faulkner LR. Electrochemical methods: fundamentals and applications. New York, NY: John Wiley & Sons; 1980.

Bhushan B. MEMS/NEMS and BioMEMS/BioNEMS: materials, devices, and biomimetics. In: Bhushan B, editor. Springer handbook of nanotechnology. Berlin: Springer-Verlag; 2010. p. 1663–740.

Bhushan B. Nanotribology, nanomechanics, and materials characterization. In: Bhushan B, editor. Springer handbook of nanotechnology. Berlin: Springer-Verlag; 2010. p. 789–856.

Bhushan B. Springer handbook of nanotechnology. Heidelberg: Springer-Verlag; 2010.

Bird G. Molecular gas dynamics and direct simulation of gas flows. Oxford: Oxford University Press; 1994.

Bowski L, Saini R, Ryu DY, Vieth WR. Kinetic modeling of hydrolysis of sucrose by invertase. Biotechnol Bio Eng 1971;13:641–56.

Burgreen D, Nakache FR. Electrokinetic flow in ultrafine capillary slits. J Phys Chem 1964;68(5):1084–91.

Burgmayer P, Murray RW. An ion gate membrane: electrochemical control of ion permeability through a membrane with an embedded electrode. J Am Chem Soc 1982;104:6139–40.

Conlisk AT. The Debye-Huckel approximation: its use in describing electroosmotic flow in micro- and nanochannels. Electrophoresis 2005;26(10):1896–912.

Conlisk AT. Essentials of micro- and nanofluidics: with applications to the biological and chemical sciences. Cambridge, UK: Cambridge University Press; 2012.

Conlisk AT, et al. Mass transfer and flow in electrically charged micro- and nano-channels. Anal Chem 2002;74(9):2139–50.

Conlisk AT, et al. Ionic and biomolecular transport in nanochannels. Nanoscale Microscale Thermophys Eng 2007;11(1-2):177–99.

Coster HGL, Chilcott TC. The characterization of membranes and membrane surfaces using impedance spectroscopy. In: Sorenson TS, editor. Surface chemistry and electrochemistry of membranes, 79. New York, NY: Marcel Dekker; 1999. p. 749–92.

Crow SC, Champagne FH. Orderly structure in jet turbulence. J Fluid Mech 1971;48 (3):547–91.

Daiguji H, et al. Nanofluidic diode and bipolar transistor. Nano Lett 2005;5(11):2274–80.

Debye P, Hückel E. Zur theorie der elektrolyte. Physikalische Zeitschrift 1923;24:185–206.

Derjaguin BV, et al. Kinetic aspects of electrochemistry of disperse systems. Adv Colloid Interface Sci 1980;13.

Dukhin SS. Non-equilibrium surface phenomena. Adv Colloid Interface Sci 1993;44.

Dukhin SS, Shilov VN. Kinetic aspects of electrochemistry of disperse systems II. Induced dipole moment and the non-equilibrium double layer of a colloid particle. Adv Colloid Interface Sci 1980;13.

Dussan VEB. On the spreading of liquids on solid surfaces: static and dynamic contact lines. Annu Rev Fluid Mech 1979;11:371–400.

Grahame DC. The electrical double layer and the theory of electrocapillarity. Chem Rev 1947;41:441–503.

Grahame DC. Electrode processes and the electrical double layer. Annu Rev Phys Chem 1955;6:6337–58.

Harris JM, editor. Poly(ethylene glycol) chemistry. New York, NY: Plenum Press; 1992.

Henrich VE, Cox PA. The surface science of metal oxides. Cambridge, UK: Cambridge University Press; 1994.

Hille B. Ionic channels of excitable membranes. Sunderland, MA: Sinauer Associates; 1992.
Israelachvili JN. Intermolecular and surface forces. Orlando, FL: Academic Press; 1985.
Karniadakis G, Beskok A, Aluru N. Microflows and nanoflows: fundamentals and simulation. Springer-Verlag; 2005.
Kemery PJ, et al. Electric field mediated transport in nanometer diameter channels. Langmuir 1998;14:2884–9.
Kirby BJ. Micro- and nanoscale fluid mechanics: transport in microfluidic devices. New York, NY: Cambridge University Press; 2010.
Lakshminarayanaiah N. Transport phenomena in membranes. New York, NY: Academic Press; 1969.
Lee S, Bhushan B. Biological molecules in therapeutic nanodevices. In: Bhushan B, editor. Spinger handbook of nanotechnology. Berlin: Springer-Verlag; 2010. p. 453–84.
Levich VG. Physicochemical hydrodynamics. Englewood Cliffs, NJ: Prentice-Hall; 1962.
Li D. Electrokinetics in microfluidics. New York, NY: Academic Press; 2004.
Lyklema J. Fundamentals of interface and colloid science. London: Academic Press; 1995.
Macdonald JR. Impedance spectroscopy: emphasizing solid materials and systems. New York, NY: John Wiley & Sons; 1987.
Madou M. Fundamentals of microfabrication. Boca Raton, FL: CRC Press; 1997.
Madou MJ. Fundamentals of microfabrication and nanotechnology. Boca Raton, FL: CRC Press; 2012.
McNair HM, Miller JM. Basic gas chromatography. New York, NY: John Wiley & Sons; 1997.
Nernst W. Theoretical chemistry. London: Macmillan; 1895.
Nernst W. Experimental and theoretical application of thermodynamics to chemistry. New York, NY: Charles Scribner's Sons; 1907.
Newman JS. Electrochemical systems. Englewood Cliffs, NJ: Prentice-Hall; 1973.
Nguyen NT, Wereley ST. Fundamentals and applications of microfluidics. Norwood, MA: Artech House; 2002.
Outhwaite CW, et al. Theory of the electric double layer using a modified Poisson–Boltzmann equation. J Chem Soc Faraday Trans 2 1980;76:1388–408.
Panton RL. Incompressible flow. Hoboken, NJ: John Wiley & Sons; 1996.
Piruska A, et al. Nanofluidics in chemical analysis. Chem Soc Rev 2010;39:1060–72.
Prakash S, et al. Theory, fabrication and applications of microfluidic and nanofluidic biosensors. Phil Trans Royal Soc A 2012;370(1967):2269–303.
Probstein RF. Physicochemical hydrodynamics: an introduction. 2nd edition Hoboken, NJ: John Wiley & Sons; 2003.
Rautenbach R, Albrecht R. Membrane processes. Frankfurt: John Wiley & Sons; 1989.
Rayleigh JWS. The explanation of certain acoustical phenomena. Nature 1878;18: 319–21.
Rice CL, Whitehead R. Electrokinetic flow in a narrow cylindrical capillary. J Phys Chem 1965;69(11):4017–24.
Sadus RJ. Molecular simulation of fluids: theory, algorithms and object-orientation. Amsterdam, Netherlands: Elsevier; 1999.
Schmickler W. Interfacial electrochemistry. New York, NY: Oxford; 1996.
Schoch RB, et al. Transport phenomena in nanofluidics. Rev Mod Phys 2008;80:839–83.

Schroeder DV. An introduction to thermal physics. San Francisco, CA: Addison Wesley Longman; 2000.

Shannon MA, et al. Science and technology for water purification in the coming decades. Nature 2008;452(20):301–10.

Sheehan PE, Whitman LJ. Detection limits for nanoscale biosensors. Nano Lett 2005;5(4):803–7.

Simon SM, et al. What drives the translocation of proteins. Proc Nat Acad Sci USA 1992;89:3770–4.

Siwy Z, Fulinski A. Fabrication of a synthetic nanopore ion-pump. Phys Rev Lett 2002;89(19): 19803-19801-19803-19804.

Sparreboom W, et al. Principles and applications of nanofluidic transport. Nat Nanotechnol 2009;4:713–20.

Sparreboom W, et al. Transport in nanofluidic systems: a review of theory and applications. J Phys 2010;12:015004.

Squires TM, Quake SR. Microfluidics: fluid physics at the nanoliter scale. Rev Mod Phys 2005;77:977–1025.

Stein D, et al. Surface charge governed ion transport in nanofluidic channels. Phys Rev Lett 2004;93(3): 035901-035901 035901-035904.

Steinle ED, et al. Ion channel mimetic micropore and nanotube membrane sensors. Anal Chem 2002;74:2416–22.

Tegenfeldt JO, et al. Micro- and nanofluidics for DNA analysis. Anal Bioanal Chem 2004;318(7):1678–92.

Torsten Vilkner DJ, Manz A. Micro total analysis systems. Recent developments. Anal Chem 2004;76(12):3373–86.

Travis KP, Gubbins KE. Poiseuille flow of Lennard-Jones fluids in narrow slit pores. J Chem Phys 2000;112(4):1984–94.

Tretheway DC, Meinhart CD. A generating mechanism for apparent fluid slip in hydrophobic microchannels. Phys Fluids 2004;16(5):1509–15.

Tulock JJ, et al. Microfluidic separation and gateable fraction collection for mass-limited samples. Anal Chem 2004;76(21):6419–25.

Tung S, et al. A cellular motor driven microfluidic system. The 12th international conference on solid-state sensors, actuators, and microsystems, Boston, MA; 2003.

Ulman A. Formation and structure of self-assembled monolayers. Chem Rev 1996;96:1533–54.

Vinogradova OI, Belyaev AV. Wetting, roughness and hydrodynamic slip. PSP Review Volume. Singapore: Pan Stanford Publishing; 2010.

Wang H, Branton D. Nanopores with a spark for single-molecule detection. Nat Biotechnol 2001;19:622–3.

Wang Y-C, et al. Million-fold preconcentration of proteins and peptides by nanofluidic filter. Anal Chem 2005;77:4293–9.

White FM. Viscous fluid flow. New York, NY: McGraw-Hill; 1991.

Zambrano HA, Pinti M, Conlisk AT, Prakash S. Electrokinetic transport in a water–chloride nanofilm in contact with a silica surface with discontinuous charged patches. Microfluid Nanofluidics 2012; doi:10.1007/S10404-012-0992-9.

CHAPTER

Interfaces in Microfluidic and Nanofluidic Systems

3

CHAPTER OUTLINE

3.1 Introduction

In a way, all μ-Na*fl* systems can be defined as interfacial systems. Consider a slit-like nanochannel of interest with a width of 1 μm, depth of 100 nm, and length of 100 μm; or, consider a nanopore with a dimension of 2 nm diameter and 30 nm length. To perform fluidic manipulations we need to access the channel or pore, and so the question is how do we facilitate physical connections at these length scales? Here we will discuss the role of interfaces, i.e., the connection between the "nanoworld" to the "microworld," and beyond to the experimentally or physically accessible length scales permitted by the "macroworld." Additional examples are provided for experimental systems throughout chapters 4, 5, and 6. Consequently, we have two levels of interfaces to think about. First, the functional interfaces inside μ-Na*fl* systems that govern fluid flow, and second, the connections and interfaces of the system to the surroundings.

S. Prakash & J. Yeom: Nanofluidics and Microfluidics. DOI: http://dx.doi.org/10.1016/B978-1-4377-4469-9.00003-2

3.2 Introduction to surfaces

Surfaces are ubiquitous in our lives. Metaphorically Oscar Wilde referring to people wrote, "We live, I regret to say, in an age of surfaces." Wilde's words implied a lack of depth in the personality of people; extending his literary definition to the world of physics, we can also apply these words to the world of surface science. Typically, a surface is considered to be the interface region between surroundings and a material of interest. In the next several pages, we will define surfaces more precisely and discuss several aspects of surfaces relevant to μ-Na*fl* systems.

In science and technology, especially μ-Na*fl* systems, the role that surfaces play cannot be emphasized enough. As discussed in Chapters 1 and 2, μ-Na*fl* systems are characterized by high *SA/V* ratios, i.e., the fluid and species interactions in a micro- or nanochannel, tube, pore, or any other micro- or nano-structure is strongly influenced by the surface area in contact with the fluid or chemical species.

Over the years, due to the many technologically important applications in catalysis, tribology, electrochemistry, corrosion, and more recently in μ-Na*fl* systems, surface science has been a subject of great scientific interest. Given the importance of surface-mediated phenomena to a wide range of applications, governed by a few underlying and unifying principles, let us begin our discussion by defining surfaces.

One way to understand a surface (or an interface between media) is to consider it mathematically. A surface can be mathematically represented as a plane. However, in most real-world applications a mathematical plane may not fit the definition of a surface. Here, we will define a surface or an interface to mean a thin region in space within which material properties can vary significantly from the bulk or remainder of the material. In addition, this thin region of space influences transport and reaction phenomena in its vicinity through interactions with the surroundings, which may be different from the bulk material. In this surface region properties such as chemical composition, refractive index, mechanical strength, conductivity, charge, etc. can significantly differ from the bulk material. The definition is depicted schematically in Figure 3.1.

In the light of this definition, surfaces provide a unique set of challenges and opportunities in μ-Na*fl* systems. Recall that in μ-Na*fl* systems *SA/V* ratios varying from about 10^3 m^{-1} to as high as 10^9 m^{-1} have already been reported with fabrication advances enabling even higher ratios.

Now that we understand the term "surface" a little better, we should also think about the term "bulk." The "bulk" refers to the main part of the material or substrate of interest. The complete region of interest or focus for transport and interactions then would be defined as a system. Therefore, here a system is defined to be a functional unit, i.e., it can be a nanoparticle or an entire lab-on-chip-type device.

In μ-Na*fl* systems the primary effect of the surface is to exert forces on the surroundings. As we discussed in Chapter 2, most μ-Na*fl* systems work with aqueous solutions. Therefore, the predominant interactions are based on interactions (and consequently the forces) between water and surfaces. Now, it should be noted here that with our definition of surfaces as part of a system, the

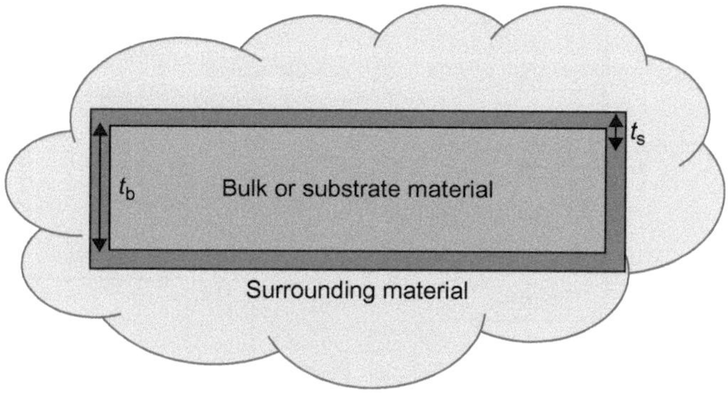

FIGURE 3.1

A schematic depicting the definition of the surface. The surface region of thickness t_s interacts with the surroundings and may be independent of the bulk material of thickness t_b. Usually, $t_s \ll t_b$. The material properties of the surface region can be different from the bulk. In μ-Na*fl* systems the surface region may dominate transport phenomena due to the interactions with surroundings governed by the high *SA*/*V* ratios.

interactions could be between particles (e.g., ions, biomolecules, nanoparticles) in the aqueous phase, or between the physical walls of the device or channels interacting with both water and particles. In short, there could be several levels of surface–system or surface–surface interactions that may influence the overall transport in the μ-Na*fl* systems.

Historically, forces have been defined by the influence region over length, i.e., if the influence of a force is "felt" for a long distance; the force was classified as long-range forces. Such a classification would include gravitational forces, electrostatic or Coulombic forces, magnetic forces, and van der Waals forces. By contrast, the short-range forces would be those that decay within a few molecular lengths. Van der Waals and Coulombic forces between strongly ionized species, and covalent bond forces between atoms, would be examples in this category. It should be noted here that van der Waals forces fall under both categories, depending on the context of the specific application. As a rule of thumb, short-range forces would be rarely considered important beyond 1 nm and long-rage forces would be considered to have an effect to nearly 100 nm from the point of origin. Therefore, it is important to relate sizes of common materials with respect to the system. Figure 3.2 presents an overview of several μ-Na*fl* systems with respect to common materials, structures, processes, and techniques.

3.3 Surface charge

In μ-Na*fl* systems, the two most common surface properties that influence flows are the surface charge and surface energy. Nearly all surfaces acquire a finite charge in contact with a liquid. In the scientific literature, two main reasons have been considered for surfaces acquiring a finite charge through (i) ionization of

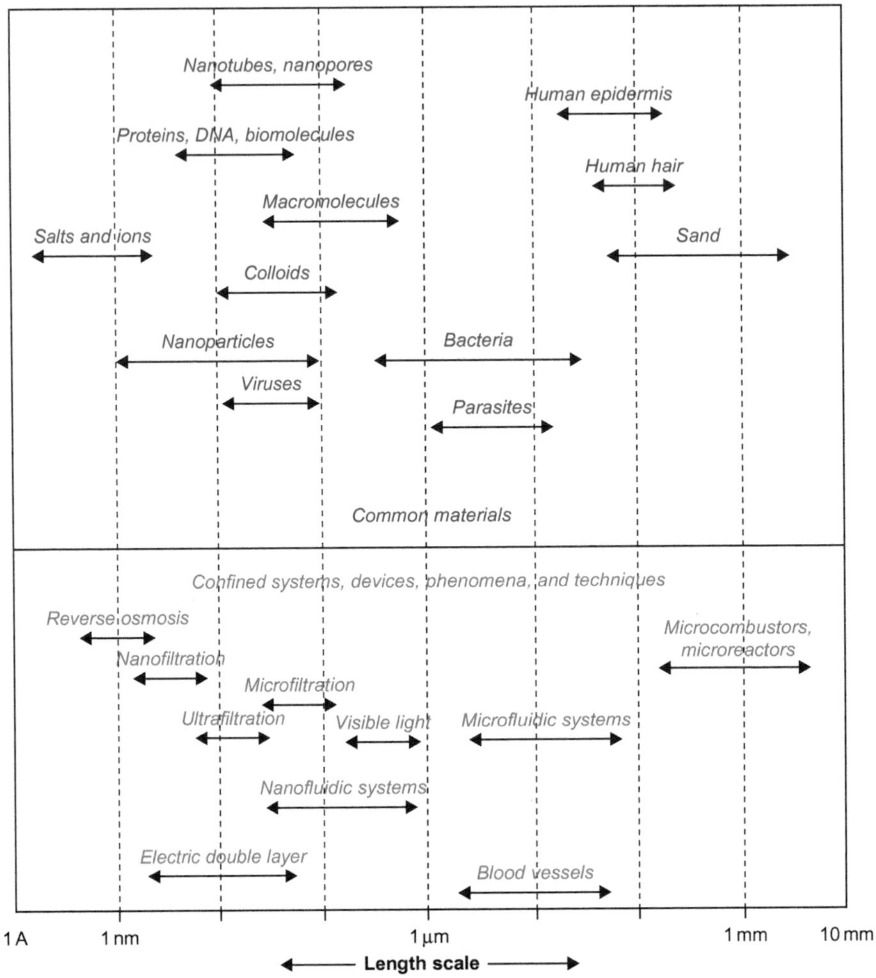

FIGURE 3.2

A description of length scales as they relate to common materials, physical objects, phenomena, and some methods relevant to micro- and nanoscale systems [1].

chemical groups on the surface or (ii) adsorption of ions from the solution to the surface. In most cases, the surface charge is expressed in terms of a surface charge density with SI units being C/m^2.

The discussions for the two reasons for surfaces acquiring charge in more detail are best understood by following an illustrative example for surface ionization. Glass or silica is one of the more common substrates used in μ-Na*fl* systems. The glass surface has silanol groups at the surface and under normal conditions, a clean glass surface would present an −OH rich surface, schematically depicted in

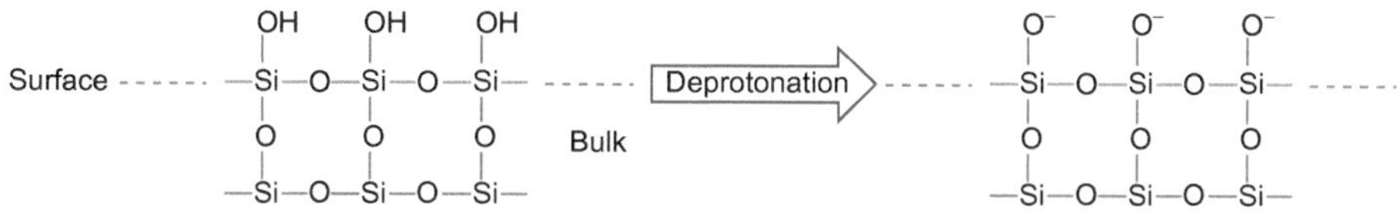

FIGURE 3.3

Schematic depiction of a silica surface undergoing deprotonation when in contact with an aqueous solution at neutral pH to acquire a negative charge.

Figure 3.3. For example, a clean silica surface presenting silanol groups to an aqueous solution at neutral pH will undergo deprotonation, i.e., the hydrogen atom (in ionic form, proton) goes into solution and leaves a negatively charged surface in contact with the aqueous solution.

A charged surface in contact with an electrolyte solution will attract ions of the opposite polarity called counterions and repel ions with the same polarity as the surface charge (called co-ions). The practical implication for the formation of a charged surface in contact with an electrolyte solution for our discussion of microfluidics and nanofluidics leads to the concept of EDLs as described in Section 2.6, in Chapter 2. Let us revisit some of the ideas from Chapter 2 with regards to the EDL. Equations 2.26–2.28, 2.30, and 2.32 present several formulations for calculating the Debye length. These results essentially imply that the effect of surface charge extends from the surface into the solution, i.e., it is a 1-D electric field from the surface to the solution. Furthermore, λ_D was noticed to increase with the characteristic thermodynamic energy, k_BT or RT with square-root dependence. The characteristic thermodynamic energy is a measure of the thermal agitation of the ions and molecules. Therefore, the direct implication is that at absolute zero, i.e., without any thermal agitation, a monolayer of counterions would perfectly shield the surface charge and the EDL thickness as correlated to our Debye length calculation would essentially collapse to a single ionic diameter.

3.4 Surface energy

In μ-Na*fl* systems the interfacial and surface energies between an interacting liquid and surface determines how the liquid will deform and/or flow in the micro- or nanochannel. The geometric angle defined when a liquid droplet interacts with a surface as shown in Figure 3.4 is called the contact angle (CA), θ_c, and is a parameter used to provide a measure of the surface energy. In Figure 3.4, a droplet is seen on a glass surface chemically modified with a fluoro-terminated monolayer to yield a hydrophobic surface or $\theta_c > 90°$. We will discuss methods to chemically modify surfaces in more detail later (see Section 3.5.2).

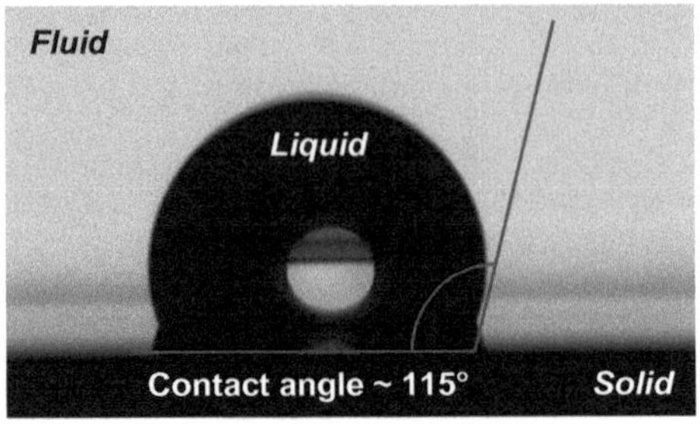

FIGURE 3.4

A digital picture of a water droplet with a contact angle of $\sim 115°$ on a fluorinated glass surface. The surface modification to yield the fluorinated surface renders the glass hydrophobic. The typical contact angle measurement is shown.

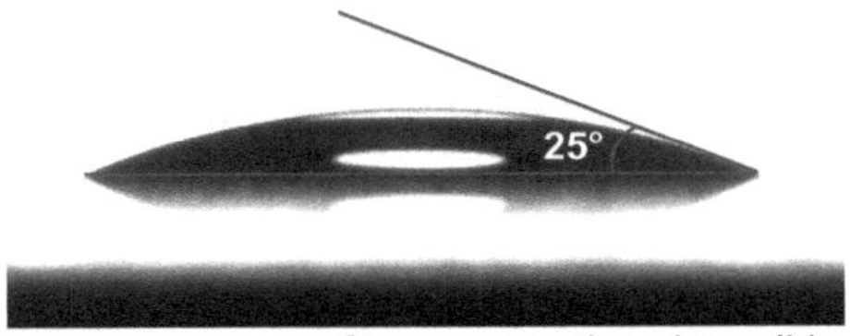

FIGURE 3.5

A digital picture of a water droplet with a contact angle of $\sim 25°$ on a clean glass surface. Typically, a clean glass surface is hydrophilic. The horizontal line shows position of the glass slide with respect to the sample mounting stage. The reflection of the water droplet is seen below the horizontal line.

By contrast, a hydrophilic surface will have $\theta_c < 90°$, and an example of a water droplet on a glass surface with a CA of 25° is shown in Figure 3.5. Contact angles are independent of surface geometry; consequently, a water droplet inside a capillary tube or micro- and/or nanochannel, or on a flat substrate will display the same θ_c but will be governed by the type of surface, liquid type, and the surrounding environment.

For the interfaces that can be seen in Figure 3.4, the surface energy terms are really interfacial energies for the different media in contact. Therefore, a set of interfacial energies for each of the contact media phases can be defined, γ_{SL} for the solid–liquid interface, γ_{SF} for the solid–fluid interface, γ_{LF} for the liquid–-fluid interface, and these terms are used to provide a governing equation connecting these interfacial energies to θ_c (Eq. (3.1) or Young's equation). Note, we have used the general term "fluid" for the surrounding region which may be vapor, air, or any other medium (usually a fluid) enveloping the substrate–droplet system.

The relation between θ_c and surface energy is usually given by Young's equation (also called Young–Laplace or Young–Dupré equation) depending on the formulation, as listed below.

$$\gamma_{SL} + \gamma_{LF}\cos\theta_c = \gamma_{SF} \tag{3.1}$$

If the droplet is considered "adhered" to the surface then an adhesion energy per unit area of contact can be defined as ΔW_{SLF} and results in the Young–Dupré equation,

$$\gamma_{SL} + \gamma_{LF}(1 + \cos\theta_c) = \Delta W_{SLF} \tag{3.2}$$

From a more mechanical point of view, if we consider the droplet formation or spreading on a surface and think of the pressure–volume (p–v) work needed to expand the curved surface we obtain the Young–Laplace equation,

$$\Delta p = \sigma\left(\frac{1}{R_1} + \frac{1}{R_2}\right) \tag{3.3}$$

Where, Δp is the pressure difference across the liquid droplet, R_1 and R_2 are the radii of curvature of the surface along any two orthogonal tangents, and σ is the surface tension. *Surface tension* is defined as the intensity of the molecular attraction per unit length along any line in the surface of the fluid in contact with some external interface, or thermodynamically, it can be understood to be the change in Helmholtz free energy per unit area change as the droplet forms (or spreads out) over a surface. In Section 3.7, we will revisit the idea of surface tension and how it can be used as a mechanism to drive fluid flow in μ-Na*fl* systems.

3.5 Thermodynamics of surfaces

Thermodynamics of surfaces and interfaces has been discussed in scientific literature for a long time, beginning with the work of Gibbs in 1878. Gibbs first presented the idea of a dividing plane to separate the "bulk" from the "surface." The dividing plane is a conceptual construct or a theoretical notion that allows one to choose the location of surface properties with respect to a hypothetical reference, similar to the mathematical notion of a surface discussed in Section 3.2. The surface properties can vary with respect to the "bulk" hypothetical reference where the properties such as energy, entropy, adsorption, etc. remain constant. In recent years, several extensive reviews and books have presented detailed analysis and review of thermodynamics of solid surfaces. In this chapter, we will review the most important ideas relevant to μ-Na*fl* systems and recommend to the reader to follow additional material from the select bibliography at the end of this chapter for more detailed information. Consequently, it is important to briefly discuss the thermodynamics of surfaces and interfaces as once a surface is modified or an interface presents to the fluid and species in μ-Na*fl* systems it often yields a change in either the surface energy or surface charge or both, which ultimately affects the interactions with the surroundings. These interactions can be better understood by considering the surface thermodynamics.

3.5.1 Formation of surface layers

The ability to control and/or change surface properties by attachment of surface coatings or to alter surface states through physical or chemical means to actively influence phenomena within micro- and nanoscale systems is an attractive and growing research area.

In most μ-Na*fl* systems, the role of walls is critical in determining subsequent surface-mediated phenomena for transport and/or reactions. Next, we present a brief discussion borrowing the ideas from colloidal sciences, and drawing upon the work published over the last century by many researchers, identifying and developing the physical laws and parameters that influence confined flows and surface-mediated phenomena.

Consider an idealized case in which a surface with some charge density, σ_s, is in contact with a solution containing only counterions (i.e., ions of polarity opposite to σ_s) with a charge density ρ_s. The surface is assumed to have a potential, ψ_s. The surface charge density can be related to the surface potential by Eq. (3.4),

$$\sigma_s = \varepsilon\varepsilon_0 \left(\frac{d\psi}{dx}\right)_s \tag{3.4}$$

Now, if the ideal solution with only counterions is replaced by an electrolyte then σ_s must account for changes in charge densities of co- and counterions in the solution adjacent to the surface and those in the bulk solution away from the surface,

$$\sigma_s^2 = 2\varepsilon\varepsilon_0 k_B T \left[\sum_i \rho_{0,i} - \sum_i \rho_{\infty,i}\right] \tag{3.5}$$

where ρ_0 is the charge density at the mid-plane between two walls (for example, think of a wide micro- or nanochannel) and ρ_∞ is the ionic concentration in the bulk solution. The subscript i denotes species i. In a more realistic case, moving from the idealized surfaces to a confined micro- or nanoscale system, the surface charge can be related to the ions in solution by the Grahame equation. Consider an electrolyte containing a mixture of NaCl and HCl in an aqueous solution within a glass device. In such a case, the modified Grahame equation relates σ_s to the ions in solution

$$\sigma_s^2 = \sqrt{8\varepsilon\varepsilon_0 k_B T}\sinh\left(\frac{e\psi_0}{2kT}\right)\{([\mathrm{Na}^+]_\infty + [\mathrm{H}^+]_\infty)(2 + e^{-(e\psi_0/kT)})\}^{1/2} \tag{3.6}$$

where ψ_0 is the mid-plane potential. It can be noted from Eq. (3.6) that as glass surfaces have a native negative charge, the surface charge density is a function of only the counterion concentration in solution. In fact, this property has been exploited in many micro- and nanoscale phenomena by manipulating the buffer concentration, i.e., the concentration of NaCl or the main salt in the electrolyte solution. Furthermore, Eq. (3.6) also shows that by changing the pH (i.e., $[H^+]$) the surface charge density can be manipulated. However, the above equations also suggest that the surface charge density can be directly manipulated through chemical modifications of the surface and directly changing σ_s to affect ψ_s.

Since confined surfaces directly play a role in surface-mediated phenomena within devices, let us explore the influence of these surfaces. The direct consequence of ion adsorption or distribution starting at the surface and extending to the bulk fluid is formation of the EDL. In Section 2.6 of Chapter 2, we discussed the structure of the EDL and the potential distributions for some existing models such as the Gouy–Chapman model and the Stern modification to the Gouy–Chapman approach that accounts for finite size of ions at the wall.

Figure 3.6 shows the concentration profile for a charged surface as the ion concentration from the wall reaches bulk value far away from the charged surface. Note that the bulk fluid is electrically neutral in microfluidics and most nanofluidics with small λ_D (correlated to an EDL thickness) leading to equilibrium between the counter- and co-ions in the bulk solution far away from the surface. Furthermore, it should be noted here that since the ion adsorption and subsequent formation of the Stern layer is a function of the surface charge, monovalent and divalent (or multivalent) ions from solutions would have different electrostatic interaction forces leading to variability in surface charge shielding as a function of z_i, the ion valence. This difference is explicitly captured in Eq. (2.28), which shows that

$$\lambda_D \propto \sqrt{\frac{1}{\sum_{i=1}^{N} z_i^2 n_{i,\infty}}} \tag{3.7}$$

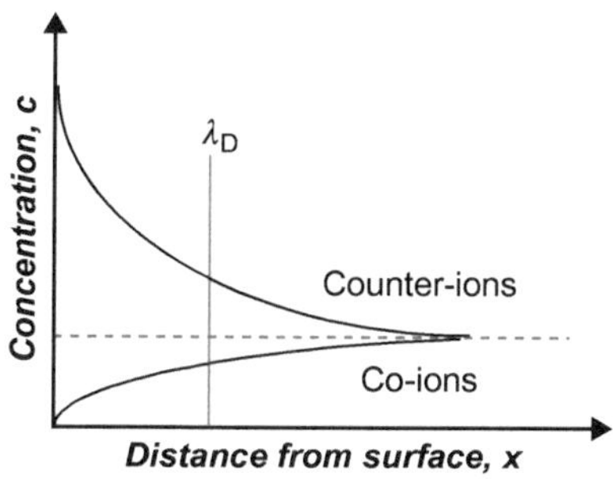

FIGURE 3.6

The schematic shows the concentration profile for any charged surface in contact with an electrolyte in terms of the trends followed by counterions and co-ions. As the distance (in the figure denoted by x) from the surface increases toward bulk solution the counter- and co-ion concentrations reach the same value. As ions are charged, equilibrium concentrations in the bulk leads to the common electroneutrality condition of the bulk fluid as discussed in Chapter 2.

Therefore, from Eq. (3.7) the importance of z_i should be noted toward the surface charge shielding and the potential evolution from the wall toward the bulk solution. Consequently, the potential at the slip plane, i.e., ζ- potential, can be different in an electrolyte solution of monovalent ions in contrast to multivalent ions at the same ionic strength. This distinction is shown in Figure 3.7 where ζ- potential for electrolytes containing mono- and multivalent ions is plotted as a function of concentration. For example, as seen in Figure 3.7A the ζ- potential for the sodium tetraborate is higher than sodium chloride solutions with both containing $MgCl_2$ at same ionic strength.

Figure 2.3 showed the structure of a typical EDL. Note that the OHP and shear or slip plane are at the same location. In most practical systems, the OHP and slip plane are identical; however, deviations from this picture may exist if there is significant wall roughness or localized areas of high charge density exist on the wall. In such cases, more detailed EDL models may be needed. Continuing with our discussion of the EDL picture as presented in Figure 2.3, note that the diffuse part of the EDL acts to store a relatively small amount of charge (typically, less than 10 $\mu C/cm^2$, with σ_s being 3–10 times or larger) leading to relatively low potentials. Therefore, analytical equations like those discussed in Chapter 2 can be used to describe the diffuse part. One other consequence of the ability to store charge is that the diffuse layer has a capacitance associated with it and is generally referred to as the differential capacitance, C_d, and is given by,

$$C_d = \frac{d\sigma_d}{d\varphi_d} \tag{3.8}$$

where σ_d and φ_d denote the charge and potential in the diffuse part of the EDL. In electrochemistry and colloidal sciences, the differential capacitance is plotted as a function of potential, as shown in Figure 3.8 for a symmetric 1:1 electrolyte following the Gouy–Chapman model.

It should be noted from Figure 3.8 that C_d is small ($\sim$150 $\mu F/cm^2$ or less) for dilute solutions at concentrations of 1 mM or less. Furthermore, the potential range is restricted to low values, typically on the order of 100–150 mV or less. The final observation to be made is that as solution concentration or potentials increase, C_d increases rapidly, reaching values $\sim$400–500 $\mu F/cm^2$ with solutions at concentrations of 100 mM or higher. Extending this discussion to another important concept, we consider the point of zero charge (pzc). The pzc would imply $\sigma_d = 0$. Under such a condition,

$$C_d = \varepsilon\varepsilon_0\kappa \tag{3.9}$$

Or, the capacitance per unit area is the same as that for a flat plate capacitor with plate distance between plates being λ_D. We could spend several more pages going over the finer details of the EDL model; however, we now have the essential tools needed for implementing most μ-Na*fl* systems. Therefore, for further reading on the discussion just presented we refer the interested reader to the detailed discussions by Lyklema [2], Israelachvili [3], Bard, [4] and Probstein [5] in their books and additional related references listed in the select bibliography.

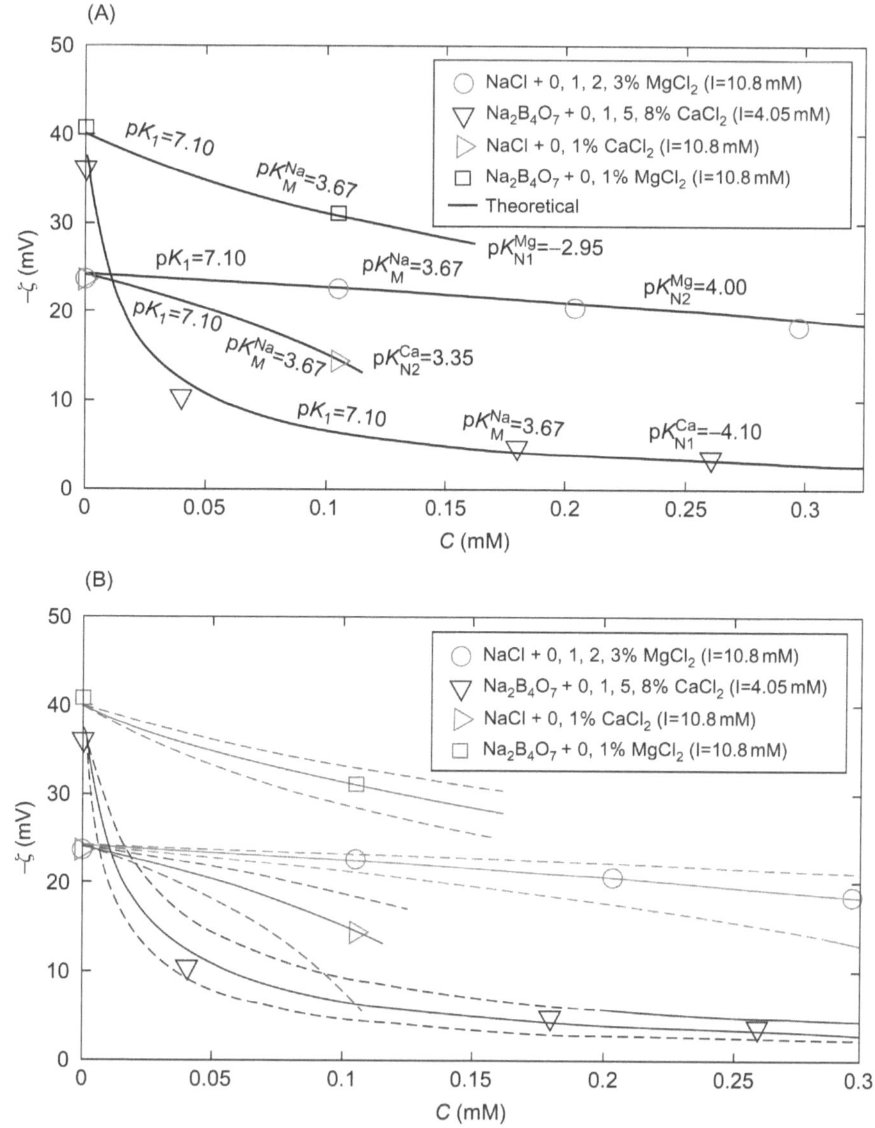

FIGURE 3.7

Effect of calcium and magnesium ions on the ζ- potential of sodium tetraborate: comparison of experimental and modeled values. The site binding model inputs are shown on the figure. (B) The theoretically predicted ζ- potential (dashed curves) when the pK value for divalent cation adsorption used in each of the solid curves of A is changed by ±5% from the values shown in A, all other inputs to the model remaining unaltered. The solid curves and corresponding set of equilibrium constants are the same as in A.

Source: *Images courtesy Prof. A.T. Conlisk and [6]*

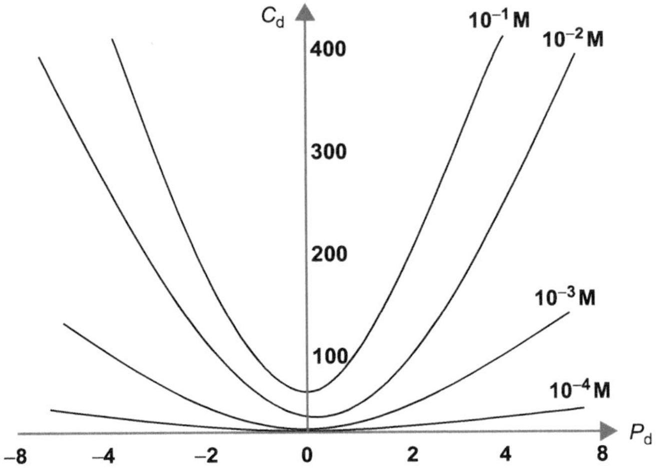

FIGURE 3.8

A plot of differential capacitance as function of dimensionless potential, P_d for a variety of 1:1 electrolyte concentrations.

Source: *[2] Lykelama J. Fundamentals od interface and colloid science, Solidliquid interface, vol. 2 and [4] Bard AJ, Faulkner LR. Electrochemical methods: fundamentals and applications.*

While essential to the working and understanding of flows ion adsorption is not the only type of surface layers that can be formed and/or attached to the wall. In the discussion around Eqs. (3.4), (3.5) and (3.6), it was suggested that chemical surface modification may be an alternate approach adhering different material layers to the physical channel wall for direct control over σ_s. The ion adsorption and formation of the EDL would then follow with respect to this new adhered layer as opposed to the underlying bulk substrate.

Formation of ultra-thin surface layers ($\sim$10 nm thick or less) either through bottom-up approaches like self-assembly or application of surface coatings through synthetic chemistry on a variety of surfaces has been a major area of study over the years. The main reason for existing and continued interest in modifying surfaces through adherent surface layers is due to the technological interest in preparing relatively defect-free surfaces with systematically engineered properties. Among the surface layers, formation of single molecule thick layers or monolayers have had a special place beginning with the early work of Langmuir on protein monolayers. In more recent times, self-assembled monolayers (SAMs) have been used as important tools in altering surface properties, building a complex array of surface functionalities based on SAMs, and as underlying building blocks for nanoscience and nanotechnology in μ-Na*fl* systems. One main distinguishing feature of SAMs in contrast to surfactant monolayers that are generally physisorbed to the surface is that SAMs are characterized by a specific interaction between the surface-binding molecule and the substrate. For instance, the specific surface–molecule interaction can lead to the formation of a covalent bond as observed in the formation of silane-based monolayers on glass or glass-like surfaces through the interactions of the silanol groups on the surface interacting with the functional silane moieties such as halogen groups in the adsorbate molecules. SAMs can be deposited from either a solution or vapor phase, and formation of SAMs on many different materials has been discussed including metals (Au, Ag, Pt, Cu, etc.), polymers (polymethylmethacrylate (PMMA), polydimethylsiloxane (PDMS), polyimide (PI), etc.), semiconductors (Si, TiO_2, etc.), and oxides (Fe_2O_3, Ta_2O_5, SiO_2, etc.). In either deposition method, the functional species adsorbs on the surface from the bulk phase near the interfacial region and may undergo surface reorganization to minimize the energy and/or entropy. Liquid- and gas-phase adsorptions are different. For example, when deposition or adsorption occurs from a liquid, the process is governed by multiple species interactions since all solution-based processes comprise more than one component in the solution (e.g., one or more solutes in a solvent).

For adsorption of a monolayer from the liquid phase, the simplest form for the adsorption process is given by the Langmuir adsorption isotherm as discussed by Probstein in his book on physiochemical hydrodynamics [5] and reviewed here,

$$\theta = \frac{c}{c + b} \tag{3.10}$$

where c is the concentration of the adsorbing species near the surface, θ is the fraction of surface sites occupied, and b is a constant related to the adsorbent. It should be noted that several isotherm models exist and Lyklema [2] in his multi-volume book explains several models in detail. For our discussions, it is pertinent to note a few key ideas: (i) most monolayer formation from liquid phases happens in a dilute solution, (ii) dilute solutions can be approximated as ideal solutions, and consequently the Langmuir isotherms for heterogeneous reactions (surface–solution interactions giving rise to reactions leading to formation of either physisorbed or chemisorbed layers) work well to provide analytical answers, and (iii) the shape of isotherm plots can give diagnostic information for several details of the adsorption process. In Figure 3.9, we see a summary of several common isotherms that have been used previously and are more common to μ-Na*fl* systems. A variety of parameters have been studied for the formation of SAMs. Some of these include solvent type, processing conditions such as immersion time, species concentration, deposition temperature, chain length, and substrate type.

A brief discussion is presented next to demonstrate how surface modification by use of SAMs or other monolayers can be a powerful tool in developing engineered surfaces within micro- and nanosystems for influencing confined phenomena. Generally, a covalently bonded SAM or other monolayers has three parts: (i) a head group, which binds to the surface and also makes linkages between adjacent molecules on the surface to form a stable covalently linked surface network, (ii) a backbone, which is often a hydrocarbon chain, and (iii) an end group, which is the functional end of the SAM and is responsible for surface properties and the interaction between the functionalized surface and the external surroundings. This SAM

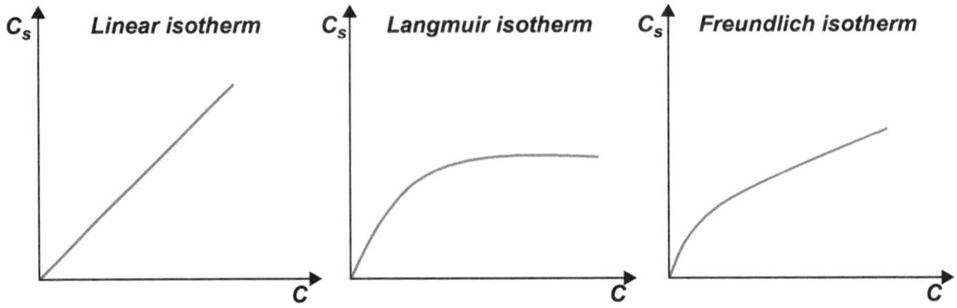

FIGURE 3.9

The three schematics show some common isotherms of relevance to adsorption processes in μ-Na*fl* systems. The linear isotherm is generally not very common unless penetration to the underlying substrate occurs. The Langmuir and Freundlich isotherms are much more common for surface–solution adsorption processes. C_S denotes the surface (also called excess) concentration, and C is the concentration of the adsorbing species in bulk solution.

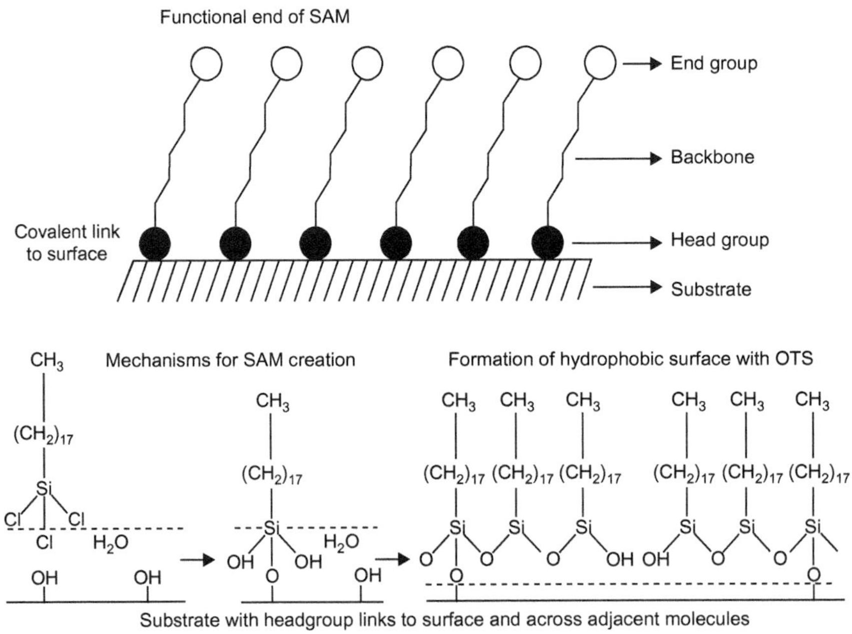

FIGURE 3.10

Schematic showing the structure for SAM and the mechanisms thought to be important for formation of the SAM. The importance of surface-bound water is also depicted in the figure.

Source : *[1] Prakash S. Karacor MB, Banerjee S. Surface modification in microsystems and nanosystems.*

structure is represented in Figure 3.10 with the mechanisms for formation of the SAMs also depicted pictorially. One of the outstanding questions in the formation of silane-based monolayers is the role played by water during the surface reactions. It is believed that a small amount (a few ppm) of water is needed to form monolayers, while excess water encourages cross-linking and formation of poor quality surface-bound monolayers.

The monolayer formation in turn leads to changes in the chemical composition of the surface in contrast to the bulk leading to an alteration of the surface chemical potential. The formation of the monolayers from the solution phase can be explained by the transient Langmuir adsorption isotherm, now expressed as a rate controlling step in an adsorption process,

$$\frac{d\theta}{dt} = \frac{k_a}{N_0}c(1-\theta) - \frac{k_d}{N_0}\theta \tag{3.11}$$

where θ is the fractional surface coverage, t is the adsorption time, k_a and k_d are the adsorption and desorption rate constants, respectively, c is the concentration of the adsorbate in the solution phase, and N_0 is the surface adsorbate concentration at full coverage. Integrating Eq. (3.11) with respect to time gives,

$$\theta = \frac{k_a c}{k_a c + k_d}\left[1 - \exp\left\{-\frac{k_a}{N_0}\left(c + \frac{k_d}{k_a}\right)\right\}t\right] \tag{3.12}$$

As discussed above, in practice most adsorption occurs from dilute solutions of the adsorbate to form defect-free monolayers, which suggests from Eqs. (3.11) and (3.12) that the rate of surface coverage should be slow for formation of high-quality surface layers. Formation of defect-free monolayers relates to allowing sufficient time for entropic processes to assist in the formation of dense, defect-free monolayers binding to a high density of available surface sites. In practice, the formation of defect-free thiol-based monolayers on gold surfaces uses this approach of allowing long deposition time for formation of monolayers from thiol solutions. Similarly, equations of state for formation of monolayers from the gas phase have also been derived and are extensively discussed in Lyklema's second volume on interfaces and colloidal science book series [2]. The equations of state depend on the chemical potentials of the surface and adsorbate and have been derived as shown previously.

Several investigations have been carried out to determine the mechanisms for formation of covalently bound monolayers on surfaces. Three main mechanisms have been identified. First, in island growth, the surface-active species adsorb to surfaces and form a vertical stack. As the growth process proceeds, these individual stacks begin to coalesce to form "islands" which in turn grow with increasing deposition time to form the surface layer. The second mechanism described in the literature is uniform growth, the surface-active molecules randomly adsorb to the surface and with increasing deposition time the molecules begin packing together. It is likely that the surface reorganization of the molecules to form the monolayer is driven by entropic minimization. Finally, the third mechanism proposed is the

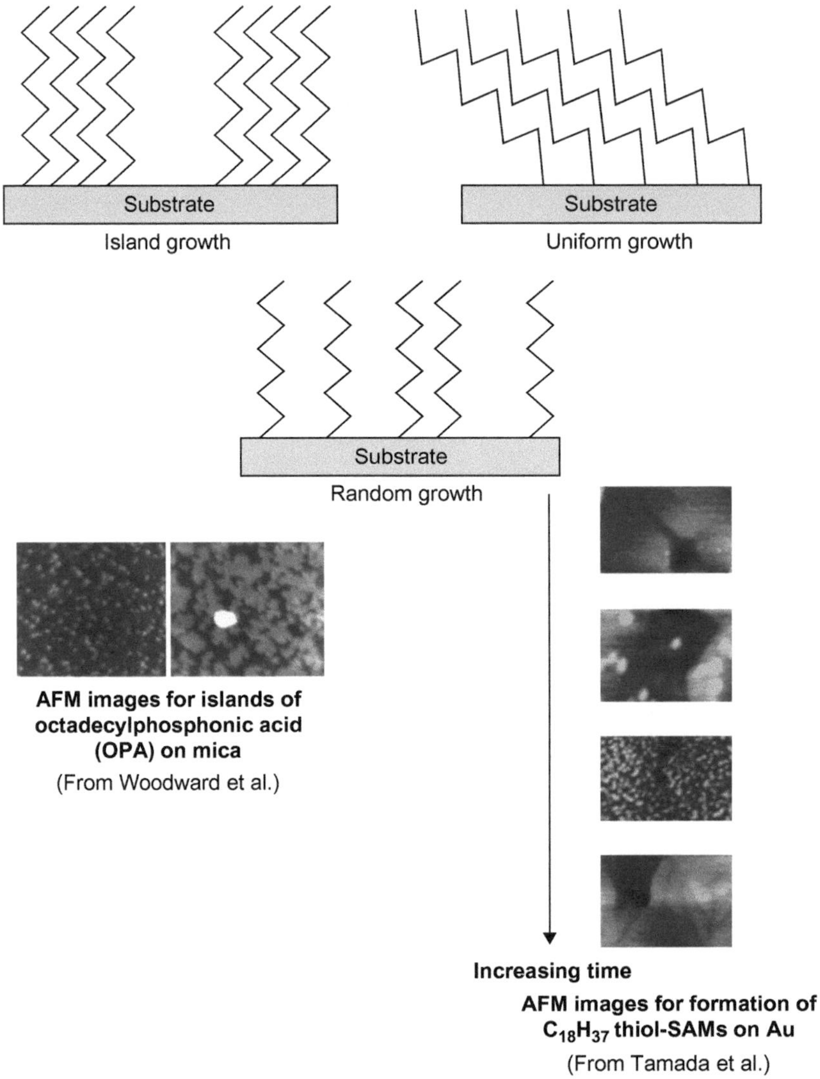

FIGURE 3.11

The figure depicts the three common mechanisms for formation of SAMs.

Source: *[2] Lyklema J.Fundamentals of interface and colloid science. Solidliquid interface, vol.2.*

random growth model in which the surface-active species randomly lay down on the surface and do not reorganize. It is likely that this form of monolayer formation leads to defects within the monolayer structure. A pictorial representation of these three mechanisms is presented in Figure 3.11.

Given the complex mechanisms of surface-layer formation and the multitude of parameters and physical conditions like substrate type, concentration of surface-active species, deposition time, and deposition temperature, it is not surprising that few well-defined protocols exist for forming high-quality surface layers. Most researchers develop their own recipes to form surface layers suited for their study. For example, choice of substrates also governs the methodology for characterizing the modified surfaces.

3.5.2 Surface modification methods

Surface modification methods can be divided in two broad categories: physical and chemical methods. The definition of these broad categories depends on how the process actually affects the surface. Physical methods, in most cases, do not change the chemical composition of the surface. These methods may change the surface roughness, grain sizes and grain boundaries, and faceting. Physical methods often relate to use of lasers, plasmas, temperature or thermal gradients, heavy metal ion beams, and polishing and grinding to alter the surface state of a material of interest. For example, Figure 3.12 shows the faceting that develops on an alumina surface due to thermal treatments under different conditions.

Chemical methods are often classified as such because these methods introduce a change in the eventual chemistry or chemical composition at the surface of a material. Consequently, the surface may possess chemical properties that are different from the bulk material as discussed in section 3.2. Among the various chemical surface modification methods, additive processes that permit formation of surface layers, either covalently bonded or physisorbed, are most common. The implication is that additional material with thickness on the order of a few angstroms (one to two atomic layers) to a few nanometers (may be a few molecular layers) is added to the bulk material or underlying substrate. Other chemical methods include treatment with UV light and reactive plasmas. These changes generally introduce a change in the eventual surface charge density or the surface energy, which provides a measure of the relative hydrophilicity or hydrophobicity of a surface. Next, we present a discussion of modifying surfaces using both physical and chemical methods by using several illustrative examples.

3.5.2.1 Physical methods

One of the oldest methods of modifying the physical characteristics of surfaces is by polishing or grinding the surface. A simple process by using an abrasive material such as sandpaper to alter the surface roughness can be found dating back to ancient times where hard objects such as small pebbles and shells were attached to bark, animal skin, or paper by using gum resin, creating ancient sandpaper. Modern advances in this simple technology appeared in the early 1900s with 3 M formally patenting and developing the various coarse- and fine-grain sandpapers commercially available today.

Another surface treatment and modification method is that of thermal treatments. Temperature gradients and thermal treatments have often been used to change surface roughness and alter the grain sizes and grain boundaries. In the past few years, thermal methods have been employed to create nanoscale features, facets,

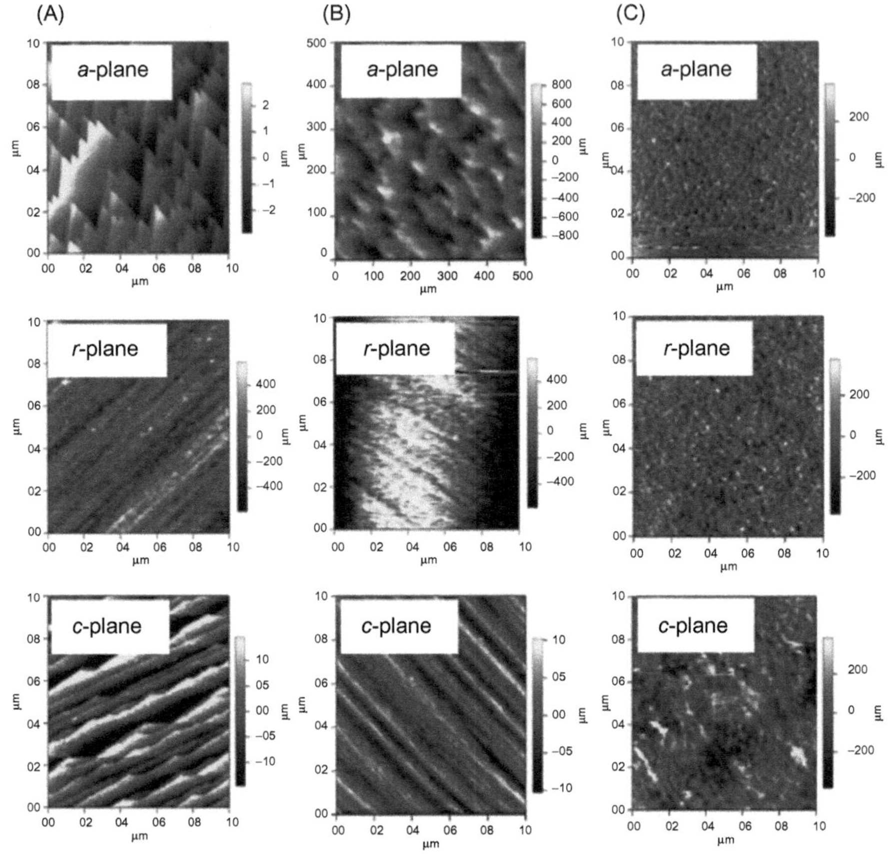

FIGURE 3.12

Tapping mode AFM images depicting structural changes to *a*, *r*, and *c* planes of α-Al_2O_3 as function of temperature and atmosphere for thermal exposure. (A) treated at 1550°C in H_2/He and O_2 mixture; (B) treated at 1550°C in Ar/O_2; and (C) treated at 1000°C in H_2O/O_2 mixture.

Source: *[7] Chandrasekharan R, et al. High temperature hydroxylation of alumina crystalline surfaces.*

textures, and nanoparticles, on a variety of material surfaces including ceramics, metals, polymers, and semiconductors sometimes in the presence of adsorbates. Thermal treatments in the presence of common gases such as oxygen or water vapor can be powerful tools for modifying existing surfaces through creation of steps or inducing other forms of nanostructures. For example, it was shown that thermal processing of crystalline α-Al_2O_3 surfaces with orientations (1 1 2 0), (1 −1 0 2), and (0 0 0 1) which were heat treated at 1500°C in Ar/O_2 and H_2/He/O_2 led to step formation and roughening as quantified through atomic force microscopy (AFM) topology images (Figure 3.12), but the effects of step formation and roughening were much less significant on samples heat treated at 1000°C in moist O_2.

3.5.2.2 Chemical methods

Altering the surface chemistry by wet or dry processes is the most common methodology used for chemical modification of surfaces. The processes are so named because of the processing methods and conditions, as discussed next. Modification schemes are governed by a wide range of parameters including sample type (polymers, metals, ceramics, etc.), stability to treatment conditions (e.g., thermal or structural), and eventual applications. For example, polymeric surfaces are often modified by photochemical methods of which UV irradiation in air, other reactive atmospheres such as ozone, combined with lasers or lamps, and grafting surface layers is fairly common. Efficacy of UV irradiation is primarily governed by the ability of a material to absorb UV light at a given wavelength, and the depth of modification is determined by the extinction coefficient of the material with the intensity variation given by the Beer–Lambert law (which predicts an exponential decay with depth). In addition, the mechanism for surface modification is due to photo-initiated cross-linking or bond scission on polymeric surfaces.

Among the dry surface modification methods use of reactive plasmas has been gaining popularity, most likely due the wide compatibility of materials and integration to microfabrication processes for device development. Many different types of gas plasmas have been cited in the literature including air, oxygen, H_2O, ammonia, and argon for modification of polymer surfaces. Figure 3.13 shows an example of treating poly(methyl methacrylate) of PMMA with water vapor plasma for activation toward formation of SAMs.

Note from Figure 3.13 that roughening due to plasma etching does not occur till nearly 150 W plasma power and increases further at 200 W with complete details discussed in the work by Long et al. [8]. Figure 3.14 shows another example of a SAM-functionalized cantilever used for sensing applications. Details on the discussions for sensors are in Chapters 5 and 6.

Plasma modification methods have their own sets of advantages and disadvantages. The biggest advantage is probably that surfaces are modified uniformly and the modification is limited to a few nanometers in depth without affecting the bulk material. The main disadvantage is probably the use of vacuum equipment with system parameters that can vary over different systems adding to cost of operation and developing optimal recipes for surface modification. A detailed discussion of the various applications and advantages and disadvantages can be found, for example, in existing literature and the reader is referred to the extensive bibliography at the end of this chapter for several references. Generally, plasma modification processes generate new chemical species on polymer surfaces. The new chemical species can arise due to surface reactions with reactive gases or due to physical sputtering (such as with Ar plasma) caused by active gas-phase species. These new surface chemical species can provide an anchor for attaching a series of different molecules that display different properties from the underlying bulk polymer. In fact, peroxides generated on polymer surfaces have been utilized for radical-based graft polymerization of methacrylates and

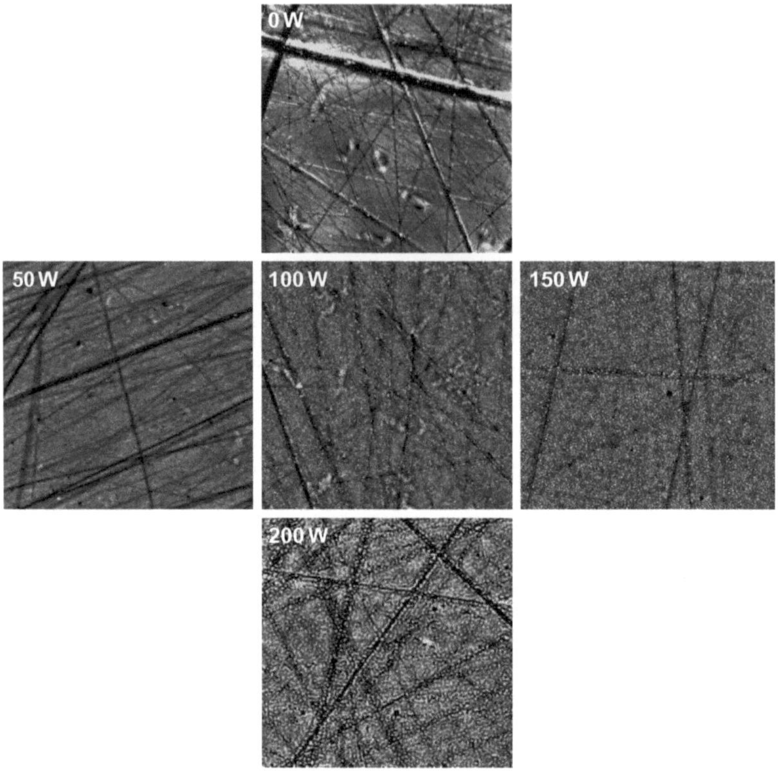

FIGURE 3.13

AFM images depicting structural changes to plasma treated PMMA as function of plasma power under treatment with water vapor plasma as a function of plasma power.

acrylamides. Furthermore, preferential hydroxylation of PMMA surfaces has been used to activate surfaces toward trichlorosilane modification and subsequent "click" chemistries to form surface scaffolds of desired functionalities. Table 3.1 lists examples of functionalized PMMA surfaces and the related contact angles.

Other methods have also been used for chemical surface modification. For example, use of reactive gases in the presence of photons through UV light and other light sources (e.g., lasers) has also been sometimes used to initiate radical polymerization on surfaces for formation of adherent polymer layers. These photochemical methods are generally clubbed with plasma methods. Photochemical surface modification has found varied applications including UV-hardening, micro- and nanofabrication, development of biosensors and biomolecule arrays, and surface grafting of various chemically functional polymeric layers.

In contrast to the "dry" methods that rely on gas-phase systems for surface modification, "wet" methods relying on "traditional beaker chemistry" have also been extensively used for surface modification. Wet chemical methods often rely on the

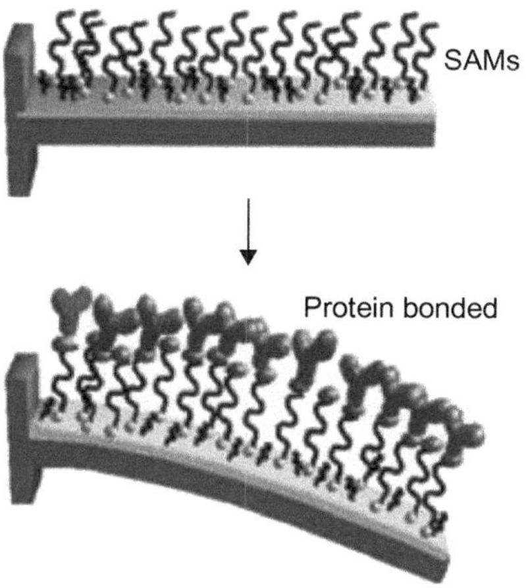

FIGURE 3.14

Functionalized microcantilevers bend when a protein is bonded to the SAM layer, allowing for label-free sensing.

Source: *[9] Serafina C ,Damiana C, Stefania C, et al. Microcantilevers and organic transistors: two promising classes of label-free biosensing devices which can be integrated in electronic circuits.*

Table 3.1 Summary of Contact Angles for Functionalized PMMA Surfaces*

Sample	Contact Angle (°)
PMMA	64
Br-PMMA	85
OTS-PMMA	104
F-PMMA	109
N_3-PMMA	75
PEG-PMMA	45

N_3 = azide group termination, PEG = polyethylene glycol, DTS = methyl terminated surfaces functionalized with octadecyltrichlorosilane, Br = bromo terminated surfaces, F = fluro terminated surfaces.

principles of organic or inorganic chemistry for formation new surface chemical functionalities. For example, a PMMA surface primarily consists of methyl ester groups, these groups can undergo reduction to alcohols with lithium aluminum hydride in ether. Alternatively, amino functionalities may also be placed on the PMMA surface through aminolysis of the ester groups by treatment with a solution of *N*-lithiodiaminopropane in cyclohexane; this aminated surface could then be reacted with a substituted isocyanate for further surface reactions. The interested reader is directed to a review article in 2009 by Prakash et al. [1] for more details and additional references.

3.6 Surface characterization methods with relevance to microfluidics and nanofluidics

Once a surface has been modified it needs to be characterized in order to (i) confirm successful and desired modification, (ii) quantify the modified surface states, and (iii) evaluate the effect of the modified surface on the interactions with the surroundings. In this section, we briefly review some of the existing surface analysis techniques.

The choice of surface analysis method is a function of the desired information about the surface state. Surface analysis and probing methods can be divided into two broad categories: (i) indirect methods and (ii) direct methods. These methods are so classified because data from indirect methods often provides a measure of a physical surface quantity that correlates to a specific surface state. For example, contact angle measurements provide a measure of the interfacial force between a liquid drop and the surface of interest. Contact angle data must be analyzed further to extract quantitative information on an actual surface state or relevant parameter such as the surface tension or surface energy. However, a technique such as X-ray photoelectron spectroscopy (XPS) provides direct verification of the chemical composition on the surface of interest. It must be noted here that most surface characterization techniques work well with flat substrates or open interface materials. Furthermore, many of the high resolution techniques were initially developed to work under high vacuum in a controlled environment. Characterization of surface states within confined micro- and nanoscale devices is a challenging proposition and continues to be an active research area.

3.6.1 Indirect methods

Contact angle (CA) measurements provide quantitative data on the interfacial energy between a liquid drop and a solid surface. Once the contact angle is measured the interfacial energies can be calculated from the Young–Laplace equation, as described in Eq. (3.1). For complete data, the advancing, receding, and static contact angles must be measured. The most common liquid used is water; however, a variety of liquids have been used to determine interfacial energies for either quantifying the modified surface state or for applications such as transfer printing. Contact angle data has also been used to obtain a measure of the surface pK_a values, which can be critical in many μ-Na*fl* systems since the surface pK_a determines the dissociation of the surface-bound chemical functionalities and subsequently the electrostatic interaction between the device walls and the confined electrolyte solutions. While CA measurements are relatively quick and can be accomplished with simple equipment, they are prone to uncertainties due to changes in water quality, temperature, and relative humidity during measurements.

The discussions presented above for CA measurements and use of Young's equation usually work well for flat, smooth surfaces. However, in most practical applications or within devices surfaces would have finite roughness (see Chapter 4 for examples of channel and surface morphology following device fabrication procedures) or may present a heterogeneous environment to the liquid droplet due to presence of multiple materials or liquid–gas phases at the interface. Under such circumstances, further analysis may be required. In a 2011 review, Bhushan and Jung discuss surfaces for superhydrophobicity, self-cleaning, low adhesion, and drag reduction with specific examples on learning from nature [11]. The underlying concepts of surface structure and subsequent interactions apply to our overall discussion as well, keeping in mind our discussions for surface charge and the related concepts in Chapter 2.

Let us begin by considering a rough surface with a surface roughness factor, $S_R > 1$. The surface roughness factor is defined as the ratio of the surface area of the actual surface to that of a flat surface projected area. The contact angle, $\theta_{c,R}$, for a rough surface can then be expressed in terms of S_R and the ideal contact angle, θ_c, and is referred to as the Wenzel equation.

$$\cos \theta_{c,R} = S_R \cos \theta_c \tag{3.13}$$

The implication from Eq. (3.13) is that for a hydrophobic surface, increasing roughness will enhance the effect of hydrophobicity and the measured contact angle will be higher than that for a flat surface of the material. Along a similar vein of discussion, if a surface is composed of two materials or phases with phase I having a fractional area A_{F1} and phase II having a fractional area A_{F2} with contact area on each phase being θ_{F1} and θ_{F2}, respectively, then the Cassie equations tell us that,

$$\cos \theta_{c,R} = A_{F1} \cos \theta_{F1} + A_{F2} \cos \theta_{F2} \tag{3.14}$$

Using the definition for S_R and combining with the Cassie equation gives the Cassie–Baxter equation,

$$\cos \theta_{c,R} = A_{F1} S_R \cos \theta_c - 1 + A_{F1} \tag{3.15}$$

where, $A_{F1} + A_{F2} = 1$. In Eq. (3.15), we inherently assumed that the composite interface is organized such that $\theta_{F1} = \theta_c$. A pictorial representation of the concepts for the Wenzel and Cassie (and Cassie–Baxter) cases is presented in Figure 3.15.

The idea of using a droplet configuration to characterize surfaces has been extended to a variety of cases. For example, surfaces modified with bioactive compounds are often characterized by colorimetric dye assays or by evaluating specific binding events to study the role of surface immobilized biological species. The dye assays are coupled to optical absorbance measurements to quantify the interaction of the solution-phase dyes with the surface-bound biological moieties. The dye assay methods are plagued with problems of sensitive surface detection as the dye interactions with surfaces are determined by electrostatic forces or surfaces presenting hierarchical structures, which can strongly affect the wetting and consequently the surface binding of molecules. Other examples have relied on using contact angle titrations to extract surface properties such as surface pK_a values.

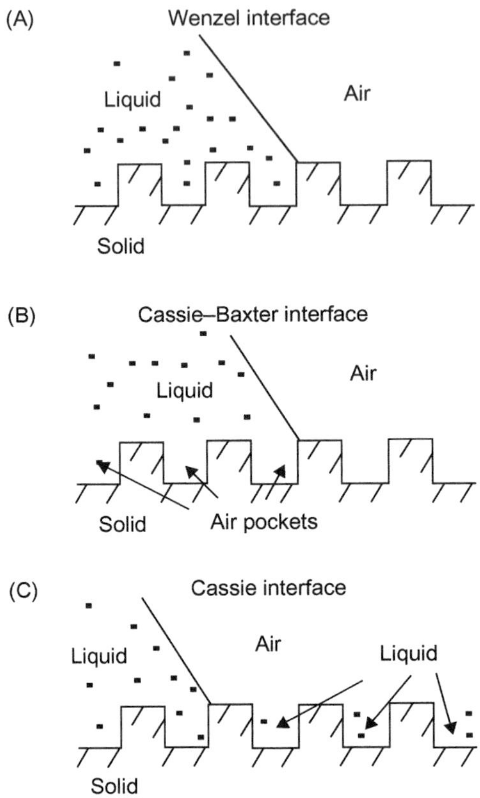

FIGURE 3.15

Schematic showing the configurations described by the (A) Wenzel equation for the homogeneous interface, (B) Cassie–Baxter equation for the composite interface with air pockets, and (C) the Cassie equation for the homogeneous interface.

Source: *[10] Bhushan B, Jung YC. Natural and biomimetic artificial surfaces for superhydrophobicity, self-cleaning, low adhesion, and drag reduction.*

The structure of surfaces has long intrigued scientists and engineers for better understanding of droplet–surface interactions and eventual control of droplet motion on surfaces. Figure 3.16 shows a schematic for how water droplets may sit on different surfaces as a function of surface structure.

Nature gives us several examples of hierarchical structures such as those on a lotus leaf, which is being studied extensively for the development of self-cleaning surfaces due to the superhydrophobic characteristics. Figure 3.17 shows a droplet on a lotus leaf with SEM images that present details of the leaf microstructure including the presence of nanotubules.

While surface measurements such as those discussed above are very useful, μ-Na*fl* systems are often characterized by flows in confined spaces. Therefore, for

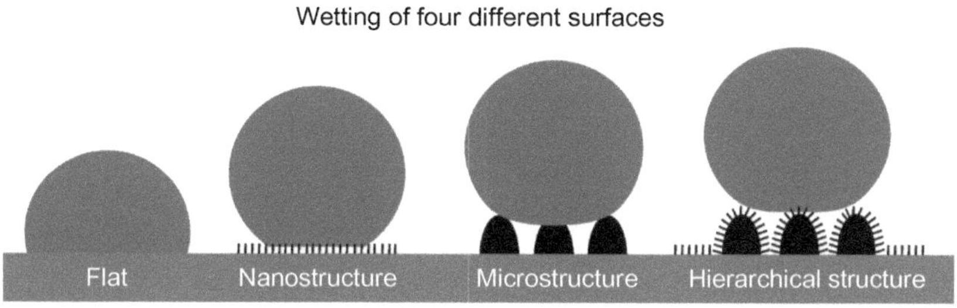

FIGURE 3.16

The schematic shows the wetting characteristics on a surface as a function of surface physical structure.

Source: *Image courtesy: Prof. B. Bhushan, reproduced with permission.*

FIGURE 3.17

The scanning electron microscopy (SEM) images show the microstructure of a lotus leaf with the presence of nanotubules. The digital photograph shows the droplet on the surface.

Source: *Image courtesy: Prof. B. Bhushan, reproduced with permission.*

such systems, one of the more important surface characterization methods with enormous practical implications for μ-Na*fl* systems relies on use of electrokinetic measurements. As discussed in Chapter 2, at a finite distance from the surface the slip plane exists, where a phenomenological quantity, the ζ (zeta) potential, is defined. ζ potential can be measured electrokinetically by either the current monitoring or the slope method in a flow driven by an applied potential. Alternatively, it can also be measured by the streaming potential method.

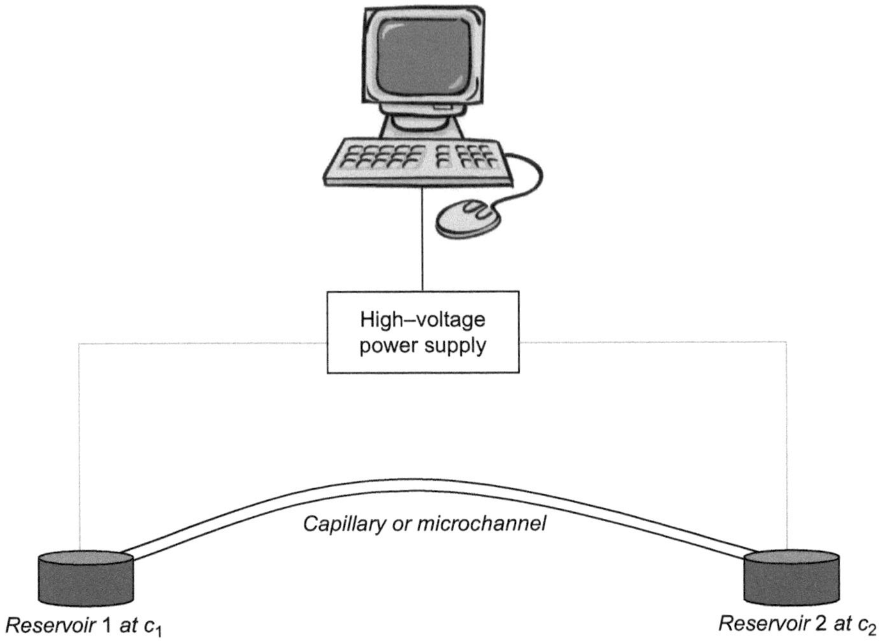

FIGURE 3.18

Schematic depiction of the current monitoring experimental setup to measure the electroosmotic flows in microchannels.

All three approaches are discussed next. In the current monitoring method, the microchannel, capillary, or nanochannel connects two reservoirs at different electrolyte concentrations. Figure 3.18 shows a schematic for the current monitoring experiments work. The channel is filled with one of the solutions and the electric potential drives the other solution to replace the solution in the channel. As electrolyte moves through the channel, a current time data set is generated and recorded, generally using a computer-controlled data acquisition setup. In our chapter available online, we discuss several tips and additional details for the current measurement method initially reported by Huang and Zare (1988; see Select bibliography) for capillaries.

In Figure 3.19 we see a schematic for the type of current time plot that may be generated. Note the differences in the ideal and realistic or measured current profiles, with the region possibly introducing errors marked in Figure 3.19B. From this data, the transition from a rising current (more conducting solution of c_1 and c_2 filling the channel) to a steady-state current (channel completely filled with the new solution) is identified within time Δt, and the electromigration velocity, u_{avg}, for thin EDLs can be calculated if the length of the channel is L,

$$u_{avg} = \frac{L}{\Delta t} \tag{3.16}$$

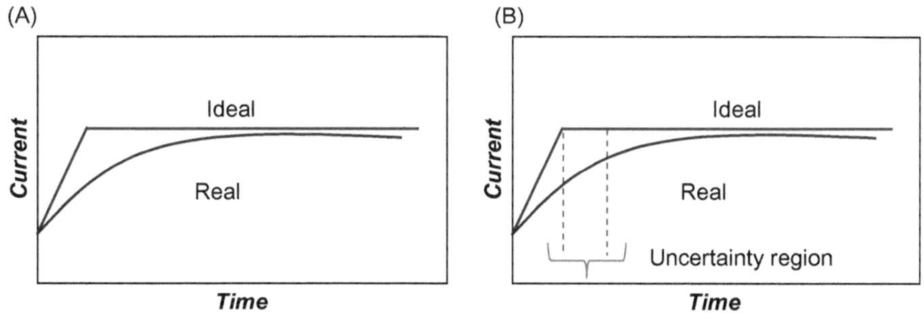

FIGURE 3.19

Schematic depiction of the current time data that is generally generated by using the current monitoring method.

Recall from Chapter 2 Eq. (2.42) that for thin EDLs under the Helmholtz–Smoluchowski approximation, the velocity is proportional to the ζ potential. Therefore, combining Eqs. (2.42) and (3.16), we get

$$\zeta = \frac{\eta L}{\varepsilon E_x \Delta t} \tag{3.17}$$

Furthermore, since it must be noted that the ζ potential is a phenomenological quantity that relates to the surface charge density for microscale systems ζ potential is a useful concept as it can be correlated to the surface charge density by the expression,

$$\sinh\left(-\frac{e\zeta}{2k_{\mathrm{B}}T}\right) = \frac{\sigma_{\mathrm{s}}\lambda_{\mathrm{D}}e}{2\varepsilon k_{\mathrm{B}}T} \tag{3.18}$$

It can be seen from Eq. (3.18) that surface charge σ_{s} and the ζ are related through a non-linear functional relationship. Furthermore, it has been shown that ζ potential can be used to quantify the isoelectric points for a surface thereby providing a means to quantify changes in ionizable groups on surfaces. Consequently, electrokinetic measurements such as the current monitoring method can provide one tool to relate surface properties to measurable flow data.

Another method that can be used is the slope method, which provides improved accuracy over the current monitoring method. This improved accuracy arises largely due to reduced noise in the ohmic region and not relying on the "mixing zone" as one electrolyte solution replaces another in the microchannel. By working within the ohmic region (as shown in Figure 3.20), effects of Joule heating can also be avoided. The slope method usually works well for small differences in c_1 and c_2 ($\sim$10% or less).

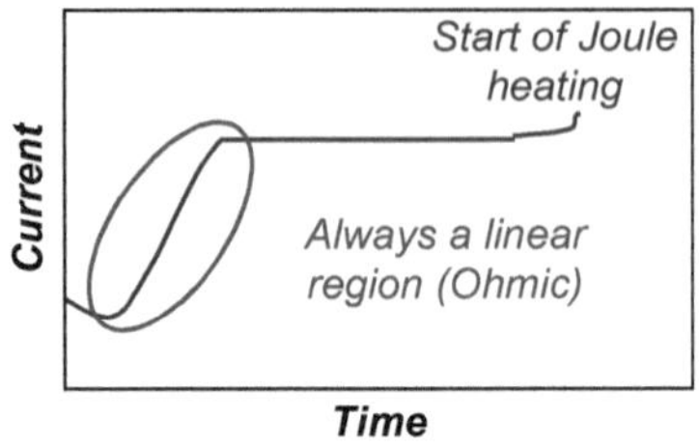

FIGURE 3.20

Schematic depiction of the current time data that is generally generated by using the slope method.

The slope, S, can be calculated based on the recorded data and is expressed as

$$S = \frac{\Delta I}{\Delta t} = u_{\text{avg}} \frac{E_x A_c (\Lambda_2 - \Lambda_1)}{L} \tag{3.19}$$

Λ_i denotes the conductivity of the electrolyte i at a given concentration and temperature.

A third method relies on the principles of streaming potential. When a pressure difference between the inlet and the outlet of the microchannel (or nanochannel) drives an electrolyte through the channel, the excess counterions (i.e., those not shielding the surface charge) in the diffuse layer are transported downstream, leading to charge transport or current flow. Consequently, a measurable potential difference sets up between the locations of the applied pressure and is called the streaming potential, while the associated current is called the streaming current. Typically, in microchannels, the wall potential is considered to be the ζ potential and therefore the streaming potential, E_s, can be related to the pressure gradient, Δp, as,

$$\frac{E_s}{\Delta p} = \frac{\varepsilon \varepsilon_0 \zeta}{\eta \Lambda_s} \tag{3.20}$$

where, Λ_s is the solution conductivity. The implication of Eq. (3.20) is that if a known pressure gradient exists across a microchannel, then the streaming potential can be measured. Furthermore, as properties of the solution are known, the ζ potential can be calculated. Note that Eq. (3.20) does not have an explicit dependence on the channel geometry. Consequently, this method has also been used to determine ζ potential of particles in a flow. In his book Li has discussed the use of a correction term, Φ, to account for the EDL. Specifically, Φ is defined as,

$$\Phi = \frac{\beta_1}{1 + ((\beta_2 \mathrm{K}^2 \varepsilon^2 \varepsilon_0^2 \zeta^2)/(\sinh^2(\mathrm{K}) a^2 \eta \Lambda_s))} \tag{3.21}$$

where, $K = a/\lambda_D$, and the parameters β_1 and β_2 are defined as,

$$\beta_1 = 1 - \frac{\cosh(K) - 1}{K\sinh(K)}; \quad \beta_2 = \frac{\sinh(K)\cosh(K)}{2K} + \frac{1}{2} \tag{3.22}$$

Despite the usefulness of ζ potential measurements one of the drawbacks is the lack of information on the number density or chemical nature of surface functional (or ionizable) groups. This drawback is due to the inherent measurement technique which is an average measurement over the entire surface of the channel being probed. Therefore, direct methods are often needed to quantify the chemical nature and surface coverage on modified or functionalized surfaces. Next, we discuss a few direct methods that have been commonly used to quantify details on the surface state.

3.6.2 Direct methods

As discussed above, direct methods are so named because they provide a direct, quantifiable measure of the surface states. Some direct methods are spectroscopic in nature and provide detailed structural and chemical information on the surface states.

The most common direct method used for surface analysis is XPS. Sometimes XPS is also called ESCA (electron spectroscopy for chemical analysis), although the use of this latter term is declining. XPS is based on the principles of the photoelectric effect. An X-ray beam is used to excite photoelectron emission from a small solid volume near the surface ($\sim$10 nm deep) (see Figure 3.21). The kinetic energy of the emitted electrons is measured and related to the binding energy. The binding energy provides a unique signature for the chemical composition of the surface. XPS can also be used for quantitative determination of the relative concentrations of chemical species in addition to the capability of modern instruments to create chemical image maps. There are two main drawbacks of XPS. First, samples must be handled carefully as the sensitive nature of the technique makes surface contamination an issue as even small levels of contamination show up in measured spectra. Second, the use of high-energy X-ray beams and ultra-high vacuum (UHV) equipment limits the type of samples that can be used, since soft samples or materials likely to out-gas pose problems in measurements conducted in ultra-high vacuum (UHV) environment. However, despite these challenges, with careful and detailed sample preparation procedures, studies have been conducted on microbial cell surfaces.

As an example, consider the surface diagnostics using an XPS for formation of azido-terminated monolayers on glass or silicon substrates with spectra shown in Figure 3.22. Note that the characteristic surface azide double-peak structure is easily detected on silicon substrates but not on glass substrates due to degradation of the surface layers under the highly energetic X-ray beams. If azido-terminated monolayers are formed on glass substrates for subsequent "click" modifications, additional characterization by Fourier transform infrared (FTIR) may be required as the azido layers undergo faster degradation on glass surfaces in contrast to silicon substrates and thus making confirmation of successful surface modification harder.

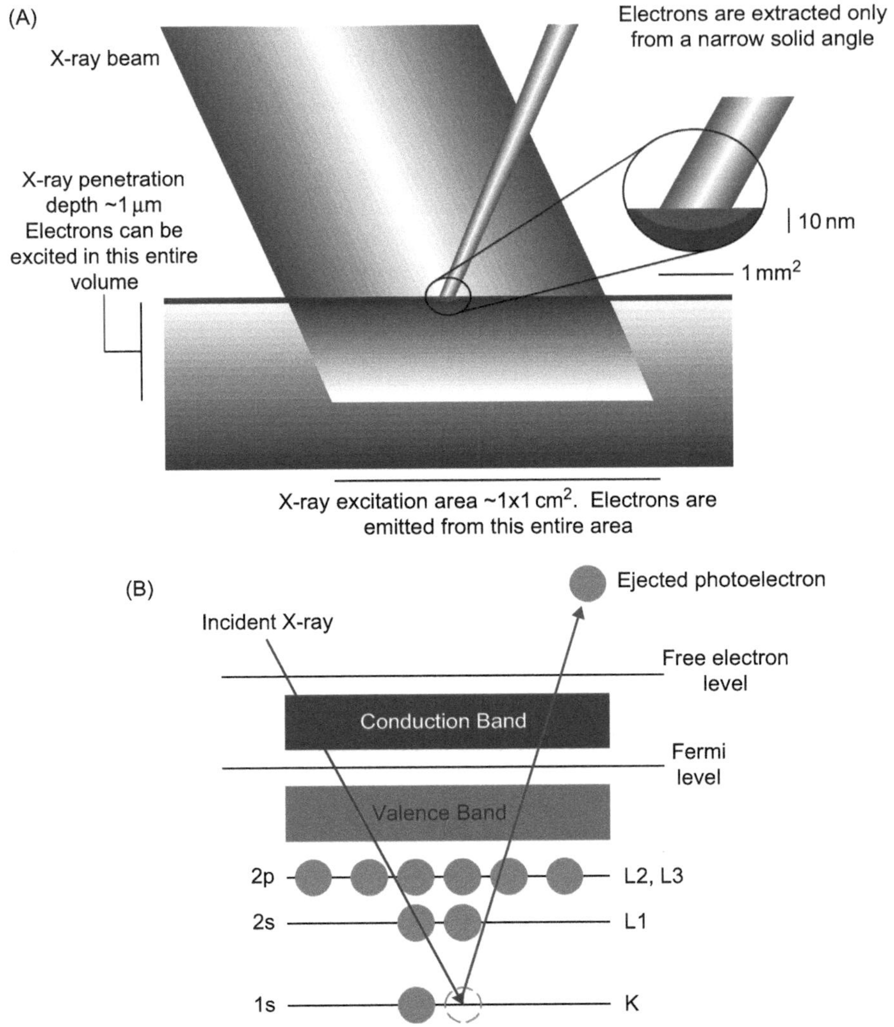

FIGURE 3.21

A schematic depiction of the working principle for XPS is shown in (A) and (B) shows the process of generation of the photoelectron. The XPS course offered through the F. Seitz Materials Research Laboratory at the University of Illinois at Urbana-Champaign is acknowledged for the content described in the figures.

A complementary technique used with XPS is FTIR-attenuated total reflection (FTIR-ATR) spectroscopy. FTIR by itself is commonly used in the transmission mode to identify the chemical bonds present in a compound. For most surface analysis IR beams can penetrate to several millimeters within the substrate and therefore the technique is not very surface sensitive. However, by using a high-

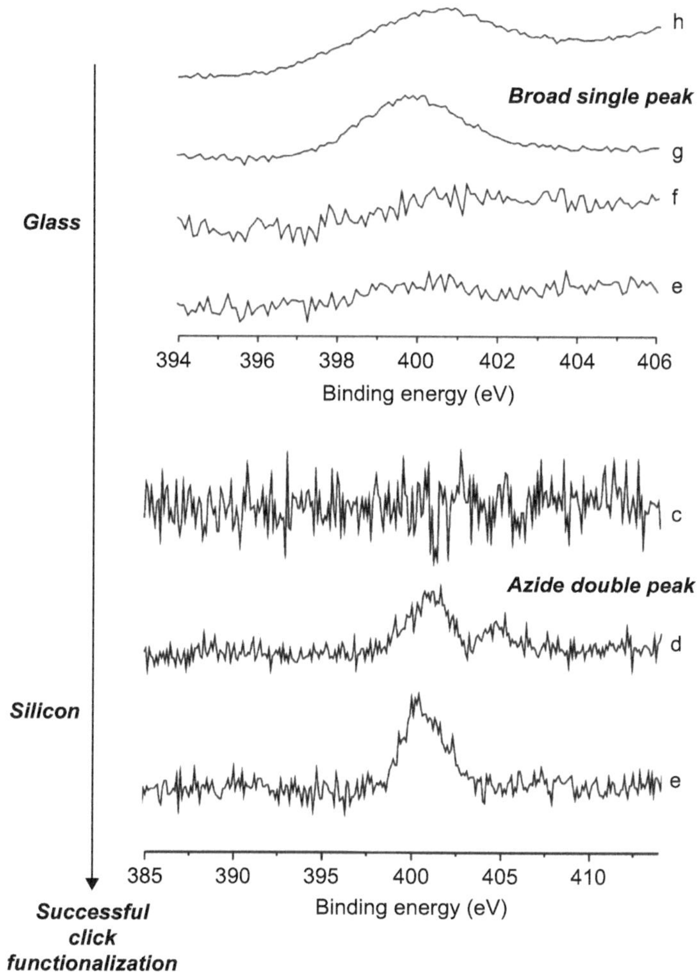

FIGURE 3.22

XPS data for surface modification for formation of different chemically functional layers. Specifically, the figure compares azido-group layers on glass and silicon surfaces for subsequent click functionalization.

Source: *[1] Prakash S, Karacor MB, Banerjee S. Surface modification in microsystems and nanosystems.*

refractive-index crystal and operating in the ATR mode, penetration of the IR beams can be limited to a few micrometers and surface states can be probed with appropriate background subtraction. Many different types of ATR crystals are available including ZnSe, diamond, and Ge. Each crystal has a specific window for IR beam transmission and therefore the resolution is limited by choice of the ATR crystal. The main advantage of FTIR-ATR over XPS is that equipment does not require UHV environments and a much broader range of samples can be used. FTIR-ATR has been used to quantify chemical bonding states and surface

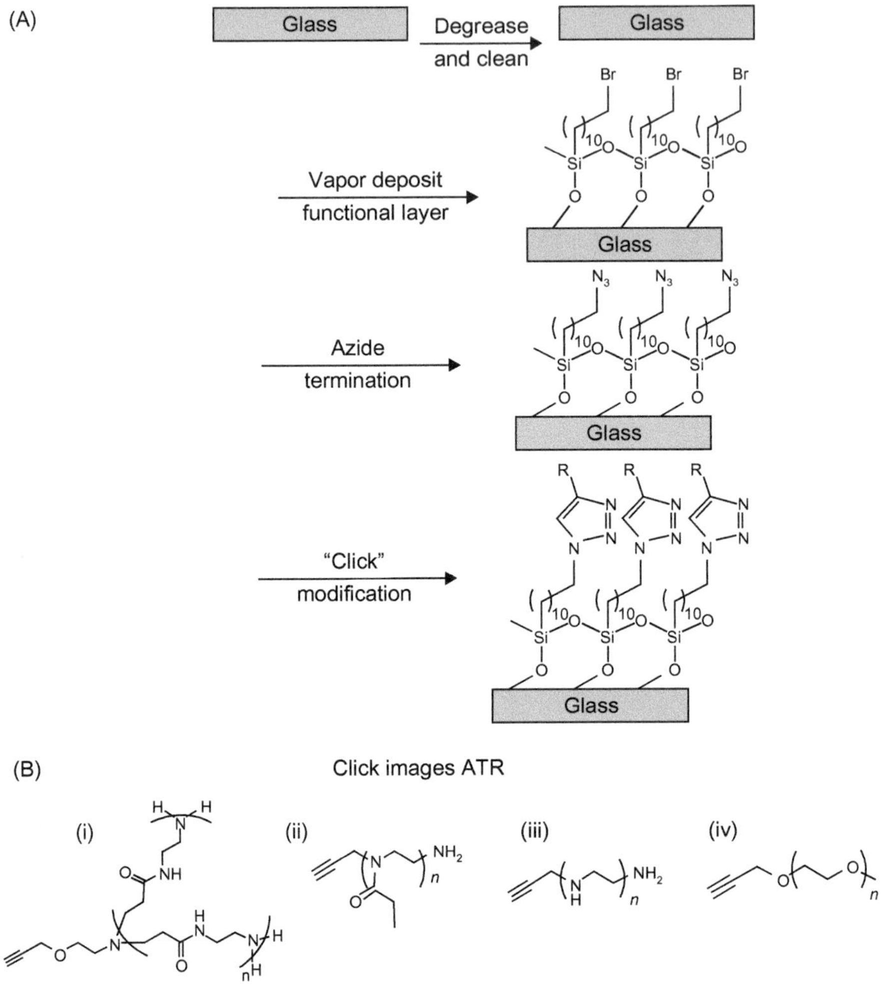

FIGURE 3.23

(A) shows the reaction scheme followed for "click" functionalization of surfaces with some key functional groups also shown. (B) shows the terminal alkynes used for click modification and (C) shows spectra for the PAMAM dendrimers after "click" functionalization of silica surfaces. Note presence of the azido peak as the generation of the dendrimers increases.

Source: *[11] Prakash S, et al. 'Click' modification of silica surfaces and glass microfluidic channels*

coverage of monolayers on silica and glass surfaces, structure of water at surfaces in the presence of an electric field, and the influence of surface defects. Several other review articles describe the theory and use of IR techniques for surface characterization in greater detail and are listed in the chapter bibliography for the interested reader. Figure 3.23A shows a schematic depiction for surfaces by

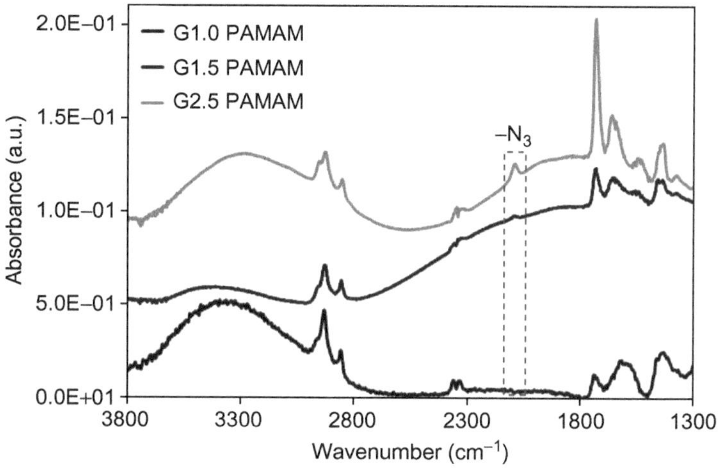

FIGURE 3.23

Continued

different "click" functionalization on glass substrates as an illustrative example of using chemical surface modification, the FTIR-ATR technique was used to characterize chemical bonding states on surfaces to characterize functional groups shown in Figure 3.23B, and the spectra are shown in Figure 3.23C.

Among other direct characterization methods, mass spectrometry is also quite popular. Several mass spectrometry techniques exist and are useful for surface information. As an example, we will discuss time-of-flight-secondary ion mass spectrometry (ToF-SIMS). ToF-SIMS is used to determine the chemical functionality and coverage or distribution of ionizable species on a surface. A high-energy ion beam (commonly Ga^+ or Au^+) is used to irradiate a surface which in turn emits secondary ions. These surface-emitted ions can be separated in a mass spectrometer according to the charge-to-mass ratio (q/m). Since the q/m ratio is unique for a given atom, ToF-SIMS provides a unique signature for the chemical functionality. In addition, ToF-SIMS can be used to create chemical images of a surface in order to identify the coverage density of the surface layers. For example, Figure 3.24 shows ToF-SIMS images for a —Br functionalized PMMA surface.

The chemical contrast in the underlying PMMA substrate and a —Br terminated monolayer helps create the chemical image. The main advantage of ToF-SIMS over XPS is the extremely high sensitivity (on order of ppb). However, the disadvantages of working with UHV and high-energy ion beams (~10 keV or higher, 30 keV is not uncommon) are similar to those for XPS.

Several other surface analysis techniques exist that rely on UHV systems and can be used to probe surface states including providing depth profile information on the chemical composition starting at the surface and toward the bulk substrate, for example, electron energy loss spectroscopy, Auger electron spectroscopy,

FIGURE 3.24

ToF-SIMS data for bromo functionalized PMMA surface. The figure suggests a high coverage density for –Br terminated SAM on a PMMA surface.

Source: *[8] Long TM ,et al. Water-vapor plasma-based surface activation for trichlorosilane modification of PMMA*

Rutherford back scattering, and dispersive X-ray methods (EDX or EDS). However, due to space constraints those are not discussed in detail here but the reader is referred to the Select bibliography for additional information.

In addition to the spectroscopic methods discussed above, there are microscopy techniques that can be used to quantify surface states. For example, atomic force microscopy (AFM) has been used to measure surface charge, surface energy, surface roughness, work of adhesion and can also be used to measure slip-lengths. The measurement of slip-lengths can be a useful parameter for the development of nanofluidic systems in order to relate the confined transport phenomena to the actual surface state. In recent work by Wu et al. [13] the dynamic AFM measurements were used to estimate the drainage velocity, which is correlated to the

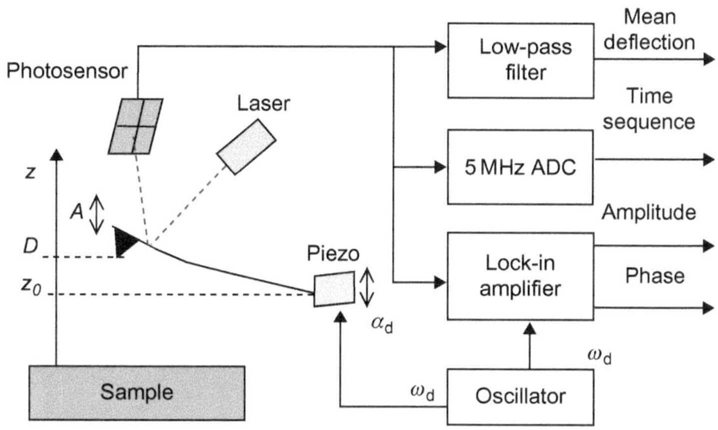

FIGURE 3.25

The figure depicts a schematic for the use of dynamic AFM method for measurements. The cantilever is oscillated at small amplitude and high frequency. The surface–AFM tip interaction alters the tip oscillation frequency and the change in phase and magnitude to the oscillation signal is correlated to the force exerted by the surface. Details of the method have been discussed by Wu et al. [12,13].

Source: *Image courtesy of Prof. Y. Wu.*

surface slip-length for both hydrophobic and hydrophilic surfaces. Figure 3.25 shows a schematic for the dynamic measurement technique and Figure 3.26 shows the measurement of the damping coefficient for a thin liquid film on silica surfaces functionalized using "click" chemistry.

Other microscopy techniques such as SEM and transmission electron microscopy (TEM) can also be used for direct imaging of surfaces. Most SEMs and TEMs can be coupled with electron or X-ray dispersion techniques (e.g., EDS) to obtain simultaneous information about the chemical composition of the surface. However, both SEM and TEM also require careful sample preparation and UHV equipment. Illustrative examples of the use of SEMs and TEMs are discussed in subsequent chapters.

3.7 Surface tension-driven flows

The origin of microchannel flows probably occurred with the use of capillaries for a variety of applications. Therefore, the term capillary motion should be defined. Capillary motion describes a broad category of flows governed due to surface tension. So, the question arises as to what is surface tension? It is defined as the intensity of the molecular attraction per unit length along any line in the surface of the fluid in contact with some external interface. Physically,

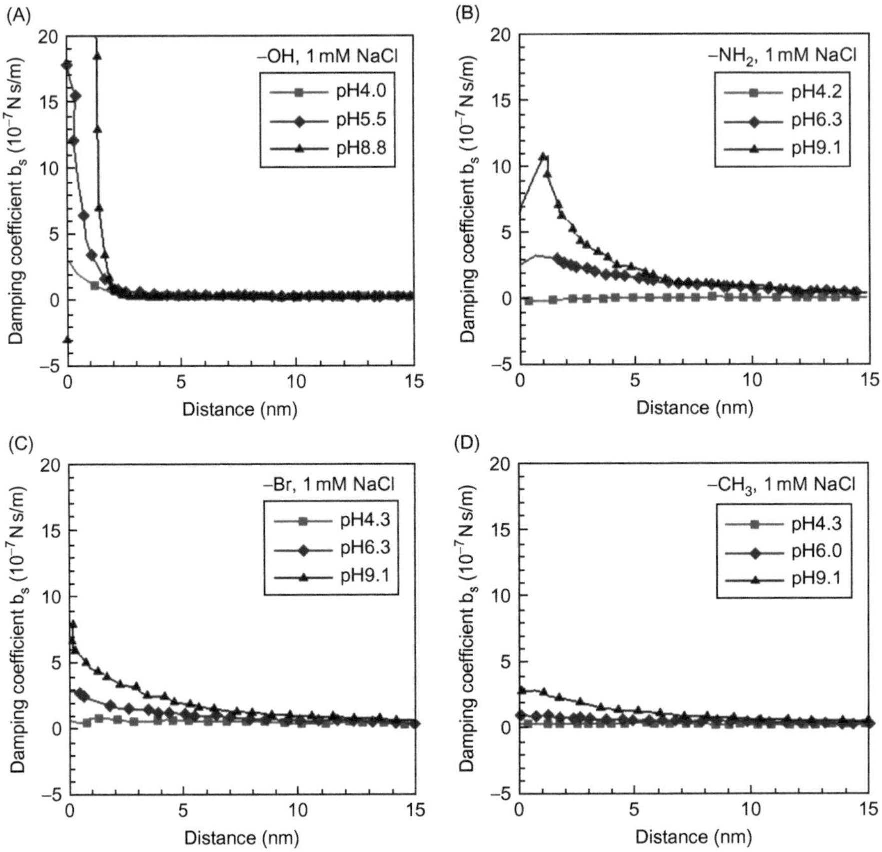

FIGURE 3.26

Sample data reported by Wu et al. [12] for the damping coefficient measured on functionalized silica surfaces as a function of solution pH by using the dynamic method [12].

consider a droplet sitting on a surface. At equilibrium, the interior of the fluid is in balance as molecules of the fluid are attracting each other while on the interface there is a net inward pulling force. This interfacial force that balances the molecular forces within the liquid is surface tension. Therefore, surface tension is a function of the liquid type, surface type, and the properties at the interface. In micro- and nanofluidics, surface tension plays an important role in manipulating fluid motion.

Spatial gradients or variations in surface tension at the liquid–vapor interface in a microchannel (or nanochannel) lead to tangential forces (or stresses when evaluated per unit area) allowing for fluid motion. The motion of the fluid due to tangential gradients in surface tension are named after Marangoni for his work in the early 1870s. However, other scientists have pointed out, as also discussed by Probstein [5], that James Thomson discussed the phenomena in detail in 1882.

So, the question that arises is, how do these gradients in surface tension arise and can they be controlled for engineering flows in micro- and nanochannels? Let us explore this question in the next few paragraphs.

Spatial gradients in surface tension can be caused by temperature changes, concentration gradients, and/or differences in electric field or surface charge. Since the flows most relevant to these changes were first observed in small, confined tubes, these are referred to as thermocapillary, diffusocapillary, and electrocapillary flows respectively. Over the years, several books and papers have been published on each of these categories and further references are listed in the Select bibliography. The key to note is that if a tangential force acts on a liquid at an interface, then

$$\vec{F}_{\mathrm{T,surf}} = \vec{\nabla}_{\mathrm{surf}}\sigma \tag{3.23}$$

where $\vec{F}_{\mathrm{T,\,surf}}$ denotes the tangential surface force with $\vec{\nabla}_{\mathrm{surf}}$ denoting the surface gradient. The liquid will have a tendency to move from a region of low to high surface tension. Several non-dimensional parameters have been defined to provide quick estimates of the relative importance of surface tension in a flow-field compared to other forces. Table 3.2 summarizes the main non-dimensional parameters and their use. Note that in Table 3.2, new terms are defined. Other terms have the meanings as discussed previously.

Surface tension flows in μ-Na*fl* systems have found several applications, specifically in the area of electrowetting leading to the development of droplet or digital microfluidics as discussed in Chapters 5 and 6. In addition, use of surface charge gradients has been used to develop micro- and nanomixers and surface-based valves for directing flows as also discussed in detail in later chapters.

3.8 Device interfaces

As discussed previously, all μ-Na*fl* systems are characterized by several interfaces. Among these, the channel interfaces between micro- and nanochannels have found special attention. This is in part due to the formation of concentration polarization or CP. CP refers to the development of concentration gradients formed at the interface. Significant CP can occur in systems independent of the driving force used, whether pressure or electrokinetically driven, especially if the interface is ion selective. The effects of CP have been imaged in electrokinetic micro- and nanoscale devices. An example of the effects of CP is presented in an experimental study and shown in Figure 3.27.

As seen in Figure 3.27, enrichment and depletion zones form at the micro/nanoscale interface. These zones were imaged with fluorescent dyes. The direct consequence of CP in micro- and nanoscale devices with electrokinetic flow is on the magnitude of electrically based sensor measurements leading to shifts in the current

Table 3.2 Summary of Some Common Non-dimensional Parameters Relevant to μ-Na*fl* Systems

Nondimensional Parameter	Ratio of Physical Quantities	Mathematical Formulation	Use and Physical Implication
Re (Reynolds number)	Inertial/viscous forces	$\frac{UL_c}{\nu}$	All μ-Na*fl* systems; $Re \ll 1$ means dominance of viscous forces
Bo (Bond number)	Gravity/surface tension	$\frac{(\rho_l - \rho_F)gL_c^2}{\sigma}$	$Bo \ll 1$ implies surface tension effects dominate; examples of insects walking on water relate to low *Bo*
Ca (Capillary number)	Viscous forces/surface tension	$\frac{\eta U}{\sigma}$	All μ-Na*fl* systems; $Ca \ll 1$ means dominance of surface tension effects with pressure drop across a droplet being small
Pr (Prandtl number)	Momentum diffusivity/thermal diffusivity	ν/α where α is the thermal diffusivity	Thermocapillary flows
Bi (Biot number)	Internal thermal resistance/boundary layer thermal resistance	hL_c/k where h is the convection coefficient and k is the thermal conductivity of the solid	Thermocapillary flows; $Bi \ll 1$ implies conduction within the solid is less important than convection across fluid boundary layer
Pe (Peclet number)	Advection/diffusion	$\frac{UL_c}{\alpha}$ or $\frac{UL_c}{D_{AB}}$ where D_{AB} is the mass diffusion coefficient	For thermocapillary flows $Pe = Re\,Pr$; For mass diffusion, $Pe = Re\,Sc$
Sc (Schmidt number)	Momentum/mass diffusion	$\frac{\nu}{D_{AB}}$	Mass diffusion in μ-Na*fl* systems
Sh (Sherwood number)	Mass transfer/diffusion	$\frac{h_m L_c}{D_{AB}}$ where h_m is the mass transfer coefficient	In μ-Na*fl* systems where mixing is important; compares convective mass transfer to diffusive mass transfer
We (Weber number)	Inertial/surface tension	$\frac{\rho U^2 L_c}{\sigma}$	Droplet-based systems or μ-Na*fl* systems with curvature
Ma (Marangoni number)	Surface tension (temperature)/viscous forces	$\frac{\sigma_T \beta h^2}{\eta\alpha}$ where β is the vertical temperature gradient in a fluid of depth h and σ_T is the surface tension gradient at the unperturbed free surface temperature	Thermocapillary; above a critical value for the given problem will lead to unstable convective flows; can also be interpreted as ratio of heat transfer by convection arising from surface tension gradients to heat transport by conduction

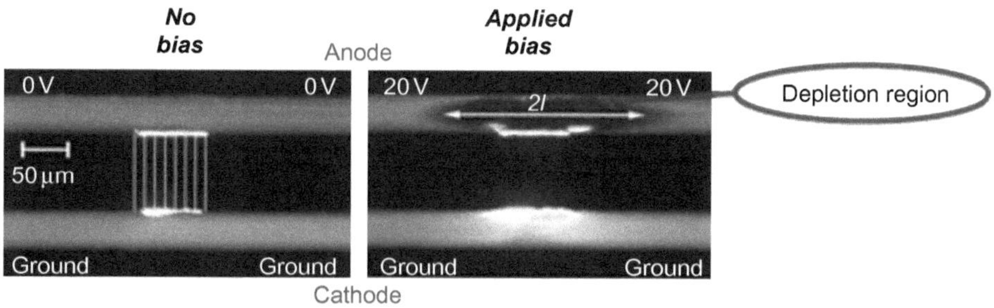

FIGURE 3.27

Enrichment/depletion zones in an electrokinetically driven micro/nanoscale device identified by increased or decreased fluorescence of the marker dye. These zones have been considered as evidence of polarization at the interface of micro- and nanochannels.

Source: *[14] Kim SJ, Wang Y-C, Lee JH, et al. Concentration polarization and nonlinear electrokinetic flow near a nanofluidic channel.*

density data recorded, thus adversely affecting the accuracy if only calibrated by assuming the micro-nanochannel device as a resistor following Ohm's law.

References

[1] Prakash S, Karacor MB, Banerjee S. Surface modification in microsystems and nanosystems. Surf Sci Rep 2009;64:233–54.

[2] Lyklema J. Fundamentals of interface and colloid science. Solidliquid interface, vol. 2. London: Academic Press; 1995.

[3] Israelachvili JN. Intermolecular and surface forces. San Diego, CA: Academic Press; 1992.

[4] Bard AJ, Faulkner LR. Electrochemical methods: fundamentals and applications. John Wiley & Sons; 2004.

[5] Probstein RF. Physicochemical hydrodynamics: an introduction. John Wiley & Sons; 2003.

[6] Datta S, Conlisk AT, Li HF, Yoda M. Effect of divalent ions on electroosmotic flow in microchannels. Mech Res Commun 2009;36:65–74.

[7] Chandrasekharan R, et al. High temperature hydroxylation of alumina crystalline surfaces. Surf Sci 2008;602(7):1466–74.

[8] Long TM, et al. Water-vapor plasma-based surface activation for trichlorosilane modification of PMMA. Langmuir 2006;22(9):4104–9.

[9] Serafina C, Damiana C, Stefania C, et al. Microcantilevers and organic transistors: two promising classes of label-free biosensing devices which can be integrated in electronic circuits. Anal Bioanal Chem 2012;402(5):1799–811.

[10] Bhushan B, Jung YC. Natural and biomimetic artificial surfaces for superhydrophobicity, self-cleaning, low adhesion, and drag reduction. Prog Mater Sci 2011;56(1): 1–108.

[11] Prakash S, et al. 'Click' modification of silica surfaces and glass microfluidic channels. Anal Chem 2007;79(4):1661–7.

[12] Wu Y, Misra S, Karacor M, et al. Dynamic response of AFM cantilevers to dissimilar functionalised silica surfaces in aqueous electrolyte solutions. Langmuir 2010; 26(22):16963–72.
[13] Wu Y, et al., Hydrophobic interaction at solid/water interface studied by dynamics force microscopy. Materials research society spring meeting. San Francisco, CA; 2009.
[14] Kim SJ, Wang Y-C, Lee JH, et al. Concentration polarization and nonlinear electrokinetic flow near a nanofluidic channel. Phys Rev Lett 2007;99:044501.

Select bibliography

Abe H, et al. Influence of NH_3/Ar plasma irradiation on physical and photocatalytic properties of TiO_2 nanopowder. J Photochem Photobiol A Chem 2006;183(1–2):171–5.

Abgrall P, Nguyen NT. Nanofluidics. Artech House; 2009.

Ahn W, et al. Electroless gold island thin films: photoluminescence and thermal transformation to nanoparticle ensembles. Langmuir 2008;24(8):4174–84.

Akamatsu Y, et al. Improvement of roller bearing fatigue life by surface roughness modification. SAE (Soc Automot Engrs) Trans 1991;100(2):44–9.

Allara DL, Nuzzo RG. Spontaneously organized molecular assemblies. 1. Formation, dynamics, and physical properties of *n*-alkanoic acids adsorbed from solution on an oxidized aluminum surface. Langmuir 1985;1:45–52.

Andryushechkin BV, et al. Direct STM observation of surface modification and growth of AgCl islands on Ag(111) upon chlorination at room temperature. Surf Sci 1999;431 (1-3):96–108.

Aumann T, et al. Rapid surface modification of polyethylene in microwave and r.f.-plasmas: comparative study. Surf Coat Technol 2001;142:169–74.

Bae B, et al. Development of a portable gas analyzer using a micro-Gas Chromatograph/Flame Ionization Detector (micro-GC/FID) for NASA's environmental missions. 42nd international conference on environmental systems (ICES). San Diego, CA, USA: American Institute of Aeronautics and Astronautics; 2012.

Bae B, et al. A five-microvalve fully integrated preconcentrator. IEEE sensors conference, Atlanta, GA; 2007.

Bae B, et al. A fully-integrated MEMS preconcentrator for rapid gas sampling. The 14th international conference on solid state sensors, actuators and microsystems:transducers '07, Lyon, France, IEEE; 2007.

Bain CD, Whitesides GM. Formation of monolayers by the coadsorption of thiols on gold: variation in the length of the alkyl chain. J Am Chem Soc 1989;111:7164–75.

Bain CD, Whitesides GM. A study by contact angle of the acid-base behavior of monolayers containing w-mercaptocarboxylic acids on adsorbed gold: an example of reactive spreading. Langmuir 1989;5:1370–8.

Bang ZL, MacDonald NC. An RIE process for submicron, silicon electromechanical structures. International conference on solid-state sensors and actuators. San Francisco, CA; 1991. p. 520–3.

Bauerle D. Laser-chemical processing: recent developments. Appl Surf Sci 1996;106:1–10.

Bard AJ, Faulkner LR. Electrochemical methods: fundamentals and applications. John Wiley & Sons; 2004.

Berdichevsky Y, et al. UV/ozone modification of poly(dimethylsiloxane) microfluidic channels. Sens Actuat B–Chem 2004;97(2–3):402–8.

Bhattacharyya A, Klapperich CM. Mechanical and chemical analysis of plasma and ultraviolet-ozone surface treatments for thermal bonding of polymeric microfluidic devices. Lab Chip 2007;7(7):876–82.

Bhushan B. MEMS/NEMS and BioMEMS/BioNEMS: materials, devices, and biomimetics. In: Bhushan B, editor. Springer handbook of nanotechnology. Berlin: Springer-Verlag; 2010. p. 1663–740.

Bhushan B. Nanotribology, nanomechanics, and materials characterization. In: Bhushan B, editor. Springer handbook of nanotechnology. Berlin: Springer-Verlag; 2010. p. 789–856.

Bhushan B. Springer handbook of nanotechnology. Heidelberg, Germany: Springer-Verlag; 2010.

Bhushan B, et al. Boundary slip study on hydrophilic, hydrophobic, and superhydrophobic surfaces with dynamic atomic force microscopy. Langmuir 2009;25(14): 8117–21.

Bhushan B, Jung YC. Natural and biomimetic artificial surfaces for superhydrophobicity, self-cleaning, low adhesion, and drag reduction. Prog Mater Sci 2011;56(1): 1–108.

Bodas D, Khan-Malek C. Formation of more stable hydrophilic surfaces of PDMS by plasma and chemical treatments. Microelectron Eng 2006;83(4-9):1277–9.

Bodas D, Khan-Malek C. Hydrophilization and hydrophobic recovery of PDMS by oxygen plasma and chemical treatment—An SEM investigation. Sens Actuat B–Chem 2007;123(1):368–73.

Bonaccurso E, et al. Hydrodynamic force measurements: boundary slip of water on hydrophilic surfaces and electrokinetic effects. Phys Rev Lett 2002;88(7):076103.

Browne WR, Feringa BL. Light switching of molecules on surfaces. Annu Rev Phys Chem 2009;60:407–28.

Bullett NA, et al. Polymer surface micropatterning by plasma and VUV-photochemical modification for controlled cell culture. Appl Surf Sci 2004;235(4):395–405.

Butt H-J, et al. Force measurements with the atomic force microscope: technique, interpretation and applications. Surf Sci Rep 2005;59:1–152.

Cai YH, et al. Photoinduced construction of a second covalently bonded organic layer on the Si(111)-7 $\times$ 7 surface. J Am Chem Soc 2007;129(27):8404–5.

Cappella B, Dietler G. Force-distance curves by atomic force microscopy. Surf Sci Rep 1999;34:1–104.

Carroll GT, et al. Photochemical micropatterning of carbohydrates on a surface. Langmuir 2006;22:2899–905.

Chabal YJ. Surface infrared spectroscopy. Surf Sci Rep 1988;8:211–357.

Chadov AV, Yakhnin ED. Colloid J Russ Acad Sci 1979;41.

Chan CM, et al. Polymer surface modification by plasmas and photons. Surf Sci Rep 1996;24(1–2):3–54.

Chandrasekharan R, et al. Change in radiative optical properties of Ta_2O_5 thin-films due to high temperature heat treatment. J Heat Transf 2007;129(1):27–36.

Chaudhury MK, Whitesides GM. Correlation between surface free energy and surface constitution. Science 1992;255:1230–2.

Chen Q, Richardson NV. Surface faceting induced by adsorbates. Progr Surf Sci 2003;73 (4–8):59–77.

Chen SH, Frank CW. Infrared and fluorescence spectroscopic studies of self-assembled *n*-alkanoic acid monolayers. Langmuir 1989;5(4):978–87.

Cheng JY, et al. Direct-write laser micromachining and universal surface modification of PMMA for device development. Sens Actuat B Chem 2004;99(1):186–96.

Chilkoti A, et al. Plasma-deposited polymeric films prepared from carbonyl-containing volatile precursors—XPS chemical derivatization and static SIMS surface characterization. Chem Mater 1991;3:51–61.

Choi S, Chae J. Surface plasmon resonance biosensor based on Vroman effect: Towards Cancer Biomarker Detection. IEEE 15th international mixed-signals, sensors, and systems test workshop. Scottsdale, Az; 2009. p. 1–3.

Conlisk AT. Essentials of micro- and nanofluidics: with applications to the biological and chemical sciences. Cambridge, UK: Cambridge University Press; 2012.

D'Agostino R, et al. In: d'Agostino R, editor. Plasma deposition, treatment, and etching of polymers. New York: Academic Press; 1990. p. 95–162.

Dorronsoro C, et al. Suitability of filofocon A and PMMA for experimental models in excimer laser ablation refractive surgery. Opt Express 2008;16(25):20955–67.

Ducker WA, et al. Measurement of forces in liquids using a force microscope. Langmuir 1992;8(7):1831–6.

Eijkel J. Liquid slip in micro- and nanofluidics: recent research and its possible implications. Lab Chip 2007;7:299–301.

Eijkel JCT, van de Berg A. Nanofluidics: what is it and what can we expect from it? Microfluid Nanofluid 2005;1:249–67.

Ferral A, et al. Chemisorption and physisorption of alkanethiols on Cu(111). A quantum mechanical investigation. Surf Sci 2003;525(1-3):85–99.

Fox HW, Zisman WA. Some advances in techniques for the study of adsorbed monolayers at the liquid–air interface. Rev Sci Instrum 1948;19(4):274.

Gao SL, et al. Photochemical surface modification of PET by excimer UV lamp irradiation. Appl Phys B—Lasers Opt 2005;81(5):681–90.

Girifalco LA, Good RJ. A theory for the estimation of surface and interfacial energies. I. Derivation and application to interfacial tension. J Phys Chem 1957;61(7):904–9.

Goddard JM, Hotchkiss JH. Polymer surface modification for the attachment of bioactive compounds. Prog Polym Sci 2007;32(7):698–725.

Godfrey SP, et al. Plasmachemical functionalization of porous polystyrene beads. Chem Mater 2001;13(2):513–8.

Goldblatt RD, et al. Characterization of water vapor plasma-modified polyimide. J Appl Polym Sci 1992;46:2189–202.

Good RJ, Girifalco LA. A theory for estimation of surface and interfacial energies. III. Estimation of surface energies of solids from contact angle data. J Phys Chem 1960;64 (5):561–5.

Graubner VM, et al. Photochemical modification of cross-linked poly(dimethylsiloxane) by irradiation at 172 nm. Macromolecules 2004;37(16):5936–43.

Groning P, et al. J Appl Phys 1994;76:887–92.

Gupta C, et al. Double transfer printing of small volumes of liquids. Langmuir 2007;23:2906–14.

Gupta C, et al. Electronic properties of a monolayer–electrolyte interface obtained from mechanistic impedance analysis. J Phys Chem C 2009;113(21):9375–91.

Haensch C, et al. Chemical surface reactions by click chemistry: coumarin dye modification of 11-bromoundecyltrichlorosilane monolayers. Nanotechnology 2008;19.

Henry AC, et al. Surface modification of poly(methyl methacrylate) used in the fabrication of microanalytical devices. Anal Chem 2000;72(21):5331–7.

Hiraoka H. In: Anpo M, Matsuura T, editors. Photochemistry on solid surfaces. Amsterdam: Elsevier; 1989. p. 448–68.

Hiraoka H, Pacansky J. J Electrochem Soc 1981;128.

Hollins P. The influence of surface defects on the infrared spectra of adsorbed species. Surf Sci Rep 1992;16:53–94.

Holmes-Farley SR, et al. Acid-base behavior of carboxylic-acid groups covalently attached at the surface of polyethylene—the usefulness of contact-angle in following the ionization of surface functionality. Langmuir 1985;1:725–40.

Holt JK, et al. Fast mass transport through sub-2-nanometer carbon nanotubes. Science 2006;312:1034–7.

Hou Z, et al. Self-assembled monolayers on electroless gold impart pH-responsive transport of ions in porous membranes. Langmuir 2000;16:2401–4.

Hozumi A, et al. Spatially defined surface modification of poly(methyl methacrylate) using 172 nm vacuum ultraviolet light. Langmuir 2002;18(23):9022–7.

Hu K, Bard AJ. Use of atomic force microscopy for the study of surface acid–base properties of carboxylic acid-terminated self-assembled monolayers. Langmuir 1997;13 (19):5114–9.

Huang X, et al. Current-monitoring method for measuring the electroosmotic flow rate in capillary zone electrophoresis. Anal Chem 1988;60:1837–8.

Ito M. Structures of water at electrified interfaces: microscopic understanding of electrode potential in electric double layers on electrode surfaces. Surf Sci Rep 2008;63: 329–89.

Johansson B-L, et al. Characterization of air plasma-treated polymer surfaces by ESCA and contact angle measurements for optimization of surface stability and cell growth. J Appl Polym Sci 2002;86:2618–25.

Kim W, et al. Acetylene molecules on the Si(001) surface: room-temperature adsorption and structural modification upon annealing. Surf Sci 2002;514(1–3):376–82.

Kim YJ, et al. Surface characterization and in vitro blood compatibility of poly(ethylene terephthalate) immobilized with insulin and/or heparin using plasma glow discharge. Biomaterials 2000;21:121–30.

King WP, et al. Atomic force microscope cantilevers for combined thermomechanical data writing and reading. Appl Phys Lett 2001;78:1300–2.

Kim SJ, Wang Y-C, Lee JH, et al. Concentration polarization and nonlinear electrokinetic flow near a nanofluidic channel. Phys Rev Lett 2007;99:044501.

Kirby BJ. Micro- and nanoscale fluid mechanics. Cambridge; 2010.

Kirby BJ, Hasselbrink Jr. EF. Zeta potential of microfluidic substrates: 1. Theory, experimental techniques, and effects on separations. Electrophoresis 2004;25:187–202.

Kiss E, et al. Surface grafting of polyethylene oxide optimized by means of ESCA. Prog Colloid Polym Sci 1987;73:113–9.

Langowski BA, Uhrich KE. Microscale plasma-initiated patterning (μPIP). Langmuir 2005;21:10509–14.

Ledbetter JW, Bowen JR. Spectrophotometric determination of critical micelle concentration of some alkyldimethylbenzylammonium chlorides using fluorescein. Anal Chem 1969;41:1345–7.

Lee JH, et al. Cell adhesion and growth on polymer surfaces with hydroxyl groups prepared by water vapour plasma treatment. Biomaterials 1991;12:443–8.

Lee KJ, et al. Photolithographic technique for direct photochemical modification and chemical micropatterning of surfaces. Langmuir 2004;20(5):1812–8.

Lee S, Bhushan B. Biological molecules in therapeutic nanodevices. In: Bhushan B, editor. Spinger handbook of nanotechnology. Berlin: Springer-Verlag; 2010. p. 453–84.

Lee SB, Martin CR. Controlling the transport properties of gold nanotubule membranes using chemisorbed thiols. Chem Mater 2001;13:3236–44.

Li D. Electrokinetics in microfluidics. New York: Academic Press; 2004.

Liang SH, et al. UV-laser treatment of PET fabrics: changes in fibre structure and dyeing behaviour. Color Technol 2003;119(1):19–23.

Liu J, et al. Surface-modified poly(methyl methacrylate) capillary electrophoresis microchips for protein and peptide analysis. Anal Chem 2004;76:6948–55.

Lockett MR, et al. Aldehyde-terminated amorphous carbon substrates for the fabrication of biomolecule arrays. Langmuir 2008;24:9198–203.

Loginova E, et al. Nanoscopic nickel aluminate spinel ($NiAl_2O_4$) formation during NiAl (111) oxidation. Surf Sci 2007;601(3):L11–4.

Loh FC, et al. X-ray photoelectron spectroscopy and secondary ion mass spectrometry studies of some surface modified hydrocarbon films. J Vacuum Sci Technol A 1994;12:2705–10.

Love JC, et al. Self-assembled monolayers of thiolates on metals as a form of nanotechnology. Chem Rev 2005;105:1103–69.

Luckham PF. Manipulating forces between surfaces: applications in colloid science and biophysics. Adv Colloid Inter Sci 2004;111(1-2):29–47.

Lummerstorfer T, Hoffman H. IR reflection spectra of monolayer films sandwiched between two high refractive index materials. Langmuir 2004;20:6542–5.

Ma Z, Zaera F. Organic chemsitry on solid surfaces. Surf Sci Rep 2006;61:229–81.

Mack GL. The determination of contact angles from measurements of the dimensions of small bubbles and drops. I. The spheroidal segment method for acute angles. J Phys Chem 1936;40:159–67.

Madou M. Fundamentals of microfabrication. Boca Raton, FL: CRC Press; 1997.

Madou MJ. Fundamentals of microfabrication and nanotechnology. CRC Press; 2012.

Mailhot B, et al. Photooxidation of blends of polystyrene and poly(vinyl methyl ether): FTIR and AFM studies. Polymer 2000;41(6):1981–8.

Mao C, et al. A photochemical method for the surface modification of poly(vinyl chloride) with *O*-butyrylchitosan to improve blood compatibility. Process Biochem 2004;39(9):1151–7.

Maoz R, Sagiv J. Penetration-controlled reactions in organized monolayer assemblies. 1. Aqueous permanganate interaction with monolayer and multilayer films of long-chain surfactants. Langmuir 1987;3.

Masliyah JH, Bhattacharjee S. Electrokinetic and colloid transport phenomena. John Wiley and Sons; 2006.

Matsuda T. Photoiniferter-driven precision surface graft microstructures for biomedical applications. Adv Polym Sci 2006;197:67–106.

McCarley RL, et al. Resist-free patterning of surface architectures in polymer-based microanalytical devices. J Am Chem Soc 2005;127:842–3.

McNair HM, Miller JM. Basic gas chromatography. New York, NY: John Wiley & Sons; 1997.

Mecomber JS, et al. Photochemical functionalization of polymer surfaces for microfabricated devices. Langmuir 2008;24:3645–53.

Mireault N, Ross GG. Modification of wetting properties of PMMA by immersion plasma ion implantation. Surf Rev Lett 2008;15:345–54.

Nguyen NT, Wereley ST. Fundamentals and applications of microfluidics. Artech House; 2002.

Oh KW, et al. In-line micro ball valve through polymer tubing. Proc microTAS 2001;407–8.

Otero E, et al. Surface modification of several steels after their exposure at high temperature to oxygen and sulfur mixtures. Surf Coat Technol 1995;76(1-3):53–60.

Pardon G, van der Wijngaart W. Electrostatic gating of ion and molecule transport through a nanochannel-array membrane. Proc IEEE solid-state sensors, actuators, microsystems conf (transducers). Beijing; 2011. p. 1610–3.

Patrito N, et al. Photolithographically patterned surface modification of poly(dimethylsiloxane) via UV-initiated graft polymerization of acrylates. Langmuir 2006;22(8): 3453–5.

Ponter AB, et al. Surface energy changes produced by irradiating polymers with ultraviolet-ozone. Polymer Eng Sci 1994;34:1233–8.

Prakash S, et al. OH concentration profiles over alumina, quartz, and platinum surfaces using laser-induced fluorescence spectroscopy in low-pressure hydrogen/oxygen flames. Combust Sci Technol 2005;177(4):793–817.

Prakash S, Karacor MB. Surface mediated flows in glass nanofluidic channels. The 13th solid-state sensors, actuators and microsystems workshop, Hilton Head Islands, SC, Transducer Research Foundation; 2010.

Prakash S, Yeom J, Jin N, Adesida I, Shannon MA. Characterization of ionic transport at the nanoscale. Proc I Mech E J Nanoeng Nanosyst 2006;220:45–52.

Prakash S, et al. Surface modification within enclosed polymer and glass microfluidic channels. The 11th international conference on miniaturized systems for chemistry and life sciences. Paris: K. Jensen; 2007.

Prakash S, et al. Nanofluidics: systems and applications. IEEE Sensors J 2008;8 (5):441–50.

Prakash S, et al. Characterization of ionic transport at the nanoscale. Proc Inst Mech Engrs Pt N J Nanoeng Nanosyst 2007;220(2):45–52.

Praschak D, et al. PET surface modifications by treatment with monochromatic excimer UV lamps. Appl Phys A—Mater Sci Process 1998;66(1):69–75.

Ren L, et al. A new method of evaluating the average electro-osmotic velocity in microchannels. J Colloid Interface Sci 2002;250:238–42.

Revathi N, et al. Synthesis and physical behaviour of In_2S_3 films. Appl Surf Sci 2008;254:5291–8.

Rochat N, et al. Attenuated total reflection spectroscopy for infrared analysis of thin layers on a semiconductor substrate. J Appl Phys 2002;91:5029–34.

Rye RR, et al. Mechanistic aspects of alkylchlorosilane coupling reactions. Langmuir 1997;13:2965–72.

Sadler DJ, et al. A new magnetically actuated microvalve for liquid and gas control applications. 10th international conference on solid-state sensors and actuators (transducers); 1999. p. 1812–5.

Sahoo NG, et al. Micro/nanoparticle design and fabrication for pharmaceutical drug preparation and delivery applications. Curr Drug Therapy 2008;3(2):78–97.

Saul G, et al. The effects of thermal treatment on the austenitic grain size and mechanical properties of 18 pct Ni maraging steels. Metall Mater Trans B 1970;1(2):383–7.
Schmickler W. Interfacial electrochemistry. Oxford, University Press, New York; 1996.
Schreiber F. Structure and growth of self-assembling monolayers. Prog Surf Sci 2000;65:151–256.
Schroder K, et al. Plasma-induced surface functionalization of polymeric biomaterials in ammonia plasma. Contrib Plasma Phys 2001;41(6):562–72.
Schroeder DV. An introduction to thermal physics. San Francisco, LA: Addison-Wesley/ Longman; 2000.
Schulze RK, et al. A variable temperature X-ray photoelectron spectroscopic study of the surface conversion of diethylaluminum azide to AlN. J Vac Sci Technol A 1990;8 (3):2338–43.
Singh RK, Fitzgerald JM. Beam interactions with materials and atoms. Nucl Instrum Meth Phys Res Sect B 1997;121:363–6.
Serafina C, Damiana C, Stefania C, et al. Microcantilevers and organic transistors: two promising classes of label-free biosensing devices which can be integrated in electronic circuits. Anal Bioanal Chem 2012;402(5):1799–811.
Sochanski J, Gatos HC. The effect of thermal desorption of water on surface states of germanium in ultrahigh vacuum. Surf Sci 1969;13:393–400.
Spiertz E, Vollenbroek. Microcircuit Eng 1984;68.
Steen ML, et al. Langmuir 2001;17:8156–66.
Steen ML, et al. Low temperature plasma treatment of asymmetric polysulfone membranes for permanent hydrophilic surface modification. J Membr Sci 2001;188:97–114.
Szczepkowicz A, et al. A comparison of adsorbate-induced faceting on flat and curved crystal surfaces. Surf Sci 2005;599(1-3):55–68.
Tamada K, et al. Surface Phase Behavior of *n*-Alkanethiol Self-Assembled Monolayers Adsorbed on Au(111): An Atomic Force Microscope Study. Langmuir 1997;13:1558.
Tiller JC, et al. Designing surfaces that kill bacteria on contact. Proc Nat Acad Sci USA 2001;98:5981–5.
Tokuhisa H, et al. Preparation and characterization of dendrimer monolayers and dendrimer-alkanethiol mixed monolayers adsorbed to gold. J Am Chem Soc 1998;120:4492–501.
Troughton EB, et al. Monolayer films prepared by the spontaneous self-assembly of symmetrical and unsymmetrical dialkyl sulfides from solution onto gold substrates: structure, properties, and reactivity of constituent functional groups. Langmuir 1988;4.
Ulbricht M. Photograft-polymer-modified microporous membranes with environment-sensitive permeabilities. React Funct Polym 1996;31(2):165–77.
Ulbricht M, et al. Photomodification of ultrafiltration membranes—Part 11—chemically and morphologically defined ultrafiltration membrane surfaces prepared by heterogeneous photo-initiated graft polymerization. Colloids Surf Physicochem Eng Aspects 1998;138(2–3):353–66.
Ulman A. Formation and structure of self-assembled monolayers. Chem Rev 1996;96:1533–54.
Uzunova-Bujnova M, et al. Effect of mechanoactivation on the structure, sorption and photocatalytic properties of titanium dioxide. Mater Chem Phys 2008;110(2–3):291–8.
Vasquez-Borucki S, et al. Hydrogen plasma treatment of poly(ethylene terephthalate) surfaces. Surf Coat Technol 2001;138(2-3):256–63.

van de Mei HC, et al. X-ray photoelectron spectroscopy for the study of microbial cell surfaces. Surf Sci Rep 2000;39:1–24.

Vinogradova OI, Belyaev AV. Wetting, roughness and hydrodynamic slip. PSP rev vol. Singapore: Pan Stanford Publishing; 2010.

Wagner CD. Sensitivity factors for XPS analysis of surface atoms. J Electron Spectrosc Relat Phenom 1983;32:99–102.

Walton J, Fairley N. Noise reduction in X-ray photoelectron spectromicroscopy by a singular value decomposition sorting procedure. J Electron Spectrosc Relat Phenom 2005;148:29–40.

Wang H, et al. Morphological evolution in oxygen-induced faceting of Re(123-1). Phys Rev B Condens Matter Mater Phys 2006;74.

Wang RS, Ong HC. Study of interfacial diffusion in Al_2O_3/ZnO and MgO/ZnO heterostructures. J Appl Phys 2008;104.

Wasserman SR, et al. Structure and reactivity of alkylsiloxane monolayers formed by reaction of alkyltrichlorosilanes on silicon substrates. Langmuir 1989;5(4):1074–87.

Wei SQ, et al. Surface morphological modification of Pt thin films induced by growth temperature. Phys Rev B 1998;58(7):3605–8.

Wei SY, et al. Photochemically patterned poly(methyl methacrylate) surfaces used in the fabrication of microanalytical devices. J Phys Chem B 2005;109(35):16988–96.

Wilson DJ, et al. Plasma modification of PTFE surfaces Part I: surfaces immediately following plasma treatment. Surf Interf Anal 2001;31(5):385–96.

Wollman EW, et al. Photosensitive self-assembled monolayers on gold: photochemistry of surface-confined aryl azide and cyclopentadienylmanganese tricarbonyl. J Am Chem Soc 1994;116:4395–404.

Woodward JT, Ulman A, Schwartz DK. Self-assembled monolayer growth of Octadecylphosphonic acid on mica. Langmuir 1996;12:3626–9.

Wu Y, et al. Effect of solution concentration, surface bias and protonation on the dynamic response of amplitude-modulated atomic force microscopy in water. Langmuir 2008;24:10817–24.

Wu Y, Misra S, Karacor M, et al. Dynamic response of AFM cantilevers to dissimilar functionalised silica surfaces in aqueous electrolyte solutions. Langmuir 2010;26(22):16963–72.

Xu H, et al. Surface modification of polytetrafluoroethylene by microwave plasma treatment of H_2O/Ar mixture at low pressure. Mater Chem Phys 2003;80(1):278–82.

Xue CY, Yang KL. Chemical modifications of inert organic monolayers with oxygen plasma for biosensor applications. Langmuir 2007;23(10):5831–5.

Yakhnin ED, Chadov AV. Colloid J Russ Acad Sci 1983;45.

Yao YX, et al. Effect of low energy ECR ion beam irradiation on micro nano scale mechanical properties of silicon. Appl Mech Mater 2008;10-12:344–7.

Yasuda HK, et al. A growth mechanism for the vacuum deposition of polymeric materials. Pure Appl Chem 1990;63.

Yeo S, et al. Solid state self-assembly of nanocheckerboards. Appl Phys Lett 2006;89.

Zangmeister RA, Tarlov MJ. UV graft polymerization of polyacrylamide hydrogel plugs in microfluidic channels. Langmuir 2003;19(17):6901–4.

Zhang G-J. MicroRNA and Cancer: Methods and Protocols; 2011.

Zhang W, et al. Anti microbial properties of copper plasma-modified polyethylene. Polymer 2006;47:7441–5.

Zhang Y, et al. Surface modification of polyimide films via plasma polymerization and deposition of allylpentafluorobenzene. Polymer 2002;43(26):7279–88.

Zhao B, et al. Surface-directed liquid flow inside microchannels. Science 2001;291:1023–6.

Zhao XL, Kopelman R. Mechanism of organosilane self-assembled monolayer formation on silica studied by second-harmonic generation. J Phys Chem 1996;100(26):11014–8.

Zharnikov M, Grunze M. Modification of thiol-derived self-assembling monolayers by electron and X-ray irradiation: scientific and lithographic aspects. J Vacuum Sci Technol B 2002;20(5):1793–807.

Zheng J, et al. Nanosized aluminum nitride hollow spheres formed through a self-templating solid–gas interface reaction. J Solid State Chem 2007;180:276–83.

Zhu Y, Granick S. Viscosity of interfacial water. Phys Rev Lett 2001;87(9):096104.

Zhu Y, Granick S. Limits of the hydrodynamic no-slip boundary condition. Phys Rev Lett 2002;88(10):106102.

Zhu Y, Granick S. No-slip boundary condition switches to partial slip when fluid contains surfactant. Langmuir 2002;18:10058–63.

CHAPTER

Advanced Fabrication Methods and Techniques

4

CHAPTER OUTLINE

S. Prakash & J. Yeom: Nanofluidics and Microfluidics. DOI: http://dx.doi.org/10.1016/B978-1-4377-4469-9.00004-4

4.1 Introduction to micro- and nanofabrication

4.1.1 Patterning—photolithography

Photolithography, also known as optical lithography, is a process of creating a prescribed pattern in a photosensitive film by selectively exposing the film to radiation and subsequently removing either the exposed or the unexposed portions of the film. Photolithography is the most widely used and perhaps most important technology in microelectronics and MEMS industries as it has been considered the only patterning technology that is scalable to mass manufacturing. In fabricating a vast majority of micro-/nanofluidic devices, photolithography is used once or more. The patterned photoresists (PRs) can not only serve as an etch mask or deposition template, as often observed in microelectronics and MEMS fabrication, but also play an intermediate role for masters and molds in casting and molding processes, which is particularly relevant to microfluidic device fabrication. Moreover, some patterned PRs can be used directly as end-use structural components or sacrificial layer supports and substrate bonding agents. The latter applications of photolithography are somewhat unique to micro-/nanofluidic device fabrications.

4.1.1.1 Photolithography overview

Figure 4.1 depicts the basic process steps in photolithographic patterning: (1) substrate preparation, (2) PR application, (3) prebaking (or softbaking), (4) alignment, (5) exposure, (6) development, (7) postbake, (8) subsequent processes using the PR as a mask film, and (9) stripping/descum. The process begins with a thorough cleaning of a substrate to remove contaminants from the previous processes or environments. In many cases, a wafer surface is primed with a PR adhesion promoter to facilitate a coating process and adhesion of the PRs. A film of PR is then coated onto the wafer either by spin- or spray coating. Once applied to the surface, the PR film is prebaked to remove the solvents and further improve the adhesion of the film. The term prebake or soft bake is coined in contrast to postbake or hard bake in the later steps. The wafer covered with PR is then conveyed to an aligner/exposure tool. The patterns in the photomask are transferred to the PR layer as latent images by selectively sensitizing the film. Depending on the type of the PRs, the exposed regions can remain or be developed away in a mild developer solution. Before development, some PRs require postbaking and/or another round of light exposure. The patterns in the PR film are then subject to visual inspection.

4.1.1.1.1 Wafer cleaning and substrate preparation

Prior to the application of priming agents and/or PRs, a wafer(s) need to be cleaned. Typical contaminants include dust particles, metal ions, and organic residues from the previous processes. The type of the cleaning procedure depends on the nature of the substrate and the deposited films. The standard cleaning procedure for silicon wafers is the RCA clean which consists of RCA-1 (or SC-1) clean for removal of the organic contaminants, RCA-2 (or SC-2) clean for removal of ionic contamination, and HF clean for removal of the native oxide layer. First, the wafers are degreased with acetone and isopropyl alcohol and soaked in deionized (DI) water. The second step is to immerse the wafers into an SC-1 solution (1:1:5 of NH_4OH, H_2O_2, and H_2O) at around 75°C for 5–10 min to remove organic residues. After a DI water bath, a brief dip into a 1:50 solution of HF and H_2O at room temperature is performed to remove the native oxide. The last step is to immerse the wafers into an SC-2 solution (1:1:6 of HCl, H_2O_2, and H_2O) which effectively removes the remaining metallic or ionic contaminants. The RCA clean procedure is recommended to be performed immediately prior to any critical steps, especially the processes involving high temperatures.

In the MEMS and microfluidics applications or in the research laboratory environments, a full set of RCA clean is hardly implemented, but an SC-1 or SC-2 cleaning is separately employed depending on the nature of the contamination. As a simpler alternative, wafers can be cleaned with the piranha solution (3:1 of H_2SO_4 and H_2O_2 or the ratio varying), a strong oxidizing agent that removes most organic matter. However, great care must be taken in preparing, handling, and utilizing the piranha solution as it is highly corrosive and dangerous (e.g.,

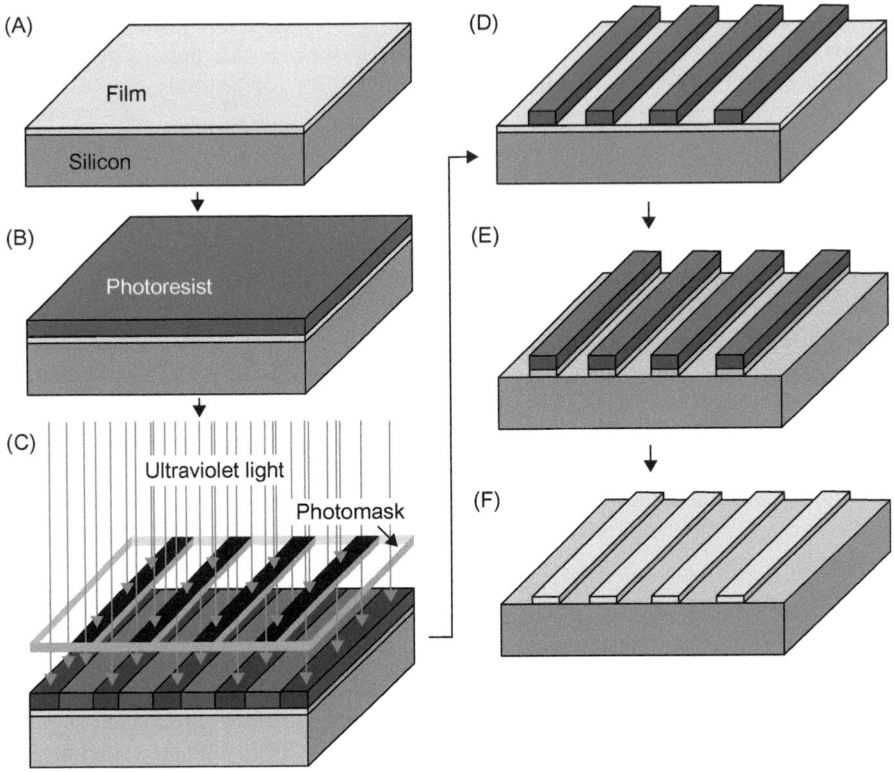

FIGURE 4.1

Schematics of the basic process sequences in photolithographic patterning: (A) thin film deposition, (B) PR application, (C) alignment and exposure, (D) development, (E) thin film etching, and (F) PR stripping.

explosive if enclosed). More specific to the semiconductor industry, plasma ashing (or O_2 descum) is used to remove the PR residues from the processed wafers. Oxygen plasma is produced by feeding a stream of oxygen gas into a vacuum chamber with the strong electromagnetic field, and the resulting reactive species etch away most of organic substances including PRs.

Note that the similar wet/dry cleaning procedures are also used to pretreat the substrates for surface modification for micro-/nanofluidic channels or some bonding processes. Aqueous-based cleaning solutions (e.g., piranha, SC-1) or oxygen plasmas terminate Si, SiO_2, or glass surfaces with the hydroxyl (−OH) groups by which silane-containing self-assembled monolayers (SAMs) can be tethered to the surface. The chemistry of the SAM formation and its applications are discussed in Chapter 3.

4.1.1.1.2 Photoresist application

After cleaning, the wafers are subject to bake-out above 100°C to remove moisture. The substrate is then primed with an adhesion promoter (see TIPS). These dehydrating and/or priming steps are critical to increase an adhesion of PR to the substrate, and otherwise the small PR patterns are often delaminated upon development. Spin coating is the standard PR application method. A wafer is first mounted on a flat vacuum chuck in a PR spinner, and a predetermined quantity (typically a few milliliters) of liquid PR is applied on the static or slowly rotating wafer. The wafer is then spun at a high speed (2000–8000 rpm) with various acceleration rates and durations. The parameters like spin speed, ramp rate, and duration are determined based on the required film thickness and viscosity of the PR solution. (Refer to Section 4.1.2.1 for further discussion of spin coating.) The thickness of the film, which is mainly controlled by spin speed, ranges between 0.5 and 2 μm for the Si-integrated circuit fabrication, but the PR film thicker than 1 cm has been reported for some MEMS structures. As a rule of thumb, the thinner the film is, the better the pattern resolution becomes. However, thinner PR films are prone to particle defects and can lead to a larger pinhole density and in turn a decreasing patterning yield. Another critical factor for the reproducible lithographic process is the uniformity of the coated film. On a planar surface, a variation of 5 nm across a wafer and from wafer to wafer is standard for a 1-μm thick PR.

TIPS—ADHESION PROMOTER

Many PRs (especially positive resists) often suffer from a poor adhesion to a substrate. The surfaces of the substrate materials get oxidized very easily, and the resulting surface oxides form long-range hydrogen bonds with water molecules adsorbed from ambient. Poor adhesion can be attributed, therefore, to the fact that PR adheres to the water molecules rather than to the surface of the substrate. In the subsequent developing or etching step, a PR layer is often delaminated from the substrate. Silicon, metals, and PSG are especially prone to the formation of these surface oxides and poor adhesion. Bis(trimethysilyl) amine (also known as hexamethyldisilazane or HMDS) is a common PR adhesion promoter, making the substrate surface hydrophobic and preventing moisture from being condensed. HMDS can be applied either by spin coating a diluted HMDS solvent or by vapor priming, that is, a concentrated HMDS vapor directly exposed to the substrate at reduced pressure. Vapor priming is known to more effectively produce a monomolecular layer of HMDS. By either way, the silyl group of HMDS readily bonds with the wafer surface while the other side (methyl and amine groups) bonds readily with the PR. Other adhesion promoters have also been developed for different types of films and substrates. For instance, amine-terminated AP3000 or AP8000 (γ-aminotriethoxysilane, Dow Chemical) adhesion promoters improve the adhesion of PR, SU-8, metal, and many polymer films on Si/glass/metal oxide substrates, and VM-651 or VM-652 (organosilane, HD MicroSystems) for polyimide. The application of these adhesion promoters also helps to increase the process yield by neutralizing the process variables such as clean room humidity and wafer storage conditions.

4.1.1.1.3 Soft bake

After spin coating, the PR film usually encompasses up to 15% of solvent and built-in stresses. Soft bake is performed at temperatures between 60°C and 120°C for a minute to tens of minutes to remove solvents and residual stresses. Once baked, the overall volume of the film is decreased and the PR film is densified. Soft bake also promotes adhesion of the PR film to the wafer substrate. If not properly cured, the PR film may stick to the photomask in contact aligner, making it difficult to separate the wafer from the mask.

4.1.1.1.4 Exposure

Once soft baked, the PR film is subject to light illumination or exposure system to transfer an image from a photomask with the alignment to the preexisting patterns or a wafer flat. The light source is often a mercury vapor lamp that delivers substantial luminescent power (10–20 mW) to the wafer in some specific wavelengths and with uniformity across the wafer. The wavelengths of the light source used in the exposure system span from 150–300 (deep UV) to 300–500 nm (near UV). While industries are using deep UV (192 nm) for patterning high-resolution features, academic and research laboratories typically use near UV light source such as the g-line (436 nm), h-line (405 nm), and i-line (365 nm).

4.1.1.1.5 Development, descum, and removal

Once exposed to UV (and, in case of negative resists, after the post exposure step), the wafers are subject to development. The exposed areas are developed away for positive PRs but remain on the substrate for negative PRs. Typical developers for positive PRs and negative PRs are alkaline aqueous solutions and organic solvents, respectively. A rule of thumb in development is a slight (~10%) overdevelopment for PR patterns to ensure the development, but depending on the resolution requirements and the sensitivity of PRs, the PRs can be underdeveloped. A descum process, a gentle oxygen plasma cleaning, is typically performed to remove residual layers (including the adhesion promoter) prior to further steps (e.g., lift-off or etching). Finally, the remaining PR layers after the pattern transfer are stripped either in wet stripping solution (PR strippers) or dry etcher.

4.1.1.2 Photomasks

Design or layout of micro-/nanofluidic elements first needs to be interfaced with a physical system, and in photolithography, such interface is established via a photomask. Numerous computer-aided drawing (CAD) tools are available to help users to lay out the design into the computer screen and convert it into a printable or machine-readable file. Photomask takes a form of a patterned opaque layer on an optically flat soda-lime glass (transparent to near UV) or fused quartz (transparent to deep UV) plates. Emulsion, chromium, and iron oxide are the common opaque materials as UV light absorbers. And the choice of the absorber materials resorts to the mask substrate materials, wavelengths of UV radiation, and mask fabrication methods.

Ways of generating patterns in photomasks from the CAD file are very different depending on the minimum feature size and its tolerance. One of the cheapest and easiest methods is to use an office laser printer and print the design as an emulsion layer onto an overhead transparency film. Due to its simplicity and low cost, research communities in micro- and nanofluidics have extensively used this type of printed photomasks [1] for large pattern sizes (>100 μm). The printed circuit board industry and academic researchers who want a better resolution utilize more advanced laser plotters and polyester transparency films. The feature size of the printed mask is directly related to the printer resolution, and the 5080, 16,000, 54,000-dpi high-resolution laser printer can produce the minimum feature size of about 30, 10, 6 μm, respectively. Polymer-based masks suffer from wear and tear and from dimensional instability. Therefore, the emulsion pattern on the transparency film is often transferred to a Cr-covered glass plate via photolithography for better handling and reproducibility.

In the integrated circuit fabrication and when a more high-resolution (below 5 μm) mask is in need, the mask pattern is generated by direct writing either using a laser or an e-beam writer. Direct beam writing is perhaps the most precise method to generate patterns—its ultimate resolution is less than 10 nm (when e-beam used), but it takes several orders of magnitude longer time to expose on a large-area substrate compared to optical lithography. Therefore, it makes sense to use electron beam writing for photomask production but not for full-scale IC fabrication. Beam physics, writing strategies, registration, and other topics related to direct beam writing can be found elsewhere.

4.1.1.3 Photoresists

PRs (also called resists) are light-sensitive materials, that is, the exposed areas get switched on or off for photochemical reactions responsible for either removing or retaining the materials. PRs consist of three key ingredients: (i) a polymer resin providing a structural support and determining the mechanical and thermal properties, (ii) a solvent diluting a resin to promote spin coating and thin film formation on a substrate, and (iii) a photoactive compound altering polymeric phase via photochemical reactions. PR is generally categorized into positive PR versus negative PR, depending on its polarity. Figure 4.2 illustrates the resultant pattern of these two different types of PR with the same mask image. In the positive resists, the exposed regions of the resists will dissolve away in the developer while the unexposed regions remain intact. Therefore, the notion of "positive" indicates that the image of the mask layout is faithfully transferred to the resist. Opposite is true for the negative PR, and the mask image is now reversed in the resulting pattern of the resist, that is, "negative."

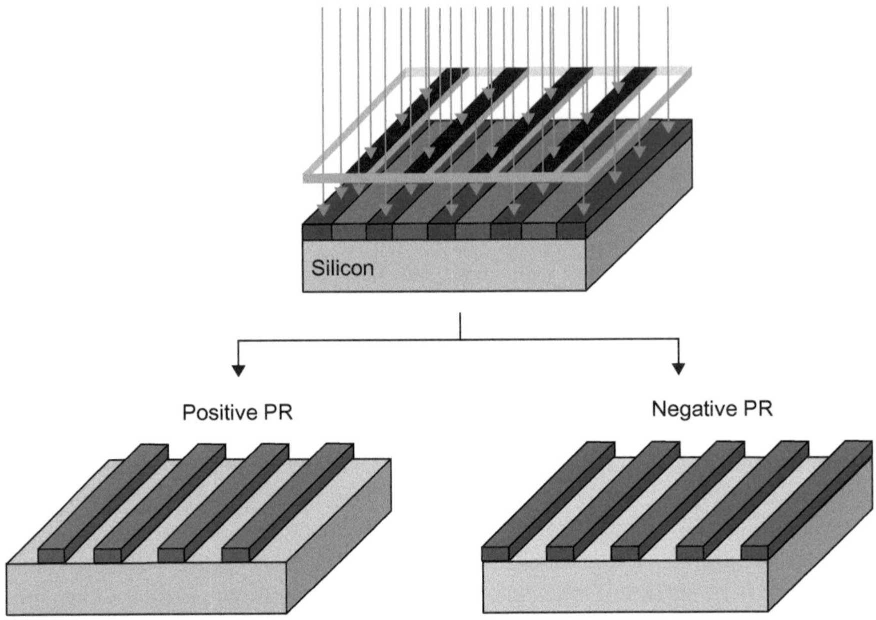

FIGURE 4.2

Schematics to illustrate the positive versus negative PR. For the positive PR, the exposed areas are dissolved away in the developer while for the negative PR the exposed regions stay intact after development.

4.1.1.3.1 Positive resists

Positive resists are dominant in IC fabrication due to their high-resolution capabilities. The most common positive resist is the DQN family, which is sensitive at a wavelength of 365 or 405 nm (called i-line or g-line resists), composed of phenolic novolak (N) as a resin and diazoquinone (DQ) ester as a sensitizer. DQ is added to the novolak resin to inhibit the dissolution in alkaline solutions, but when the resist is exposed to UV light, the dissolution rate of the exposed area is increased by more than 10-fold and therefore selectively removed in the developer. Figure 4.3 illustrates the sequence of chemical reactions that the DQN resist undergoes during exposure. Another common positive resist is poly(methyl methacrylate) (PMMA), widely used as a high-resolution resist for direct write e-beam lithography as well as X-ray and deep UV lithography. Upon illumination (by UV or e-beam), the exposed region in PMMA becomes soluble in the developer through chain scission. Doses for e-beam and DUV are 50–500 $\mu C/cm^2$ with 20–50 kV of the beam energy and >500 mJ/cm^2 at 248 nm, respectively. Typical developer solutions for positive resists are alkaline ones including potassium hydroxide (KOH) or sodium hydroxide (NaOH) diluted with water.

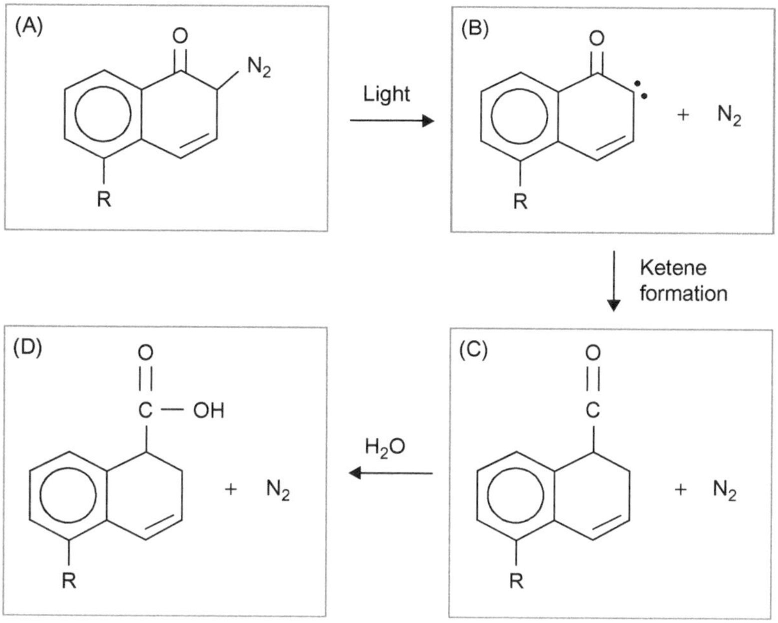

FIGURE 4.3

The underlying chemical reactions for the DQN-based positive resist. (A) and (B): when a diazoquinone (DQ), the base insoluble photoactive compound, is exposed to light, it produces a carbine, which then undergoes the Wolff rearrangement to form a ketene. Water in the film interacts with the ketene to form a base-soluble carboxylic acid product.

4.1.1.3.2 Negative resists

Negative resists are of special interest to the micro-/nanofluidics community as the resists themselves often serve as a structure for fluidic elements and/or their supports. The exposed regions of the negative resists are rendered insoluble in the developer as crosslinking of main chain or side chain takes place by photolysis. The common sensitizer in negative resist is an azide compound—for example, bis (aryl)azide rubber, which promotes crosslinking once photochemically activated. The unexposed regions are removed (or developed) with solvents like xylene, toluene, and methylene chloride. The advantages and disadvantages of positive and negative resists are compared in Table 4.1.

4.1.1.4 Resist application

Spin coating works well to create a thin, uniform film from the PR solution when a substrate is flat and smooth. But in many MEMS and micro-/nanofluidics applications, we often encounter the substrates that are not flat but rough, etched, or curved. Spin coating over these nonconventional substrates is challenging as depicted in Figure 4.4—as an example of a planar substrate with severely etched

Table 4.1 Comparison of Positive and Negative Resists

Positive Resists	Negative Resists
Better for smaller feature sizes (<0.5 μm)	Better for larger feature sizes (±2 μm)
Higher contrast	Lower contrast
Poor/fair adhesion to Si	Excellent adhesion to Si
Developer attacking both exposed and unexposed regions but at different rates	Swelling of crosslinked resist (by absorption) during development, limiting resolution
More expensive	Less expensive
Easy to remove—good for lift-off	Difficult to remove—not suitable for lift-off
Medium resistance to wet chemical	Excellent resistance to wet chemical
Aqueous developer	Organic developer
AZ series, SPR series, PBS, PMMA, etc.	SU-8, COP, Kodak, KMPR, etc.

or protruded topography, grooves and crevasses will be filled with liquid PR (due to higher surface tension) resulting in highly nonuniform film thickness (thinning of the resist at convex corners and accumulation inside the cavity). The uneven surface with the flat photomask also poses an issue of light diffraction during the exposure step making it difficult to resolve finer features. When spinning and spraying are not feasible, a dry film PR can be laminated onto some substrates including etched substrates, printing circuit boards, and flexible plastic films.

4.1.1.5 Alignment and exposure

Alignment is a major criterion in all lithography techniques. Patterns on a photomask need to be positioned with respect to existing features on a wafer, allowing features on the multiple layers to be aligned correctly with respect to each other. Photomasks have alignment (or fiducial) marks—typically lines, rectangles, or cross patterns situated on both far ends of the mask. In photolithography, UV exposure with appropriate alignment is performed in an aligner. And aligners are typically equipped with two sets of microscope/cameras on both ends to view the alignment marks. The stages on which the masks and wafers are mounted can move relative to each other (x, y, and rotation) until the alignment marks on both ends of the mask are well aligned with the target features on the wafer. Two types of exposure system are contact (or proximity) aligner or stepper (also called projection). Photomasks are in a 1:1 scale for the contact or proximity lithography or 5:1 (or 10:1) scale for projection lithography (for the latter, the mask is also called reticles).

4.1.1.5.1 Contact/proximity lithography

In the contact/proximity aligner, the photomask is in physical contact with or in close proximity to the PR film on the wafer. The minimum feature size is mainly

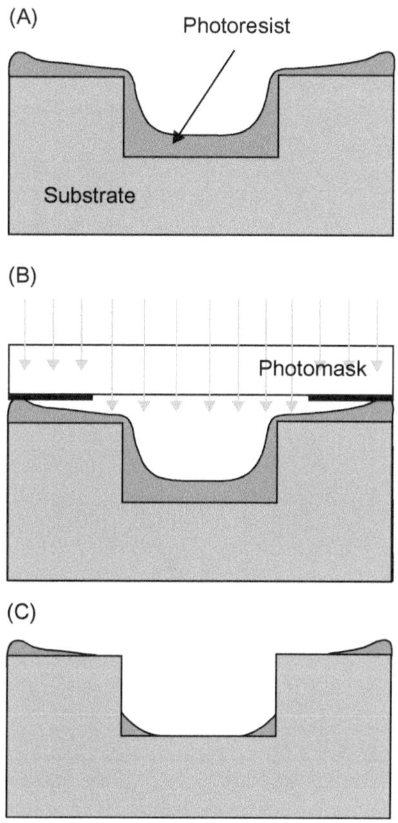

FIGURE 4.4

Schematics of the cross-sectional views of (A) nonuniformly coated PR film after spin coating on the etched substrate, (B) uneven PR film with the larger gap between the photomask and the PR film posing an issue of light diffraction, and (C) the resulting incomplete pattern transfer after development.

determined by diffraction through the gap between the mask and the PR film. Ideally, this gap can be flushed by vacuum or hard contact mode, but in reality, when two flat, right surfaces are brought together, it is very difficult to reduce the gap to less than 1 μm over the large area, making the practical resolution of contact mode lithography around 0.5 μm for the typical UV source (i-line, g-line, h-line).

Contact mode can (i) damage the photomasks and shorten their service life due to the mechanical wear and (ii) result in a high density of defects in both the PR film (lowering the lithographic yield) and the mask (again reducing its durability). A schematic of the contact aligner is shown in Figure 4.5A,B. In a

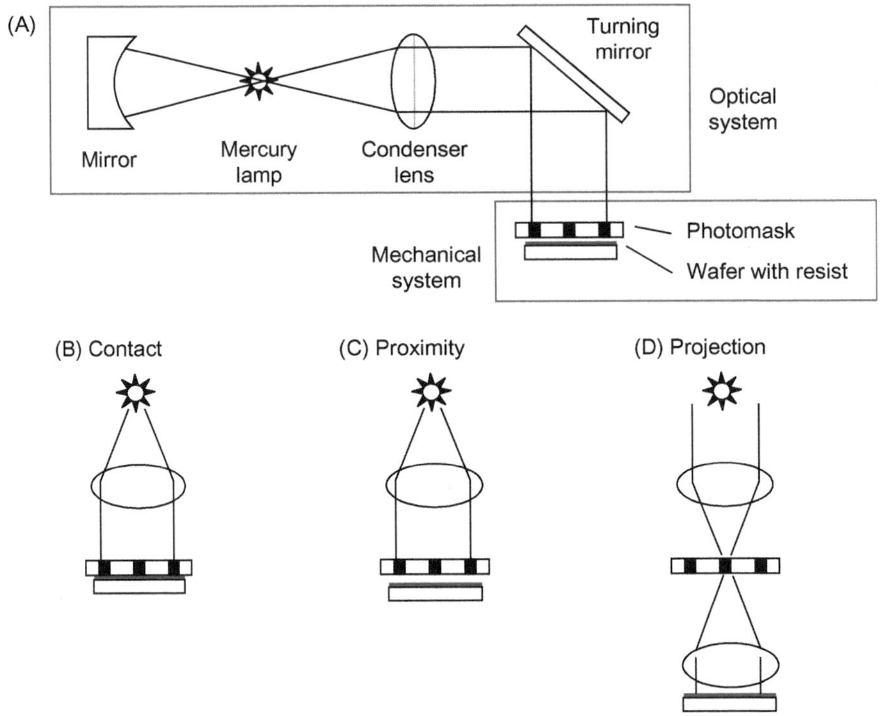

FIGURE 4.5

(A) A schematic of optical lithography system. Both photomask and wafer are on the movable stages enabling alignment before exposure, (B) contact mode (zero gap between the mask and the resist, (C) proximity mode (gap of 3–50 μm), and (D) projection mode (stepper)—additional optics between the mask and the wafer enabling the size reduction of the pattern image.

modification to contact lithography, proximity mode overcomes the issues associated with the mask damage by introducing a gap between the mask and the wafer (Figure 4.5C). A small gap (3–50 μm), however, limits the lithographic resolution. The resolution, the ability to separate two adjacent line patterns, is the criterion used for patterning accuracy and is determined by the minimum resolvable period, $2b_{\mathrm{min}}$. For contact/proximity lithography, the diffraction of the transmitted light through the gap, g, is approximated by the Fresnel formula:

$$2b_{\mathrm{min}} = 3\sqrt{\frac{\lambda}{n} \times \left(g + \frac{t}{2}\right)} \tag{4.1}$$

where λ is the wavelength of light, n is the refractive index of the resist, and t is the resist film thickness. In a typical situation of i-line exposure ($\lambda = 0.365$ μm,

$n \approx 1.6$, $t \approx 1$ μm), the resolution would be approximately 1 μm for contact lithography ($g \approx 0$) and 3.3 μm for proximity lithography ($g \approx 5$ μm), respectively.

4.1.1.5.2 Projection lithography

In the stepper mode, a complex lens system is installed between the mask and the PR, and a demagnified image of the photomask is projected and exposed repeatedly across the PR film on the wafer in a grid (Figure 4.5D). Because of the reduction lens system, the features on the PR can be considerably smaller than those on the mask. Unlike the contact aligner, the mask (or called reticle) never makes a contact with the PR film in the stepper, protecting the mask surface (reducing the defects) and prolonging its usage life. Alignment and registration is also more straightforward in the stepper. Therefore, almost all semiconductor industries use projection lithography and the stepper for IC fabrication. The resolution of the resulting pattern in projection lithography is limited by the wavelength of light used, the ability of the lens to capture light at wider angles (related to numerical aperture), and other equipment-related parameters. This relationship (Rayleigh formula) is summarized as follows:

$$R = k_1 \frac{\lambda}{\mathrm{NA}} \tag{4.2}$$

where R is the resolution or critical dimension, that is, the finest line that can be resolved, λ is the wavelength of light in the projection system, NA is the numerical aperture, and k_1 is a coefficient depending on the equipment and process conditions. Therefore, a shorter wavelength or a higher numerical aperture of the lens system would improve the overall resolution. The trend in microelectronics has been a use of smaller λ, which is the main driving force behind Moore's law—the number of the transistors on integrated circuits is doubling approximately every two years. The state-of-the-art stepper uses deep UV light from excimer lasers with wavelengths of 248 (KrF laser) and 193 nm (ArF laser), allowing the pattern resolution down to 30 nm.

One should note, however, that this strategy of decreasing the wavelength and increasing the numerical aperture encounters the competing constraint, called the depth of focus (DOF)—the spatial depth over which the image of the mask is faithfully created. The DOF is also a function of the wavelength and the numerical aperture:

$$\mathrm{DOF} = 0.5 k_2 \frac{\lambda}{\mathrm{NA}^2} \tag{4.3}$$

where k_2 is another coefficient depending on the aberration of the optical system. A higher DOF would be desirable in order to have the projected image stay in focus through the thickness of the PR films. This also means that NA should be lower for a given λ and undermine the resolution. Therefore, significant efforts have been made to reduce k_1 for the smaller resolvable features and to increase k_2 for the larger DOF, including the phase-shift masks [2,3], optical proximity correction [4], and off-axis illumination [5]. Recently, IC manufacturers introduced [6] immersion lithography to achieve the resolution below 45 nm half-pitch node. Immersion lithography is a resolution enhancement technique that replaces the usual air gap between the final lens element and the PR surface with a liquid medium with a refractive index (n) greater than 1 [7-9]. For example, the higher n of purified water ($n = 1.44$ at 193 nm) can not only increase the DOF of exposure by a factor of 2 but also enables the lens designs with NA greater than 1 (note: the NA of an air exposure system cannot exceed 1). As of 2012, Intel and other IC manufacturers use 193-nm immersion lithography for 22-nm node and expect to extend it to the next 14-nm node for the CMOS fabrication.

4.1.1.5.3 Double-sided alignment

In many MEMS and micro-/nanofluidic device fabrication, lithographic patterns on both sides of a wafer need to be aligned with respect to each other with high accuracy. This front-to-back side alignment, known as double-sided alignment, can be performed in commercial aligners. Figure 4.6 illustrates the operation of the SÜSS MA-6 system to align crosshair alignment marks on the mask to crosshair marks on the back side of the wafer [10]. First, the alignment marks on the stationary masks are viewed from the microscope objectives that are located underneath the mask, and the image of the alignment mark is stored in the computer. Then, a wafer with the already-made alignment marks on one side is loaded in the aligner with its alignment mark side facing bottom toward the microscopes, and the moving stage of the wafer is translated and rotated until these alignment marks on the wafer are aligned to the electronically stored image. The typical alignment accuracy is 2 μm.

4.1.2 Additive techniques

Additive techniques in micro- and nanofabrication refer to the processes that deposit or coat thin films on the substrate. These thin films can be metallic, semiconducting, or insulating, and serve as (i) electrical components (electrodes, interconnecting wires, insulators, resistors, capacitors, etc.), (ii) mechanical components (membranes, beams, cantilevers, etc.), and (iii) process-assisting components (etch masks, sacrificial layers, protective films, etc.). For many MEMS and micro-/nanofluidic devices, therefore, thin films are essential to provide a wide variety of functionalities. Numerous deposition/coating techniques are available to prepare thin films on substrates: physical vapor deposition (PVD) (evaporation, sputtering, etc.), chemical vapor deposition (CVD), electroplating, oxidation, liquid-phase deposition (dip coating, drop casting, spin coating, sol–gel, Langmuir–Blodgett, etc.), and so on.

The characterization of thin films deposited/grown on the substrate is often performed electrically, optically, mechanically, and spectroscopically to monitor their material properties including thickness, resistivity, breakdown voltage, index

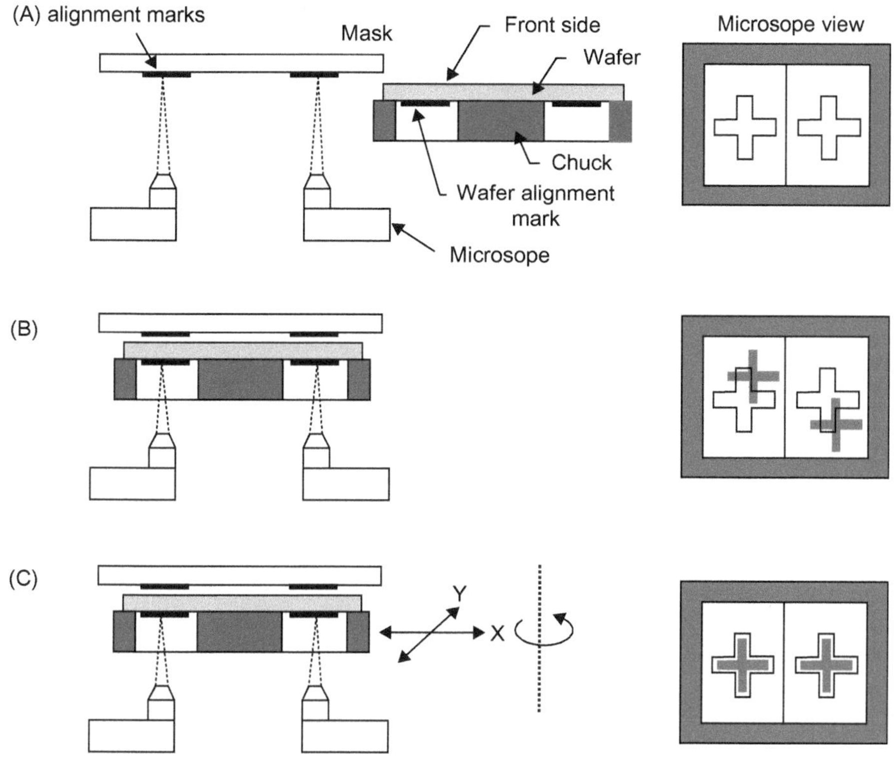

FIGURE 4.6

Double-sided alignment scheme for the SÜSS MA-6 alignment system: (A) the image of mask alignment marks is electronically stored, (B) the alignment marks on the back side of the wafer are brought in focus, and (C) the position of the wafer is adjusted by translation and rotation to align the marks to the stored image. The right-hand side illustrates the view on the computer screen as the targets are brought into alignment.
Source: *[10] Maluf N, Williams K. An introduction to microelectromechanical systems engineering.*

of refraction, transmittance/reflectance, hardness, adhesion to the substrate, and residual stress. These properties of the thin films can be adjusted to meet the requirements of each application using different deposition techniques, changing process conditions, or implementing multilayer structures. One important aspect of the additive processes is the resulting profile of the added materials: conformal versus nonconformal. Depending on the nature of the deposition/coating techniques, a film deposited over a trench could have a uniform thickness across the entire topography of the substrate or a nonuniform thickness from the processes' directionality (Figure 4.7). Both conformal and nonconformal deposition techniques are very useful because the former provides the material continuity especially useful for fabricating electrical components and the latter provides

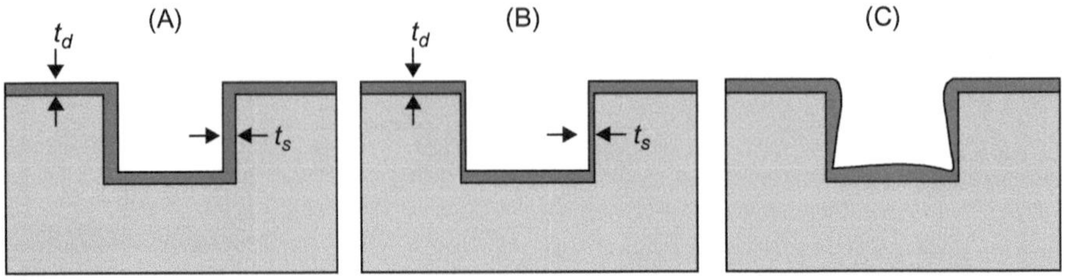

FIGURE 4.7

Schematics of the cross-sectional profiles of the deposited film on the trench: (A) conformal and uniform ($t_d \approx t_s$), (B) nonconformal ($t_d \gg t_s$), and (C) nonuniform.

a route to creating anisotropic and asymmetric features by tilting the substrates. More detailed accounts are given in the following for only the additive processes relevant to the micro-/nanofluidic device fabrication.

4.1.2.1 Spin coating

Spin coating is perhaps the most widely used method to deposit a thin film from the solution-borne materials. In the semiconductor industry, materials including PRs, polymers, and spin-on-glass are primarily deposited by spin coating. A spin-coater (or spinner) typically consists of a vacuum chuck (a flat perforated metal plate connected to a vacuum line) and a rotor that can rotate the stage up to a maximum speed of 7500–10,000 rpm at a controlled acceleration rate (Figure 4.8A). Here, we take an example of spin coating PRs on a wafer. After the wafer is mounted on the center of the chuck, a drop of PR with a known quantity is dispensed at the wafer center (Figure 4.8B). The liquid puddle is spread out to the edge of the wafer in the initial acceleration step during which more than a half of the solvent can evaporate as shown in Figure 4.8C. The thickness h of the resulting film is generally uniform across the substrate and controlled by the solution viscosity, solvent evaporation rate, and spin speed. The empirical expression is

$$h = kC\left(\frac{\mu}{\omega^2}\right)^{1/3} \tag{4.4}$$

where C is the initial concentration of polymers in solution, k is a constant, ω is the angular rotational velocity, and μ is the viscosity. Depending on the rheological behavior of the polymer, it may take a few minutes to reach an equilibrium thickness. Various factors may cause nonuniformity in the resulting film. Since viscosity increases in the acceleration stage, the acceleration rate can change the radial nonuniformity in thickness. A flow pattern around the spinner and ambient humidity influence the evaporation rate, and in turn, film thickness. Due to the higher surface tension at the edge of the wafer, the residual resist materials (called edge bead) are built up at the wafer edge (Figure 4.8D). In contact lithography, it is necessary to have intimate contact between the mask and the resist for the best

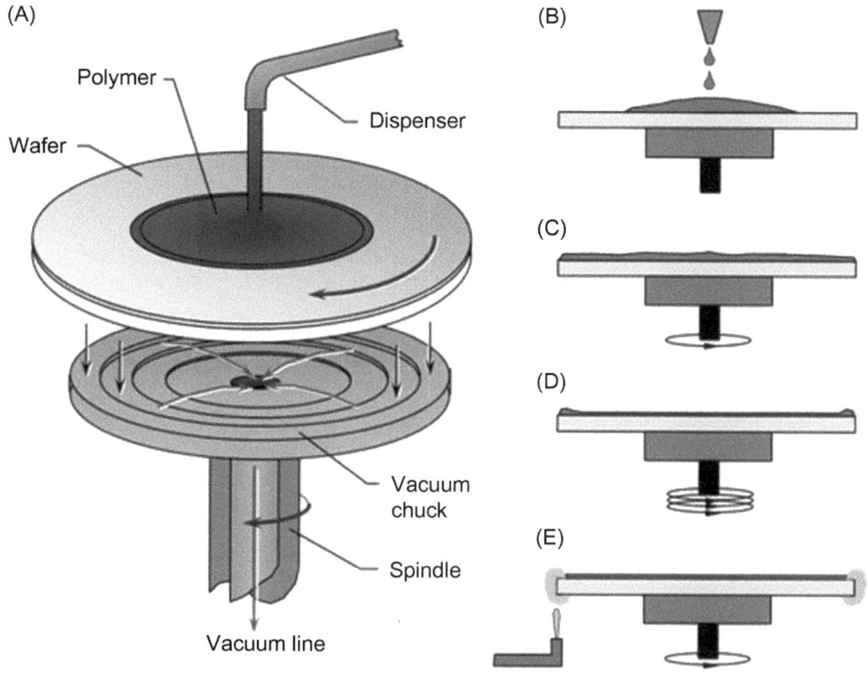

FIGURE 4.8

(A) A schematic of the spinner configuration with the step-by-step procedures of spin coating of the resist, (B) dispensing the resist, (C) spreading the resist at the acceleration step, (D) forming the edge bead after the final spinning, and (E) the edge bead removal.

resolution. Therefore, usually the edge bead must be removed prior to pattern exposure. Two general approaches to the edge bead removal are taken. In industrial facilities, the resist at the edge is selectively removed by solvent spraying within the spinner station—the edge bead removal integrated in the overall photolithography step (Figure 4.8E). When this capability is not available, the edge bead area can be selectively exposed with a larger dose and developed away prior to the main pattern exposure.

4.1.2.2 Thermal oxidation

One of the critical reasons that silicon has been a dominant material in semiconductor industry is its ability to form a high-quality oxide film. The properties and utility of a SiO_2 film grown on Si substrate are discussed in Section 4.3.1. An oxide film can be grown by reacting the silicon substrate with oxygen or water vapor (steam) at a temperature typically in a range of 1000–1200°C. The reactions for dry and steam oxidation are as follows:

Dry oxidation: $Si + O_2 \rightarrow SiO_2$

Steam oxidation: $Si + 2H_2O \rightarrow SiO_2 + 2H_2$

Typically, wet oxidation is referred to as a reaction with a mixture of O_2 and H_2O. Note that Si from the substrate is consumed during the reaction, decreasing the thickness of silicon. But the overall volume expands, increasing the total thickness of the Si/SiO_2 stack.

The generally accepted model to predict the thickness of the oxide film was developed by B.E. Deal and A.S. Grove in the 1960s [11]. The reaction can be divided into three steps occurring in succession: Step 1. Diffusion of O_2 through the gas film boundary to the surface of the Si surface, Step 2. Penetration and diffusion of O_2 through the grown SiO_2 film, Step 3. Reaction of O_2 with Si at the Si/SiO_2 interface. The resistance associated with each step usually varies as the reaction progresses, and the step with the highest resistance is considered to be rate controlling. One can develop the growth rate equation for the case in which Steps 1, 2, and 3 are rate controlling in turn and later combine the three resistances. The growth rate, da/dt, can be written as

$$\frac{da}{dt} = \frac{N^*/N_{O_2}}{(1/h) + (a(t)/D) + (1/k_s)} \tag{4.5}$$

where N^* is the concentration of source O_2, N_{O_2} is the concentration of O_2 just outside of the oxide film, h is the mass transfer coefficient, $a(t)$ is the film thickness at time t, D is the diffusion coefficient of O_2 in SiO_2, and k_s the reaction rate constant. da/dt, h, $D/a(t)$, and k_s are in units of velocity. The solution to Eq. (4.4) is

$$a(t) = \frac{A}{2}\left[\sqrt{1 + \frac{4B(t+\tau)}{A^2}} - 1\right] \tag{4.6}$$

where A and B are constants that depend on temperature and oxidation conditions: $B = C_1 \cdot \exp(-b_1/T)$ and $B/A = C_2 \cdot \exp(-b_2/T)$, T in kelvin. τ is an effective initial time corresponding to an initial oxide thickness, a_0. Table 4.2 shows some values of these parameters in different oxidation conditions. Two limiting cases can simplify Eq. (4.6). For small t, we can assume $a(t) \sim 0$, that is, $da/dt = 0$. This means $a(t)$ is linear with time and independent of D. For large t, $a(t)/D$ is dominant and thus $da/dt \propto 1/a(t)$, that is, $a(t) \propto \sqrt{t}$.

Table 4.2 Parameters in Deal–Grove Model Solution for Thermal Oxidation [11]

Type	C_1 (cm^2/s)	C_2 (cm/s)	b_1 (K)	b_2 (K)
Dry	2.144×10^{-9}	0.173	1.427×10^4	2.320×10^4
Wet	5.940×10^{-10}	2.490	8.237×10^3	2.320×10^4
Steam	1.070×10^{-9}	4.530	9.049×10^3	2.378×10^4

The oxidation rate is also dependent on the dopant type, dopant concentration, and crystal orientation. For example, the oxidation rate of the (111) plane is slower than that of the (100) plane because of the (111) plane's denser atomic arrangement. The thickness of the oxide can be measured using the color chart or more accurately with ellipsometry.

TIPS—NATIVE OXIDE

Like many metals, a thin oxide layer is formed on top of the silicon surface relatively quickly ($\sim$10 min) at ambient conditions. This 1–2 nm-thick oxide is called native oxide, and its quality is much poorer than thermally or CVD-grown counterparts. In order to obtain a good quality of the thermal oxide, this native oxide should be removed prior to oxidation in a tube furnace. It has been known that the PR adhesion to the Si substrate gets improved when a native oxide is removed prior to the application of the adhesion promoter and PR.

4.1.2.3 Physical vapor deposition

The thin films of various materials can be formed by a solid, liquid, and vapor (or gas). In PVD, solid materials are vaporized and transported to the target substrate by heat or by an energetic beam of electrons, photons (laser), or positive ions (sputtering). The PVD processes are typically performed in high vacuum, where molecules travel in straight lines ($K_n > 1$, free molecular flow regime) and the line-of-sight deposition takes place. Therefore, the uniformity of the deposition rate is often determined by source-to-target geometry. The PVD processes include evaporation, sputtering, laser ablation, and hybrid methods. Although chemical reactions in the gas phase and at the substrate are generally absent, simple metal oxides and nitrides can be grown by bleeding oxygen or nitrogen gas during the deposition process. The resulting films from the PVD processes are typically polycrystalline, and the grain size can be controlled via the deposition conditions and post-treatments.

The nonconformal nature of the line-of-sight deposition from the PVD processes can be beneficial in creating nanochannels [12]. The trenches created by nanoimprint lithography (NIL) cause local shadowing effects, leading to nonuniform deposition that can reduce the original size of the channel and seal them off on the top as shown in Figure 4.9.

4.1.2.3.1 Evaporation

Thermal evaporation utilizes heat to vaporize a source material that travels through vacuum and condenses on a substrate surface. The energy for evaporating the materials is provided by passing a high current through a highly refractory metal container (resistive heating) or by focusing a high-intensity electron beam gun for local heating (Figure 4.10). The recessed containers holding the target materials are called boats or crucibles and typically made of tungsten, tantalum, molybdenum, graphite, alumina, quartz, boron nitride, or other high-temperature (refractory) metals or alloys. In the case of e-beam evaporation (preferred method), crucibles are water-cooled to prevent them from getting heated, and

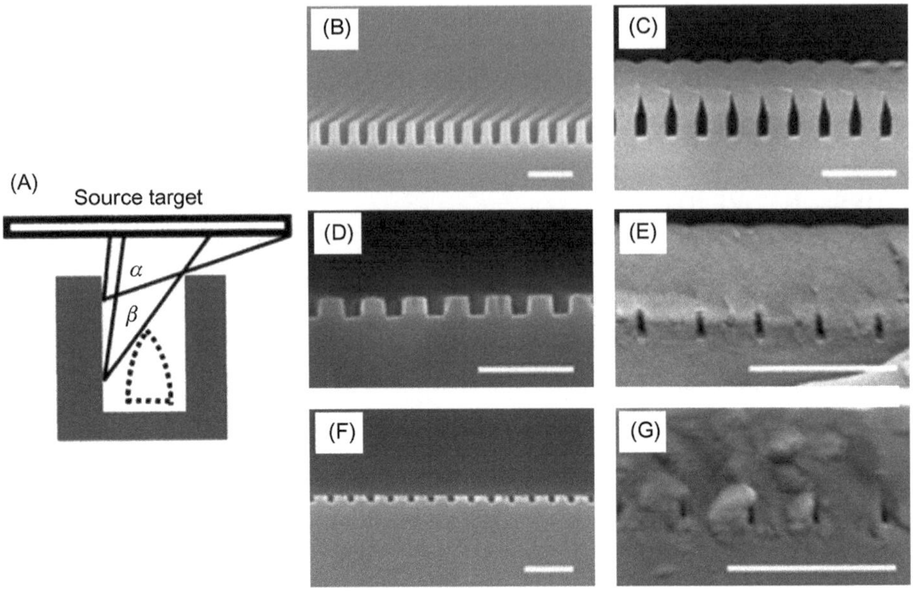

FIGURE 4.9

(A) a schematic of inclined sputtering deposition and SEM images of (B), (D), and (F) arrays of trenches fabricated by NIL and (C), (E), and (G) nanofluidic channels sealed by a single step SiO_2 sputtering at an angle. The scale bars are all 500 nm.

Source: *[12] Cao H, Yu Z, Wang J, Tegenfeldt JO, Austin RH, Chen E, et al. Fabrication of 10 nm enclosed nanofluidic channels.*

with this and localized heating by electromagnetic alignment and focusing, the sublimation of the crucible and the resulting contamination of the deposited films can be minimized. One disadvantage for e-beam evaporation is possible radiation damage to the substrate. For compounds or alloys, each component has different vapor pressure and deposition rates, and therefore the stoichiometry in the resulting film from evaporation would be different from that of the source materials. Sputtering deposition offers better stoichiometric films for compounds and alloys. The comparison of two PVD methods is summarized in Table 4.3.

4.1.2.3.2 Sputtering

Sputtering refers to the removal of atoms from the target substrate under bombardment by energetic ions. If the energy of the incident ions is greater than a certain threshold, sufficient energy can be coupled to atoms on the surface layers of the target materials, which would be ejected [13,14]. Therefore, regardless of melting temperature or heat of vaporization, sputtering can be utilized for almost any material. Unlike evaporation, sputtered atoms have the large energy flux to the substrate, knocking out the loosely bound impurities on the surface and improving the adhesion to the substrate (also making the

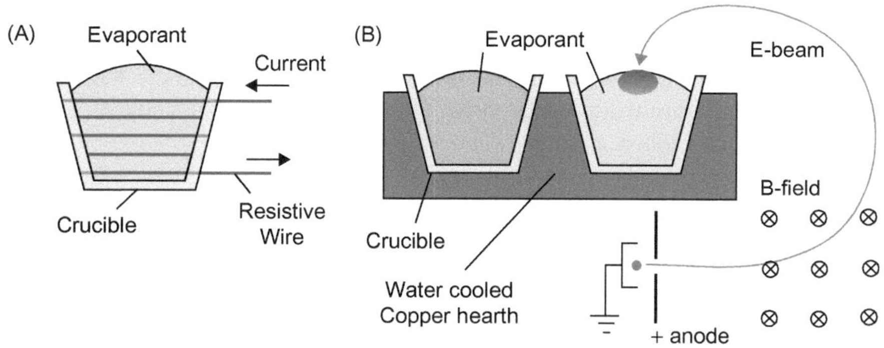

FIGURE 4.10

Diagrams for thermal evaporation by heating methods: (A) resistive (Joule heating) evaporation and (B) e-beam evaporation.

Table 4.3 Comparison Between Evaporation and Sputtering Deposition

Evaporation	Sputtering
Thermal energy	Kinetic energy (momentum)
Low energy (~0.1 eV) atoms	High-energy (1–10 eV) atoms/ions
More porous	Denser and smaller grain size
Poor adhesion	Better adhesion
High vacuum	Low vacuum
Poor step coverage (good for lift-off)	Better step coverage
Lower impurity incorporated	Gas molecules incorporated
Point source—poor uniformity	Parallel plate—better uniformity
Poor stoichiometry for alloys	Preserving stoichiometry

film denser). But the delicate substrate can be damaged due to the high-energy ion bombardment.

The energy efficiency of the sputtering process is rather low, with only highly energized atoms near the surface being ejected [15,16]. The energy transmitted deeper into the target will get lost as heat or substrate damage. However, the sputtering phenomena have been utilized primarily not in etching but in the film deposition. The sputtering rate S of a material is proportional to the ion current density (J_i), the sputtering yield (Y), and the ratio of atomic mass (M_t) to density (ρ) as shown in the following equation.

$$S = k\frac{J_i M_t Y}{\rho} \tag{4.7}$$

where k is a proportional constant. The sputtering yield is defined as the number of atoms ejected from the target material per incident ion and is generally on the order of unity. The sputtering rates of some selected materials under the standard condition (500 eV normally incident Ar^+ ion with a current density of 1.0 mA/cm^2) are listed in Table 4.4.

4.1.2.4 Chemical vapor deposition

CVD forms thin films on the surface of a substrate by thermal decomposition and/or reaction of gaseous compounds (Table 4.5). CVD is different from PVD in

Table 4.4 Measured Sputter Rates for $E = 500$ eV Argon Ion at 1 mA/cm^2 (Normal Ion-Beam Incidence) [17]

Target Material	*A* (min)	Target Material	*A* (min)
Be	159	Dy	1040
C	40	Er	881
Al	630	Hf	590
Si	310	Ta	380
Ti	336	W	340
V	337	Re	470
Cr	530	Os	440
Mn	874	Ir	540
Fe	429	Pt	792
Co	510	Au	1553
Ni	570	Pb	3073
Cu	818	Bi	8798
Ge	900	Th	740
Rb	4000	U	660
Y	837		
Zr	570	CdS	2100
Nb	390	GaAs (110)	1500
Mo	421	GaP (111)	1400
Ru	580	GaSb (111)	1700
Rh	650	InSb	1300
Pd	1150	$LiNbO_3$ (y-cut)	400
Ag	1833	PbTe (111)	3300
Sn	1217	SiC	320
Sb	3238	SiO2	400
Sm	990	Al_2O_3	129
Gd	1027	$Bi_{12}GeO_{20}$	1290

Table 4.5 The Main CVD Reactions and Their Precursors (note that temperature cited for thermally activated CVD. PECVD will have lower deposition temperature) [18]

CVD Reaction	Precursors	Examples	T_{dep} (°C)
Thermal decomposition (pyrolysis)	Halides	$TiI_4 \rightarrow Ti + 2I_2$	1200
	Hydrides	$SiH_4 \rightarrow Si + 2H_2$	600–1150
	Metal carbonyl	$Fe(CO)_5 \rightarrow Fe + 5CO$	370
	Metal-organic	$(C_8H_{10})_2Cr \rightarrow Cr + 2C_5H_{10} + 6C$	500
Reduction (precursors + reducing agents)	Halides	$SiCl_4 + 2H_2 \Leftrightarrow Si + 4HCl$	900–1200
		$WF_6 + 3H_2 \Leftrightarrow W + 6HF$	550–800
Oxidation (precursors + oxidizing agents)	Halides	$TiCl_4 + O_2 \rightarrow TiO_2 + 2Cl_2$	
	Hydrides	$SiH_4 + 2O_2 \rightarrow SiO_2 + 2H_2O$	350
	Metal-organic	$Zn(C_2H_5)_2 + 4O_2 \rightarrow ZnO + 5H_2O + 2CO$	250–500
Hydrolysis (precursors + hydrolyzing agents)	Halides	$2AlCl_3 + 3CO_2 + 3H_2 \rightarrow Al_2O_3 + 6HCl + 3CO$	800–1150
Nitridation (precursor + nitriding agents)	Halides	$TiCl_4 + \frac{1}{2}N_2 + 2H_2 \rightarrow TiN + 4HCl$	1200
	Hydrides	$3SiH_4 + 2N_2H_4 \rightarrow Si_3N_4 + 10H_2$	800
	Halohydrides		
Disproportionation	Halides	$2GeI_2 \Leftrightarrow Ge + GeI_4$	
		$3GaCl \Leftrightarrow 2Ga + GaCl_3$	
		$2SiI_2 \Leftrightarrow Si + SiI_4$	
Synthesis	Two or more gaseous precursors	$TiCl_4 + 2BCl_3 + 5H_2 \rightarrow TiB_2 + 10HCl$	1100
	Metal-organic	$Ga(CH_3)_3 + AsH_3 \rightarrow GaAs + 3CH_4$	

a way that gases, evaporated liquids, or chemically gasified solids are utilized as source materials. The desired material is deposited directly from the gas phase onto the surface, and the binding of the film to the substrate is chemical in nature, resulting in better adhesion.

In addition, the films with precise stoichiometry can be deposited from CVD, widening the choice of the materials. Almost all technologically relevant materials can be deposited as thin films via CVD. Unlike PVD, the behavior of molecules in CVD chambers is governed by the diffusion and surface migration ($K_n < 0.01$), and therefore the CVD films have much better step coverage (more conformal) than the films deposited by PVD.

The important factors in determining the film quality are gas ratio, temperature, and pressure. Precursor gas streams, CVD chamber, and substrates are typically heated over 300°C to promote reaction kinetics and ensure the reasonable deposition rates. Therefore, most polymeric substrates having low melting temperature would be restricted in the CVD process. The CVD-grown films are widely used as etch masks, insulator films, ferroelectrics, and contacts in integrated circuits and also as a variety of MEMS structures such as membranes and cantilevers. In addition, many nanomaterials (e.g., nanotubes and nanowires) are synthesized based on the CVD processes. The common configuration of CVD deposition apparatus is a horizontal furnace tube similar to high-temperature furnace for oxidation and doping. As wafers are stacked vertically in the tube, high-throughput and batch processes can be achieved. Many variants of CVD are commercially available including low-pressure CVD (LPCVD), plasma-enhanced CVD (PECVD), and atomic layer deposition (ALD). The common reaction types and examples of the deposited films are listed in Table 4.5.

4.1.2.4.1 Low-pressure CVD

LPCVD is one of the most popular CVD processes, and subatmospheric pressure (0.1−2 Torr) helps to reduce undesirable chemical reactions and improve film quality and uniformity across the wafer. The reactor configuration is very similar to a high-temperature furnace that is coupled with a vacuum pump on one end and precursor vapor sources bleeding from the other end. LPCVD typically operates at relatively high temperature (over 800°C), promoting the transport of reactant species and surface migration and therefore leading to conformal films. However, it is not suitable for the substrates with metal or polymeric films. Polysilicon, silicon nitride, low-temperature oxide (LTO), diamond films are some examples of the films grown from LPCVD. LPCVD-grown silicon nitride films are known to be an effective etch mask for wet anisotropic etching due to their pinhole-free high quality. In addition, the stress of the silicon nitride films can be controlled by the operating pressure and stoichiometric ratio of the precursors.

4.1.2.4.2 Plasma-enhanced CVD

The high-temperature requirement for LPCVD can be lifted when plasma source is used to supply the energy needed for chemical reactions to occur. Substantial deposition rates (tens of nanometers per minute) with a wider range of substrate materials (some polymers, metals, etc.) can be achieved with low operating temperature (typically 200–400°C, but sometimes at room temperature). Films deposited at low temperatures are typically associated with fewer stress issues. The main drawback of the PECVD-deposited films compared to the LPCVD ones is poorer quality—less pure due to the incorporation of foreign molecules. For example, hydrogen is often incorporated in PECVD-grown silicon nitride films, resulting in possible device instability. Examples of the films deposited by PECVD include silicon oxide, silicon nitride, silicon oxynitride, amorphous silicon, and diamond-like carbon.

4.1.2.4.3 Atomic layer deposition

ALD is a type of CVD process and used for depositing ultrathin films. ALD deserves a special attention as it is known to provide a truly atomistic control over dimension and may find relevant applications in nanofluidics. Unlike more common CVD processes, the unique feature of ALD is that the deposition rate is self-limited and therefore the total film thickness depends on the number of ALD cycles rather than the exposure time. This layer-by-layer deposition is achieved by pulsing source vapors into the reactor alternately, one at a time, separated by purging or evacuation periods, thereby breaking down the overall chemical reaction into two or more self-limited reactions [19,20]. In case of Al_2O_3 (aluminum oxide), trimethylaluminum (TMA) and water are used as precursors. TMA is first pulsed into the reactor chamber and forms a single monolayer on the surface of a substrate. Then, an inert gas such as nitrogen or argon is injected to purge the reaction byproducts and unreacted TMA. Subsequently, a wafer vapor is pulsed into the chamber to react with the first TMA layer to form an alumina film. The chamber is purged with an inert gas once again to remove residual gases and complete one cycle. The cycles can be repeated as many times as needed to achieve the desired film thickness. Since the deposition rate is precisely controlled by the number of the cycles and the diffusion of the reactants takes place in sufficient time, uniform coating of highly nonplanar features (nanoholes and nanowires) is possible. In microelectronics, ALD is considered to be the state-of-the-art technology for depositing high-k gate oxide and dielectrics. Intel, for example, in 2005 implemented the ALD process to deposit HfO_2 (high-k dielectric) for the 45 nm technology [21]. Because of the slow deposition rate, attaining thick (>1 μm) films may not be practical. However, thanks to the accurate control over the thickness; nanochannels with the exact dimension can be created using surface machining with ALD.

4.1.2.5 Electrochemical deposition

Electrochemical deposition, or electrodeposition (also known as electroplating), is a process of depositing conducting/semiconducting materials onto a substrate (often conducting) using an electric field and redox reaction. An electrical current is used to reduce cations of a target material from an electrolyte and coat on the surface of the electrode (cathode). The anode (counter electrode) that is used to close the circuit by injecting electrons into the electrolyte is typically made of platinum. Electrodeposition is beneficial because it enables to grow a conductive film of up to several hundreds of microns. Note that the metal films deposited by PVD or CVD methods are limited to a few microns. This technology is relevant to microfabricated devices as the electroplated materials can be selectively deposited on the substrate using the photopatterned film as a mask. In the early MEMS development era, many microscale metal structures of complex geometry were fabricated by electroplating metals on a thick PMMA mold patterned by the LIGA (Lithographie, Galvanoformung, Abformung) process [22]. LIGA, however, entails X-ray illumination, and its capital cost for X-ray facility (synchrotron source) and X-ray masks are very high, which deprived much attention from the field. The UV-version of the LIGA process with very thick SU-8 is capable of producing similar high-aspect-ratio (HAR) microstructures [23] and gains more popularity due to its compatibility with the mainstream microelectronic and MEMS fabrication.

4.1.3 Subtractive techniques

Patterns defined in lithography are transferred or transcribed into the substrate using the subtractive processes. Materials can be removed either by physical sputtering, denoted as physical etching, or by chemical reaction, denoted as chemical etching. While the physical erosion by laser ablation is the example of the former, wet etching is the common example of the latter. In some cases, physical and chemical etching can take place simultaneously as in the case of reactive ion etching (RIE). Etching techniques can also be classified as wet or dry etching techniques depending on whether the etch medium is a liquid or a vapor. While wet etching is primarily done chemically by nature, dry etching makes use of physical, chemical, or physical/chemical etching. In addition, etching can be categorized as isotropic and anisotropic etching in regard to resulting etching profiles and directionality. Figure 4.11 shows the representative etching profiles from isotropic (uniform in all directions) and anisotropic etching (preferential in one direction over others). The removal mechanism (physical vs. chemical etching) often has to do with the etching directionality. Physical etching relies on either momentum or energy transfer of ions or photons, which tend to be more directional. In the direction of highest momentum or energy density, highest removal of materials takes place, resulting in anisotropic etching. On the other hand, chemical etching is governed by thermodynamics and reaction kinetics and tends to be isotropic.

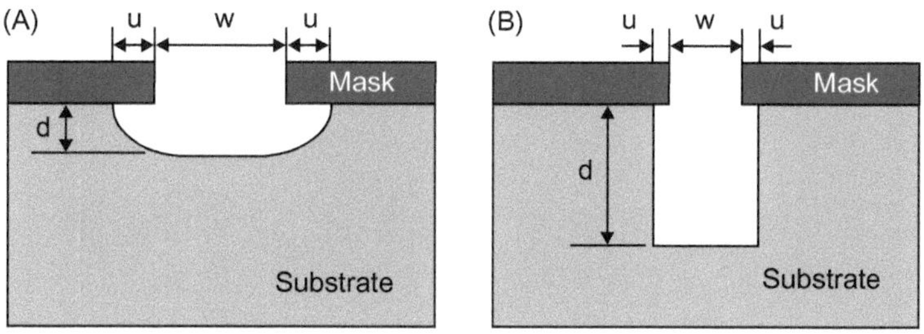

FIGURE 4.11

Typical etching profiles from (A) isotropic etching and (B) anisotropic etching. *d* is the etching depth, *u* is the undercut, and *w* is the width of the etch window.

A notable exception is the etch rate dependence on the crystal orientation or composition of the materials.

An etch rate, R, is defined by a total etched depth, d, divided by a total etching time, t, i.e., $R = d/t$. Accounting for an amount of the undercut, we can quantify the degree of anisotropy, A, as follows:

$$A = 1 - \frac{R_L}{R_V} = 1 - \frac{u/t}{d/t} \tag{4.8}$$

where R_L is the lateral etch rate (u/t) and R_V is the vertical etch rate (d/t). In isotropic etching, an etching depth is similar to an undercut distance ($u \approx d$), i.e., $A \approx 0$. In anisotropic etching, $u \ll d$, i.e., $A \approx 1$. Another important factor in a subtractive process is *selectivity*, S, which is defined as the ratio of material removal rates between the target and masking materials: $S =$ Etch Rate (for etched material) /Etch Rate (for mask material). Ideally, a subtractive process removes the target material without degrading the masking material. But often selectivity needs to be considered for material compatibility and etching duration. In general, wet etching provides a higher selectivity than dry etching. In this section, we will cover the wet and dry etching with an emphasis on etching of the three most common substrates for micro- and nanofluidic devices, namely, silicon, glass, and polymers.

Excellent reviews on the common etchants and their etch rates are available for a wide range of materials including metals and ceramics [24,25]. Table 4.6 summarizes wet and dry etchants for the common thin films and corresponding etch rates.

4.1.3.1 Wet etching

In wet etching, a solid material is reacted with liquid etchant producing soluble byproducts. Wet etching often exhibits faster etching rate than dry etching and also accommodates more wafers per batch and is suitable for mass manufacturing.

Table 4.6 Type of Etchants and Etch Rates for Wet and Dry Etching Processes of the Common Materials [10,24]

Target Substrate	Wet Etchants (Aqueous Solution)	Etch Rate (nm/min)	Dry Etchants (Plasma or Vapor)	Etch Rate (nm/min)
silicon	$HF + HNO_3 + CH_3COOH$ (isotropic)	1–20	SF_6 (RIE) (varies)	0.1–0.5
	KOH (anisotropic)	0.5–3	SF_6/C_4F_8 (DRIE) (anisotropic)	1–15
	EDP (anisotropic)	0.75	XeF_2 (isotropic)	0.1–10
	TMAH (anisotropic)	0.5–1.5		
silicon oxide (thermal)	HF	2300	$CHF_3 + O_2$	5–150
	$5NH_4F + 1NF$ (buffered HF)	100	$CHF_3 + CF_4 + He$	25–600
			HF vapor	66
silicon nitride (LPCVD)	Hot H_3PO_4	5	SF_6	150–250
			$CHF_3 + CF_4 + He$	200–600
aluminum	$H_3PO_4 + HNO_3 + CH_3COOH$	530	$Cl_2 + SiCl_4$	100–150
	HF	4	$Cl_2 + BCl_3 + CHCl_3$	200–600
gold	$KI + I_2$	660		
titanium	$HF + H_2O_2$	110–880	SF_6	100–150
tungsten	H_2O_2	150	SF_6	300–400
	$K_3Fe(CN)_6 + KOH + KH_2PO_4$	34		
chromium	$Ce(NH4)_2(NO_3)_6 + CH_3COOH$	93	Cl_2	5
PR	$H_2SO_4 + H_2O_2$ acetone	>100,000 >100,000	O_2	350

Etch rate can be controlled by the reactant concentration and temperature, and higher temperature and greater concentration would increase etch rate.

4.1.3.1.1 Isotropic wet etching

The primary etchant for silicon oxide or glass substrates is hydrofluoric acid (HF) [26] or buffered oxide etch (BOE) [27], which are extremely popular in creating glass microfluidic chips.

$$SiO_2 + 6\,HF \rightarrow H_2SiF_6 + H_2O \tag{4.9}$$

At elevated temperature, SiO_2 dissolves in concentrated alkali.

$$SiO_2 + 2NaOH \rightarrow Na_2SiO_3 + H_2O \tag{4.10}$$

HF-based wet etching gives an isotropic etching profile, i.e., undercutting the masking layer and resulting in semi-circular sidewalls or sidewalls that taper downward (Figure 4.11). The depth of the etched features becomes coupled with their lateral dimension, making it inappropriate to generate high-resolution and high-density features. Therefore, when microchannels in glass-based microfluidic chips are designed, the desired opening size (channel width) should be determined by $w + 2u$, in case of perfectly isotropic etching, $w + 2d$ (in Figure 4.11). Another challenge in HF-based wet etching is that continued exposure to HF required for etching deeper features can result in PR delamination. For this reason metal masking layers of aluminum or gold with an intermediate chrome adhesion layer are used to protect the substrate [28]. Meanwhile, a buffered solution can be prepared by adding a related salt like ammonium fluoride (NH_4F) to HF. Buffering in BOE is beneficial in that it helps not only to keep the etch solution at its maximum strength over use, time, and dilution by preventing depletion of fluoride ions but also reduces etching and delamination of the PR mask film. Figure 4.12 shows the example of the nanochannels and microchannels that were etched by HF and BOE solutions.

Isotropic etching of Si can be conducted in a mixture of nitric acid (HNO_3), HF, and acetic acid (CH_3COOH), known as HNA.

$Si + HNO_3 + 6\,HF \rightarrow H_2SiF_{6(aq)} + HNO_2 + H_2O + H_{2(g)}$

HNO_3 first oxidizes silicon, and at the same time, the resulting silicon oxide gets etched by HF. Acetic acid is preferred as a diluent compared to water because it decreases the dissociation of the nitric acid and thus preserves its oxidizing power. The relative concentration of HF and HNO_3 in water determines not only the resulting etch rate and surface roughness but also anisotropy in the etched profile. A faster etch rate tends to result in a rougher etched surface and a more isotropic etch profile. The common mask for HNA is SiO_2 with the selectivity of around 80 (i.e., the etch rate of Si is 80 times faster than SiO_2 in HNA).

4.1.3.1.2 Anisotropic wet etching

When a single-crystalline silicon wafer is exposed to some alkaline solutions such as KOH, NaOH, and LiOH, different etching rates are observed with respect to the crystal orientation of the silicon substrate (see Section 4.3.1 for discussion of the Si properties). Due to the highest atomic density along the (111) planes, the etch rate is slowest compared to a much faster etch rate along the (100) or (110) planes. Although the atomic bond density of the (111) plane is approximately 15% higher than the other two planes, it does not explain the stark difference in etch rate—the etch rate in the (111) plane is two to three orders of magnitude smaller than that of the (100) or (110) planes. Along with the bond density, the number of the dangling bonds and the type of bonds (e.g., sp, sp2, sp3) also

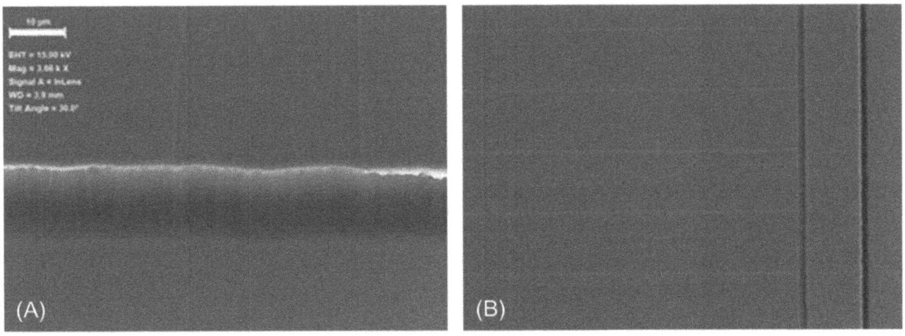

FIGURE 4.12

(A) An SEM image of a micro-/nanochannel interface. (B) An optical micrograph image of a micro-/nanochannel interface for the image in (A). The channels are etched into glass using 49% hydrofluoric acid for the microchannels and 10:1 buffered oxide etch for the nanochannels. The microchannels are ~100 μm wide and 9.5 ± 0.4 μm deep while the nanochannels are 38 ± 1.5 μm wide and 50 ± 6 nm deep.

Source: *[29] Prakash S, Pinti M, Bhushan B. Theory, fabrication, and applications of microfluidics and anofluidic biosensors.*

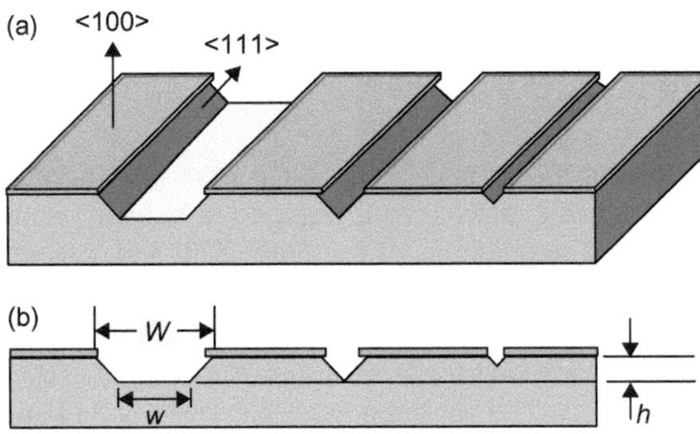

FIGURE 4.13

(a) A schematic of the anisotropically etched microchannels on silicon and the relative crystalline plane, (b) the cross-section view of (a).

contribute to the different etch rate. Figure 4.13 shows a schematic of the anisotropic etching of (100) and (110) Si wafers along with the relationship of each plane and the resulting dimensions.

$$W = w + \frac{2h}{\tan(54.74°)} \tag{4.11}$$

The angle between (111) planes and (100) planes is deterministically defined to be at 54.74°, introducing the concept of the geometrical etch stop. When a rectangular or square window pattern with each side parallel along the (100) plane on a (100) Si wafer is continually etched, etching will stop at a line or a point regardless of etching duration and the etching depth will be determined purely by the size of the mask opening. The resulting etched profile would be a channel with an isosceles triangle or an inverted pyramid. If one desires to terminate etching prior to the geometrical etch stop, a layer of silicon oxide, silicon nitride, or silicon heavily doped with boron can be used as an etch stop because of their high resistance in many alkaline etchants. For the same reason, silicon oxide and silicon nitride are the most frequently used mask materials in wet anisotropic etching. Although anisotropic wet etching theoretically allows for the production of an atomically smooth surface formed from terraces, the production of hydrogen bubbles during silicon etching by KOH can cause local fluctuations in concentration of reactants, leading to inhomogeneity in etch rate and rough surfaces. Etching at moderately elevated temperatures and with rigorous stirring reduces local roughness to below 20 nm.

The selection of a proper anisotropic etchant is typically based on the type of the mask material available and the selectivity and etch rate of the target Si planes. Potassium hydroxide (KOH) is one of the most popular etchants due to its high etch rate and high anisotropy (i.e., etch rate in (100) planes is more than 500 times faster than the other planes). Two noticeable disadvantages of KOH are its CMOS incompatibility and aggressiveness. KOH is highly corrosive and attacks almost all the materials, making it difficult to find a suitable mask material. In fact, an LPCVD silicon nitride is considered to be the only viable mask that can survive for extended etching processes. The PECVD silicon nitride also exhibits high selectivity, but a large number of pinholes in the film allow the etchant to penetrate the mask layer and attack the underlying substrates and components. A thermally grown silicon oxide layer can be used as a mask for a shorter etching time. Potassium ions in the etchant are a major source of ionic contamination leading to a failure in semiconductor devices. Ethylene diamine pyrocathecol (EDP) is an alternative to KOH and is known to be CMOS compatible. It has a relatively high etch rate while a layer of SiO_2, Si_3N_4, or metal can be used as a mask material. EDP, however, is very toxic, degrades quickly when contacting with air (rarely reused), and produces less anisotropic profiles compared to KOH. Finally, tetramethyl ammonium hydroxide (TMAH) is CMOS compatible and easy to handle and generates higher anisotropy than EDP. Moreover, both SiO_2 and Si_3N_4 can be robustly used as an etch mask. However, the etch rate of TMAH is slower than KOH or EDP.

4.1.3.1.3 Metal-assisted chemical etching

Metal-assisted chemical etching (MACE) is an electrodeless chemical etching technique to produce submicron or nanometer anisotropic structures in semiconductor substrates. When a silicon surface partly covered by a thin layer of noble metal (e.g., Au, Pt, Ag) is subject to an etchant composed of HF and an oxidant agent (mostly H_2O_2), the Si beneath the noble metal gets oxidized and etched away anisotropically, generating pores or wires in the substrate. After its first demonstration [32], the MACE technique gained increasing attention in the research community due to the fact that it is truly a low-cost and facile method to fabricate various Si nanostructures with the ability to tune the geometry, dopant type and level, and crystalline orientation of Si. Various 1D, 2D, and 3D structures in silicon include straight and zigzag silicon nanowires, vertical and helical pores, vertical or inclined channels, and spiral structures [33–37]. Porous Si and Si nanowires have been fabricated using MACE for the applications in antireflectance coating [38], photovoltaic [39], Li-ion battery [40], thermoelectric device, and gas sensors [41,42]. Figure 4.14 shows nanometer scale pores created by a single catalyst particle, hinting the possibility of constructing nanochannels by depositing narrow metal lines and performing the MACE process. More detailed information in etching mechanism and characterization can be found in the recent review paper [43].

4.1.3.2 Dry etching

Dry etching utilizes a vapor or plasma to remove materials and relies on one or more of the following three mechanisms: physical etching (thermal or sputtering), reactive etching (by chemical reaction), and photodissociation (photon energy breaking bonds). Many dry etching techniques result in anisotropic etching

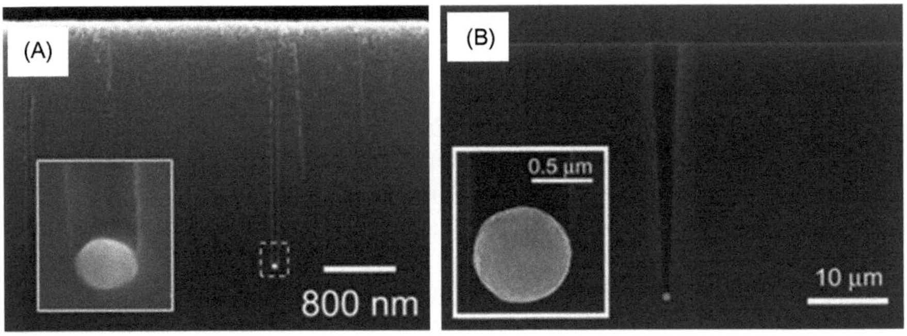

FIGURE 4.14

Cross-sectional SEM images of the pores etched in a Si(100) wafer with (A) Ag particles in an aqueous solution containing 5.3 M HF and 1.8 M H_2O_2 for 1 min [30] and (B) an Au particle in 2.6 M HF and 8.1 M H_2O_2 for 1 h.

Source: *[31] Lee C-L, Tsujino K, Kanda Y, Ikeda S, Matsumura M. Pore formation in silicon by wet etching using micrometre-sized metal particles as catalysts.*

profiles, but some (e.g., HF vapor etching of glass or XeF_2 etching of silicon) are capable of producing isotropic etching.

4.1.3.2.1 Ion-beam etching (ion milling)

Ion-beam etching is a nonreactive etching process by which material is removed through physical sputtering. The discussion of sputtering can be found in Section 4.1.2.3. Because the etch rates for different materials rarely differ by more than a factor of 2 (Table 4.4), the etching selectivity of the ion milling process is very poor. Recall that the selectivity achieved by RIE is often an order of 10:1 to 100:1. Therefore, a modification of ion milling, called reactive-ion-beam etching, was developed in which sputtering takes place in an atmosphere of reactive gases such as oxygen or chlorine [44]. In case of oxygen, the reaction does not act to increase the etch rate of the target but rather helps to decrease the etch rate of the mask material.

4.1.3.2.2 Reactive ion etching

Reactive ion etching (RIE) systems are the most common dry etchers using RF excited plasmas for etching of various materials [45]. Source gases introduced in the chamber undergo a series of chemical reactions producing plasma species, and the resulting reactants, charged or uncharged, are transported to the substrate that is to be etched. The central theme in RIE is that the dissociated ions in plasma are accelerated onto the substrate and this ion bombardment assists the etching reaction that is otherwise too slow to be effective. The etched profile tends to be anisotropic, independent of the crystal orientation of the substrate (in case of silicon), and the etch selectivity is reasonably high.

Numerous RIE processes have been developed to etch silicon. A key etchant for silicon is SF_6, sulfur hexafluoride, which is a clean, nontoxic processing gas and relatively affordable. The proposed reaction is

$$Si + 2SF_{6(g)} \rightarrow SiF_{4(g)} + 2SF_{4(g)}$$

The reaction does not, however, go forward at modest temperature (unlike XeF_2 which makes the reaction go forward at low temperature), and therefore ionization via plasma mode helps to create more reactive species like F^* (radicals) and F^- (ions) and facilitates the reaction. The common mask materials for silicon etching include PRs, silicon oxide, and metal films. Etching of the silicon nitride (SiN_x) film is also typically performed using a SF_6 discharge. In case of etching glass or silicon oxide (SiO_2), CF_4, or CHF_3 (Freon) is typically used for etching [45].

4.1.3.2.3 Deep reactive ion etching

As many Si-based MEMS or microfluidic devices entail bulk etching of Si with a HAR and/or a fast etch rate, deep reactive ion etching (DRIE) has gained its popularity in academia and industries alike. Some unique features in DRIE enable us to create 3D structures and highly anisotropic features from HAR channels to thru holes to an array of pillars, which is not possible with wet bulk etching dependent on the crystal orientation of the Si substrate. Figure 4.15 shows the scanning electron microscope (SEM) images of some exemplary DRIE-etched microfluidic elements.

In conventional RIE, the plasma density is limited by the method of coupling RF energy into the plasma. Since the identical RF power source is used to create the plasma and to accelerate ions, it is impossible to control the plasma density and its ion energy separately. In low-temperature RIE, the production of sidewall passivation layers occurs concurrently with etching of the substrate. This coupling makes process difficult to be controlled, limiting the achievable aspect ratio and etching speed.

On the other hand, the DRIE system uses two separate RF power sources: one to power the inductive coils to generate a bulk plasma and the other dedicated to accelerate ions vertically as an independent control of the substrate bias. This decoupling of plasma power and ion energy makes it possible to achieve high-density plasma (10^{11}–10^{12} cm^{-3}) in the low-pressure regime (<10–20 mTorr) and thus the directionality of etching. Another key technology for DRIE is the

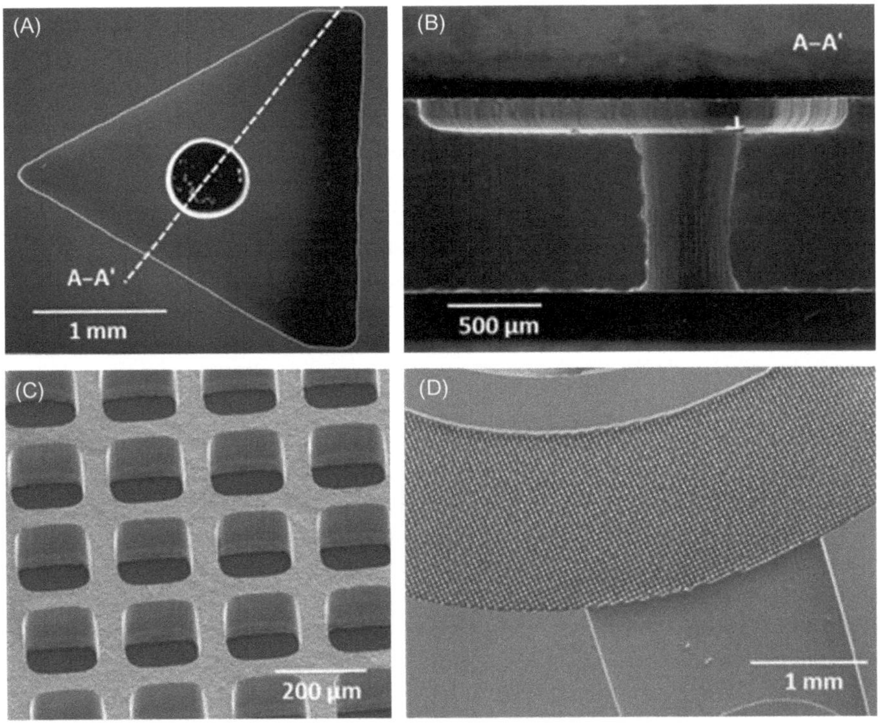

FIGURE 4.15

SEM images of the DRIE-etched microfluidic elements: (A) and (B) via hole for the microvalve, (C) a micromesh structure for the micro fuel cell, and (D) an array of micro pillars for evaporation-enhanced structures.

employment of the Bosch process, a time-multiplexed process of alternating sidewall passivation deposition and fast etching, enabling the HAR etching of the devices (Figure 4.16). The success of Bosch process hinges upon the deposition of the inhibiting films in the sidewall passivation phase, where C_4F_8 discharges

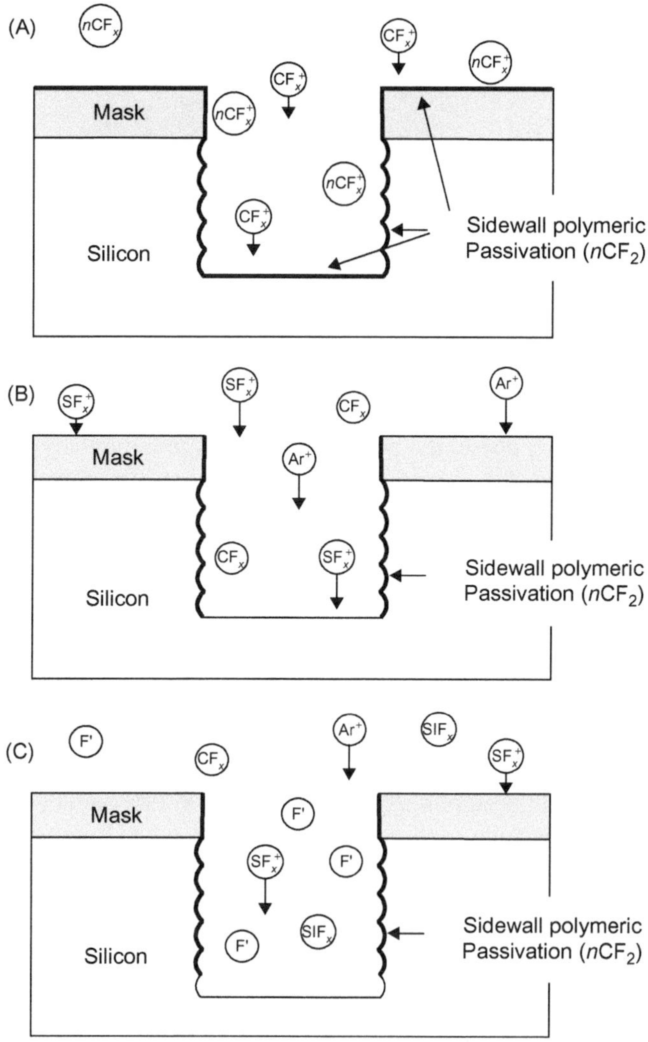

FIGURE 4.16

A schematic of the time-multiplexed Bosch process in DRIE: (A) deposition phase: Teflon-like® passivation film (nCF_2 polymer) deposited on all surfaces, (B) etch phase A: vertically directed ions removing the passivation layer on the line of sight, and (C) etch phase B: atomic fluorine reacting and etching the exposed Si.

Source: *[46] Yeom J, Wu Y, Selby JC, Mark A, Shannon J. Maximum achievable aspect ratio in deep reactive ion etching of silicon due to aspect ratio dependent transport and the microloading effect.*

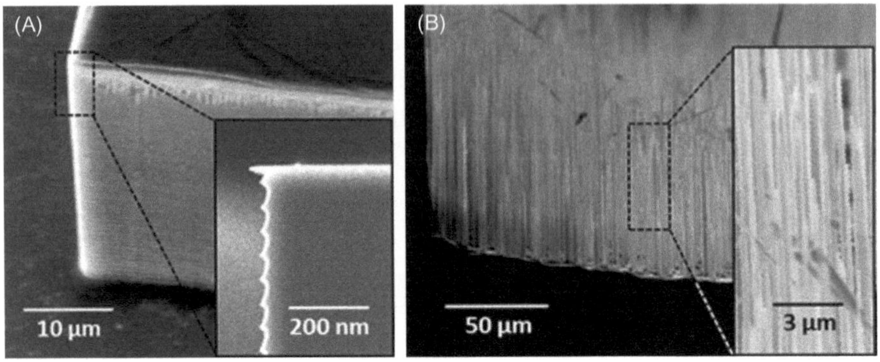

FIGURE 4.17

(A) SEM images of the DRIE-etched sidewall showing the scallops of a few hundred nanometers and (B) SEM images of the micrograss formation due to aluminum redeposition.

contain a variety of radicals that form Teflon-like® films on all surfaces. During the subsequent etching phase, the passivation film is preferentially removed from the bottom of the trenches due to ion bombardment, while preventing the sidewalls from being etched by SF_6 discharges. The duration of etching and passivation phases, applied electrode and coil powers, and gas flow rates can be prescribed independently for each phase and tuned for the desired etch rate, mask selectivity, and sidewall profiles. The nature of the cyclic process is well manifested by the nanoscale scallops seen in the sidewall profiles of the etched channel as shown in Figure 4.17A.

4.1.3.2.4 Dry etching masks

One of the important factors in plasma etching is the selectivity, which is of particular interest to those in need of deep etching. Thin film PRs are not typically resistant to harsh plasma etching conditions, and therefore, a handful of plasma-etching resistant PRs have been developed and are available in market. For example, AZ 4000 series and OGC 825 PRs are widely used as an etch mask in plasma etching. Increasing the solid content in these PRs improves their resistant to plasma but leads to higher viscosity (thus thicker films) and to more light scattering, thus reducing the photolithographic resolution. Since they are chemically inert in the plasma, PRs are in general etched away by physical sputtering or evaporation. Double (or multi) coating or spinning at a lower speed are some techniques that can generate 20 μm or even thicker film for the special applications [47].

Silicon dioxide (SiO_2) is another common choice of mask. Both CVD-deposited or thermally grown oxide can be used. While a thermally grown SiO_2 has better control over impurity incorporation than PECVD grown, the former is much harder to obtain thick films. SiN_x is less attractive as a mask than SiO_2,

Table 4.7 Comparison of DRIE Masks

Type of Mask	Deposition Method	Selectivity	Comments on Processing
PR AZ4620	Spin coating	50:1 ~ higher	Easy to coat and remove Difficult to achieve fine resolution
Silicon Oxide	PECVD	200:1 ~ 250:1	Hard to grow thick films (2 ~ 3 μm)
Silicon Nitride	LPCVD	50:1 ~ 70:1	Etched chemically, hard to grow
Aluminum	Sputtering	1000:1 or higher	Easy to deposit and remove Last very long
Chromium	Sputtering	1000:1 or higher	Easy to deposit and remove Undesired surface roughness
Titanium	Evaporation	less than 50:1	Etched chemically by C_4F_8
Nickel	Evaporation	1000:1 or higher	Micrograss on etched plane
Polyimide 2721	Spin coating	100:1 ~ 150:1	Difficult to remove after etch

especially when it is not stoichiometric because it can be etched chemically [48]. Rather, SF_6 discharge is used to etch SiN_x film with PR mask layers. Various metal layers are viable etch masks for RIE etching because of their exceptional resistance to a plasma, but the sputtering and redeposition of metal masks are known to cause the formation of micrograss on the etching plane in the case of deep silicon etching (Figure 4.17B). Table 4.7 compares the various DRIE mask materials.

4.1.4 Lift-off

Lift-off is a process used to pattern thin metal films when a proper wet etchant or dry etching chemistry is not available. Lift-off takes advantage of the fact that the step coverage of most metal thin film deposition methods is quite poor, i.e., the metal film cannot overcoat steep or undercut steps. First, a PR layer is patterned with a "negative" image (noting that the image is reversed after lift-off) and the edge of the PR layer is engineered to have reentrant profiles, promoting a discontinuous coverage of the subsequent metal coating. After the metal deposition, the "sacrificial" PR layer is dissolved away in a solvent such as acetone or PR strippers (e.g., 1165, NMP, 400T, etc.), lifting off the unwanted metal regions. A short oxygen plasma descumming is usually preferred before the metal deposition to ensure the removal of the residual PR layer on the substrate and to enhance the adhesion between the metal and substrate.

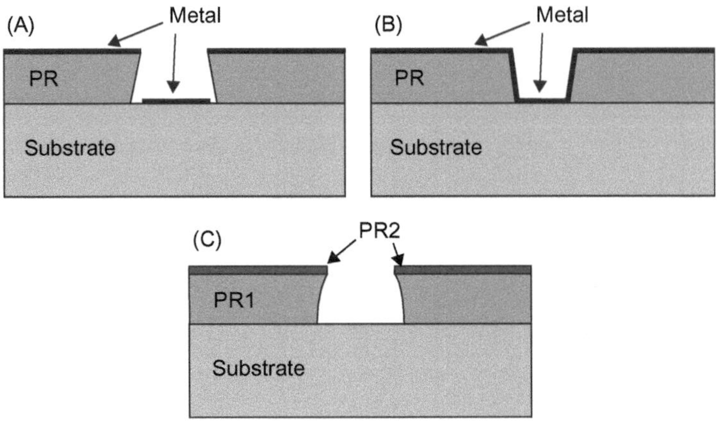

FIGURE 4.18

Lift-off patterning of metal thin film: (A) PR with a negative slope facilitating lift-off, (B) PR with a positive slope hindering lift-off, and (C) a double-layer PR structure for the ideal lift-off profile.

Figure 4.18 illustrates two possible scenarios in the lift-off process—the success of the lift-off process hinges upon the PR sidewall profile. The reentrant (or negative slope) profile in Figure 4.18A helps to shadow the deposited metal on the substrate, allowing the solvent to attack the PR layer and lift off the metal on top of the PR with ease. On the contrary, when the PR slopes outward (positive slope in Figure 4.18B), the metal layer tends to be congruous forming attachments between the substrate and top of the PR, thereby preventing clean lift-off from taking place. The thickness of the metal deposited, the step coverage of the different metal deposition methods used, and migration and ductility of the metal film are also important for the effectiveness of the lift-off process. A thicker film and more ductile metal (e.g., Au) make it more difficult to lift-off than a thin and brittle film. Evaporation is known to create the worst step coverage making it most suitable for the lift-off process.

The decent undercut profile (Figure 4.18C) can be engineered to facilitate the lift-off process. A double-layer PR structure is prepared by spin coating two layers of PRs—the top layer being thinner and less photochemically sensitive and the bottom layer being thicker and more sensitive. Amounts of light exposure and overdevelopment would result in a controllable undercut profile.

4.1.5 Bonding

Once device features have been fabricated on a substrate, the features must be isolated from the outside environment to protect the sample of interest from contamination or interferences toward sensing. Therefore, the features are often sealed using a cover layer with well-established bonding methods. Many of the wafer bonding technologies developed for IC and MEMS device packaging and silicon-on-insulator fabrication can be utilized to create the micro-/nanofluidic devices.

The recent review on the various wafer bonding techniques is available by Niklaus et al. [49], and the bonding conditions and advantages/disadvantages of each bonding method are summarized in Table 4.8. Four bonding techniques (fusion, anodic, eutectic, and adhesive bonding) will be covered in more detail in this section.

In fabricating micro- and nanofluidic devices, the substrate bonding is especially important because most patterning and pattern transfer techniques (etching, molding, printing, etc.) generate the one-sided open-cavity structures. It is the bonding technique that seals those cavities and creates enclosed channels or other fluidic elements. Figure 4.19 (p. 128) shows the cross-sectional schematic of the bonding interface for each technique. Note that fusion bonding and anodic bonding seal a channel by joining two surfaces and without altering the intended channel dimension. This is no longer valid in eutectic or adhesive bonding because the intermediate layer (metal or adhesive) has a finite thickness, which can change the eventual channel height. The thickness of the metal or adhesive layer is on the order of a few hundred nanometers, and therefore they may not be important in fabricating microchannels. However, when fabricating nanochannels, the thickness of these intermediate layers (also called spacer layers) will be the important design considerations and ultimately limit the minimum achievable nanochannel dimension.

Bonding also serves to connect one fluidic element to another. While joining two substrates may be enough to create micro-/nanochannels, the device with more complex functionality may require multilayer structures, in which a flow becomes 3D. For example, micropumps and microvalves often utilize a deformable membrane, which needs to be bonded to the other substrates to direct the flow in a desirable fashion. More 3D microfluidic devices and their applications are covered in later chapters, and the significance of the bonding techniques will become more evident.

4.1.5.1 Fusion (direct) bonding

Fusion bonding refers to a wafer bonding process with no additional intermediate layers. When two mirror-polished surfaces are brought together either spontaneously or forcibly, those two are irreversibly bonded. Fusion bonding (also called direct bonding) is typically used for glass-to-glass or silicon-to-silicon bonding. The substrates need to be sufficiently clean, flat, and smooth for bonding to take place, and the bonding quality is highly susceptible to particle defects. In silicon direct bonding, thin oxide layers including native oxide must be present on both silicon wafers. When two oxide-coated surfaces are brought together, the condensation reaction takes place to merge the interface (see Figure 4.20, p. 129). The typical annealing temperature for silicon direct bonding ranges between 600 and 1200°C. Thanks to its high bonding strength, direct bonding is frequently

Table 4.8 Summary of Wafer Bonding Techniques [49]

Bonding Technique	Bonding Conditions (Temperature, Pressure, Voltage, etc.)	Pros and Cons	Application Areas
Direct (fusion) bonding	600–1200°C small or no pressure	+ High bond strength + Hermetic seal + Resistant to high temperature and harsh chemicals − High surface flatness required − High temperature not compatible with electronics	SOI wafer fabrication
Anodic bonding	150–500°C, 200–1500 V no or small pressure	+ High bond strength + Hermetic seal + Resistant to high temperature and chemicals − Temperature and voltage requirements may not be compatible with electronics	Sensor packaging
Solder bonding	150–450°C low pressure	+ High bond strength + Hermetic seal + Compatible with electronics − Need for solder flux	Bump and flip-chip bonding
Eutectic bonding	200–400°C low to moderate pressure	+ High bond strength + Hermetic seal + Compatible with electronics − Sensitive to native oxides at surfaces	Hermetic packaging Bump and flip-chip bonding
Thermocompression bonding and direct metal-to-metal bonding	350–600°C 100–800 MPa (high pressure)	+ Hermetic seal + Compatible with electronics − Very high force required − High surface flatness required	Wire bonding Bump and flip-chip bonding

(*Continued*)

Table 4.8 (Continued)

Bonding Technique	Bonding Conditions (Temperature, Pressure, Voltage, etc.)	Pros and Cons	Application Areas
Ultrasonic bonding	Room temperature up to 250°C high pressure	+ Compatible with electronic wafers − Only demonstrated for small bond areas	Wire bonding
Low-temperature melting glass bonding	400–1100°C low to moderate pressure	+ High bond strength + Hermatic seal − May not be compatible with electronics	Sensor packaging
Adhesive bonding	Room temperature up to 400°C low to moderate bond pressure	+ High bond strength + Low bond temperature + Works practically with any substrate materials − No hermetic seal − Limited temperature stability	MEMS, sensor packaging, 3D-Ics, temporary bonds

used for SOI (silicon-on-insulator) technology. In glass-to-glass fusion bonding, glass is heated to a temperature near to the glass transition temperature (~550°C) and put under pressure to create a seal. The temperature and aspect ratio of the channels are the main factors that determine the survival of the channels. If the bonding temperature is too high, the channels tend to deform or collapse due to reflow of glass. If the temperature is too low, bonding will not occur [50]. A sealed channel with a depth of 6 nm was achieved in a Pyrex substrate [51]. Small aspect ratios down to 0.0004 for 25 nm deep channels in Pyrex have been achieved.

4.1.5.2 Anodic bonding

Anodic bonding is used most often to bond silicon to glass in MEMS packaging and microfluidic devices [52]. Typically, silicon devices are bonded with glass top covers because of the optical transparency of glass which is preferred for device testing and characterization. The silicon substrate and glass top cover are brought into physical contact and bonded by applying heat (temperatures on the order of 400°C), pressure (~100–500 kPa), and a voltage (~several kilovolts) across the device. Under the high electric field, sodium ions (Na^+) leached out

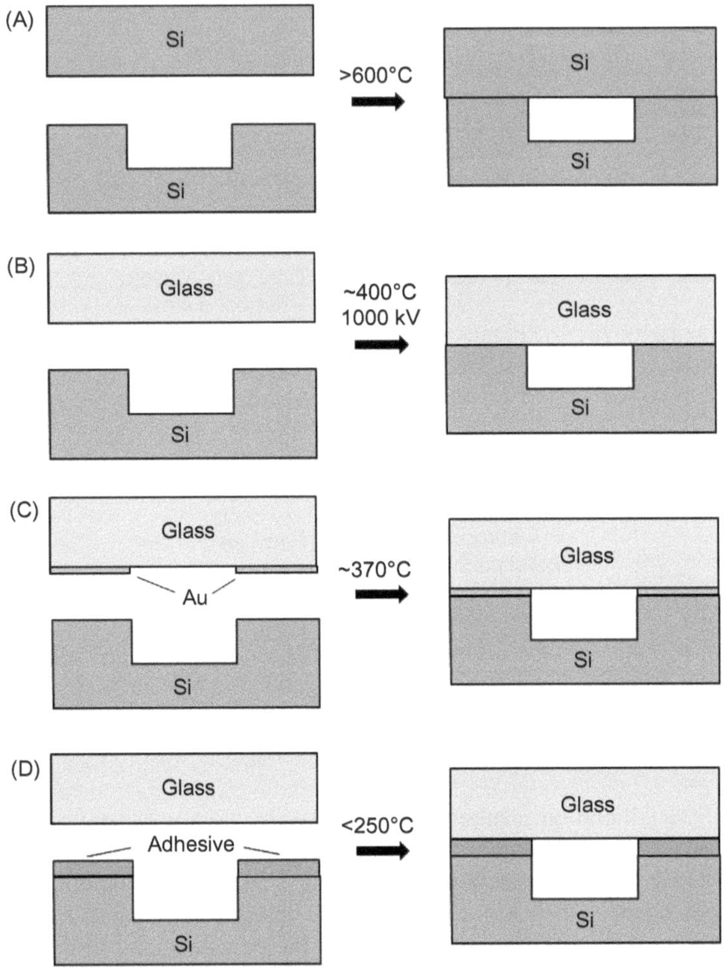

FIGURE 4.19

Schematics of bonding techniques for silicon-to-silicon or silicon-to-glass microfluidic device: (A) fusion, (B) anodic, (C) eutectic, and (D) adhesive bonding. Note that the finite thickness of intermediate layers may limit the applicability of eutectic and adhesive bonding in fabricating nanochannels.

from glass migrate into the silicon–glass interface while oxygen ions (O^-) diffuse toward the cathode. (Note that the cathode is placed on a glass or quartz substrate while the silicon is on the anode side.) The presence of oxygen ions at the silicon–glass interface leads to an oxidation reaction and strong bond between two substrates. Anodic bonding is less susceptible to particle defects compared to fusion bonding because if a particle defect is present on the surface, glass materials would flow around the particle and heal it. In order to reduce the thermally induced stress, the thermal expansion coefficients of two substrates should be carefully matched. The strong potential difference applied across the device can cause the channels to bow. If the aspect ratio for the channels is very low, the silicon and glass may come into contact in the center of the channel causing bonding due to elastic deformation of either the glass or the silicon. Once the surfaces come in contact, the bond expands quickly down the channel resulting in completely sealed off channels. Aspect ratios as low as 0.004 (channel depth of 20 nm and width of 5 μm) have been fabricated by this method [50].

4.1.5.3 Eutectic bonding

Eutectic bonding is a wafer bonding technique in which an intermediate metal layer creates a eutectic composition with most frequently silicon substrate. Since the eutectic temperature is typically much lower than the melting temperature of the substrates, it is considered to be low-temperature bonding. The most established eutectic system is between silicon and gold, and its eutectic temperature is known to be 363°C (eutectic composition being Au/Si = 97.15/2.85 wt%). A gold layer can be deposited on any substrate, and as long as the substrate is stable at the eutectic temperature, a variety of substrate pairs can be bonded. Eutectic bonding can produce strong and hermetic bonds at low temperature and is used to create a hermetic sealing of micromachined transducers [49]. One of the

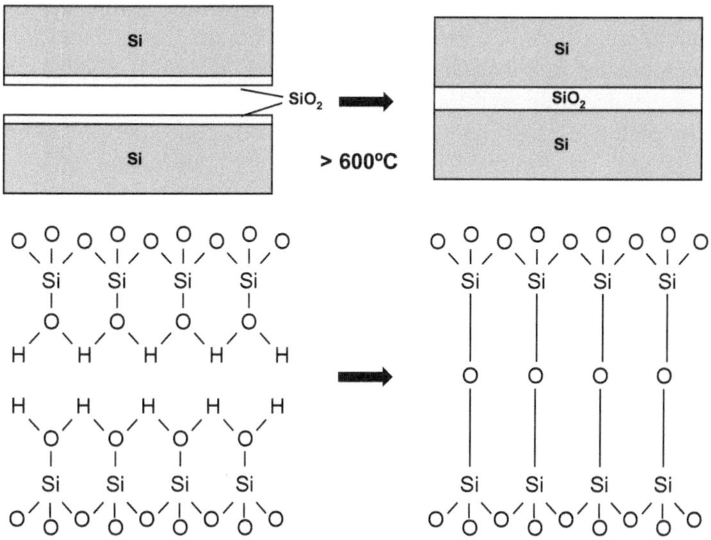

FIGURE 4.20

Schematic of the silicon fusion bonding requiring a thin oxide layer; the condensation reaction taking place at bonding interface.

notable drawbacks in eutectic bonding is a high residual stress at the bond interface due to local volumetric expansion during the eutectic formation. The high residual stress can eventually cause a bond failure especially with cyclic thermal loading. Recently, Radadia et al. [53] reported a modified eutectic bonding technique combining fusion and gold eutectic bonding—gold eutectic bonding at 410°C followed by high temperature annealing at 1100°C. This so-called gold diffusion eutectic bonding was used to fabricate Si-to-Si microcolumns with the bonding strength comparable to that of the anodic bonding and capable of withstanding abrupt temperature-programming conditions.

4.1.5.4 Adhesive bonding

In adhesive bonding, an intermediate layer is used to bond the substrates of the same and/or different materials. The common adhesive materials include SU-8, benzocyclobutene (BCB), and other commercial glues/adhesives. The advantages of adhesive bonding include (i) low bonding temperature (typically below 250°C) permitting the wider range of the substrate materials (such as polymers) and (ii) less susceptibility to particulate defects on the surface as the defects would be embedded in a relatively thick adhesive layer. But several problems of adhesive bonding should be mentioned in light of the micro- and nanofluidic device fabrication: (i) tight dimensional control would be extremely difficult due to compliance of the adhesive materials, (ii) adhesive materials may not be stable with solvents or chemicals and possible interaction could result in undesirable outgassing or byproducts, (iii) adhesive materials are typically permeable to gas and moisture, hampering hermetic seal, and (iv) the bonding would not be stable in the high-temperature applications. Some adhesive materials such as SU-8 and BCB are photopatternable, and therefore, the microfluidic components can be lithographically patterned and bonded with each other or onto other substrates. Agirregabiria and co-workers [54,55] reported the fabrication of a multilayer microstructures for microfluidic applications using successive photopatterning and bonding of SU-8 layers (Figure 4.21).

4.2 Advanced patterning techniques

4.2.1 Remarks on nanoscale patterning

As nanofluidics deals with motion of fluids below 100-nm scale, we need the fabrication technology that can produce key features of less than 100 nm in at least one dimension. The simplest and perhaps most commercially important nanostructures with only one of their dimensions being nanometer scale are thin films of 1–100 nm thickness. Thin film growth processes such as PVD, CVD, ALD, and MBE developed by semiconductor manufacturers have long been showing atomistic control over the thickness of the film on planar substrates.

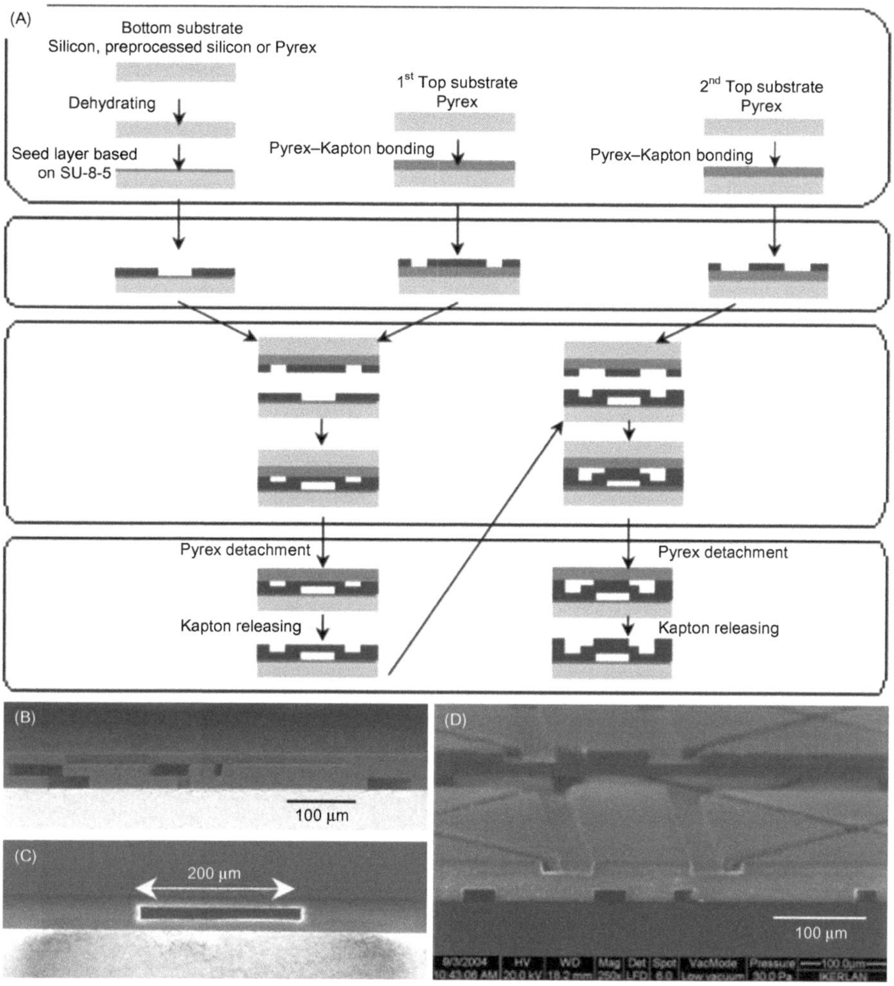

FIGURE 4.21

(A) A diagram of the fabrication process: substrate preparation, SU-8 photolithography, bonding of the photopatterned layers, and Kapton releasing. SEM images of (B) interconnected microfluidic devices made of three SU-8 layers bonded on a Pyrex™ wafer, (C) embedded microchannel of three SU-8 layers, and (D) three SU-8 layers bonded together on a silicon wafer.

Source: *[55] Agirregabiria M, Blanco FJ, Berganzo J, Arroyo MT, Fullaondo A, Mayora K, e tal. Fabrication of SU-8 multilayer microstructures based on successive CMOS compatible adhesive bonding and releasing steps.*

In Section 4.1, we have shown the examples of the micro/nanochannels fabricated by these well-established thin film techniques together with the conventional lithography. The other extreme would be 0D (e.g., nanoparticles) and 1D (e.g., nanotubes and nanowires) self-supporting nanostructures. Nanoparticles bear nanometer dimensions in all direction, and nanowires and nanotubes have nanometer scale in at least two dimensions. They have been produced mostly by chemical synthesis. Large-scale manufacturing processes are already put in practice for a variety of nanoparticles while carbon nanotubes are close to being produced at industrial scale (up to kilograms of nanotubes are commercially available now).

Different from thin film processes and inorganic syntheses methods, nanofabrication is referred to as a technique to create on substrates 2D or quasi 3D nanostructures, which are building blocks to form various functional nanodevices and systems [56]. This definition of nanofabrication is particularly relevant in our discussions because almost all nanofluidic elements of microfluidic devices are built on 2D substrates. It is arguable to include the chemical synthesis of nanomaterials, which have been well known to chemists and material scientists even before the inception of the "nanotechnology" boom, as a category of nanofabrication. However, nanofabrication indisputably encompasses techniques to assemble nanomaterials or integrate them on 2D substrates, irrespective of whether functional nanomaterials are grown directly on the substrates or suspended in solution first and deposited to the substrates.

Some of the additive/subtractive processes discussed in Section 4.1 continue to work for transferring the nanoscale pattern features to the functional materials or substrates. However, other processes will face serious challenges as the feature size is reduced to nanometer scale. While the experimental verification of fundamental nanofluidic theories and building nanofluidics devices per se entails nanofabrication at some points, it is important to note that better understanding of fluid motion inside nanoscale passages and/or around the nanoscale objects can in turn help to improve a yield of nanofabrication processes and may devise new fabrication methods.

4.2.2 Electron beam lithography

Electron beam lithography creates nanoscale patterns by directly writing on resists with focused electron beams. A typical e-beam resist is PMMA of which the solubility changes upon irradiation of e-beam. The electrons induce local chain scission, and the exposed area becomes more soluble in a developer making PMMA a positive resist. Since the wavelength of high-energy electrons is typically less than 0.01 nm (e.g., 0.005 nm for 50 keV), e-beam lithography is capable of producing ultrahigh-resolution patterns. The resolution of e-beam lithography is determined by the scattering of primary and secondary electrons in the resist film and the substrate [57]. When electrons hit resist with high energy, they scatter forward. Despite the very small beam spot size (<0.5 nm) focused on top of the resist surface, the exposed area in resist becomes much larger as scattering broadens the beam inside the resist. Some electrons are bounced off from the substrate and backscatter (large angle scattering) with micrometer range. Therefore, if the neighboring features are tightly spaced, the exposure dose should account for the backscattered electrons (called proximity effect). The optimization of resist thickness, substrate reflectivity, and dose would be necessary to resolve sub-10-nm patterns. Features of as small as 4 nm in lateral dimension have been fabricated [58]. E-beam lithography is a serial process just as any other beam-based writing techniques (ion beam and laser), and the sequential nature of the process makes writing very time consuming and impractical for mass production. Therefore, it is widely used for R&D or pilot production and photomask production for optical lithography.

Since earlier days of research in nanofluidics, e-beam lithography has been the workhorse for generating nanofluidic components with well-defined geometries and dimensions. In one of the examples, Perry et al. [59] utilize e-beam lithography to create molds for in-plane nanochannels with asymmetric features like funnels. The funnel pattern with the neck dimension of 80 nm was written in a negative resist SU-8 layer as a replica master, and high-modulus PDMS was cast to form a straight channel with a funnel (Figure 4.22A,B). They studied ionic transport across the asymmetric nanochannel and observed rectified ionic currents dependent on the taper angle. Another example is to use e-beam patternable sacrificial polymers for generating an array of nanochannels [60]. Polynorborene-based polymer was first patterned by e-beam lithography into an array of lines, capped by PECVD silicon oxide, and thermally decomposed to leave nanochannels behind (Figure 4.22C).

4.2.3 Focused ion-beam lithography

The instrumentation of the focused ion beam (FIB) bears a resemblance to the scanning electron microscope (SEM), but in FIB, a focused beam of ions is used to image and micromachine the samples rather than a focused beam of electrons in SEM. The FIB is extensively used in cross-sectional imaging of semiconductor devices, modification of the electrical routing on semiconductor devices, failure analysis, sample preparation in transmission electron microscope (TEM) and other analytical tools, mask repair, and micromachining [61,62]. Gallium is the most popular liquid metal ion sources as it is a liquid metal at room temperature and compatible with high vacuum and has a long life (up to 1500 h). Ionization and field emission of Ga atoms is induced via Taylor cone instability at the tip where a high electric field is applied. Unlike electron microscopes (i.e., SEM), FIB is inherently destructive to the samples. When bombarding the specimen, high-energy Ga ions will sputter the surface of the interrogated sample. Therefore, most FIB instruments are nowadays equipped with a combination of SEM and FIB columns (known as dual beam or cross beam), in which the electron beam is used for imaging and the ion beam used only for milling. This strategy permits nondestructive imaging at higher magnification and with better resolution as well as more accurate control over the milling process.

4.2.4 Nanoimprint lithography

Nonimprint lithography (NIL) is a method of patterning a thermoplastic material—a pattern of a stamp is transferred (more accurately embossed) onto a moldable film on a substrate through mechanical contact [63]. The resulting 3D pattern on the surface of the film is then transferred to the underlying wafer by common semiconductor processing techniques such as RIE etching. The fundamental principle of NIL is similar to that of hot embossing. Figure 4.23A illustrates the process

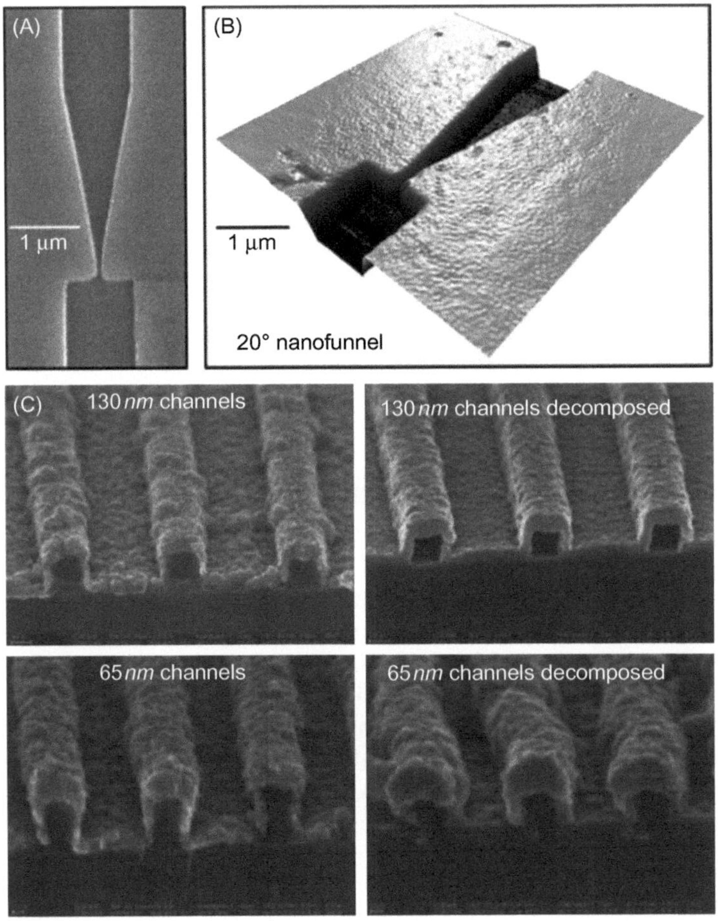

FIGURE 4.22

Examples of the nanofluidic elements fabricated by e-beam lithography: (A) an SEM image of nanofunnels with 20° taper angle and (B) an AFM image of (A). (C) SEM images of the array of nanochannels before and after decomposition of polynorbornene.

Source: *[59] Perry JM, Zhou K, Harms ZD, Jacobson SC. Ion transport in nanofluidic funnels and [60] Devlin NR, Brown DK, Kohl PA. Patterning decomposable polynorbornene with electron beam lithography to create nanochannels.*

sequence of NIL. First, a stiff mold with nanoscale relief structures is fabricated using other high-resolution lithography techniques such as e-beam lithography [64]. The stiffness of the mold patterns and the flexibility of the substrates are important for high-quality pattern replication because conformal and uniform contact is entailed over a large area. The mold is then pressed onto a soft film—typically a PR film heated above its glass transition temperature—with a range of pressure, temperature, and contact time. The resulting topography in terms of the local thickness

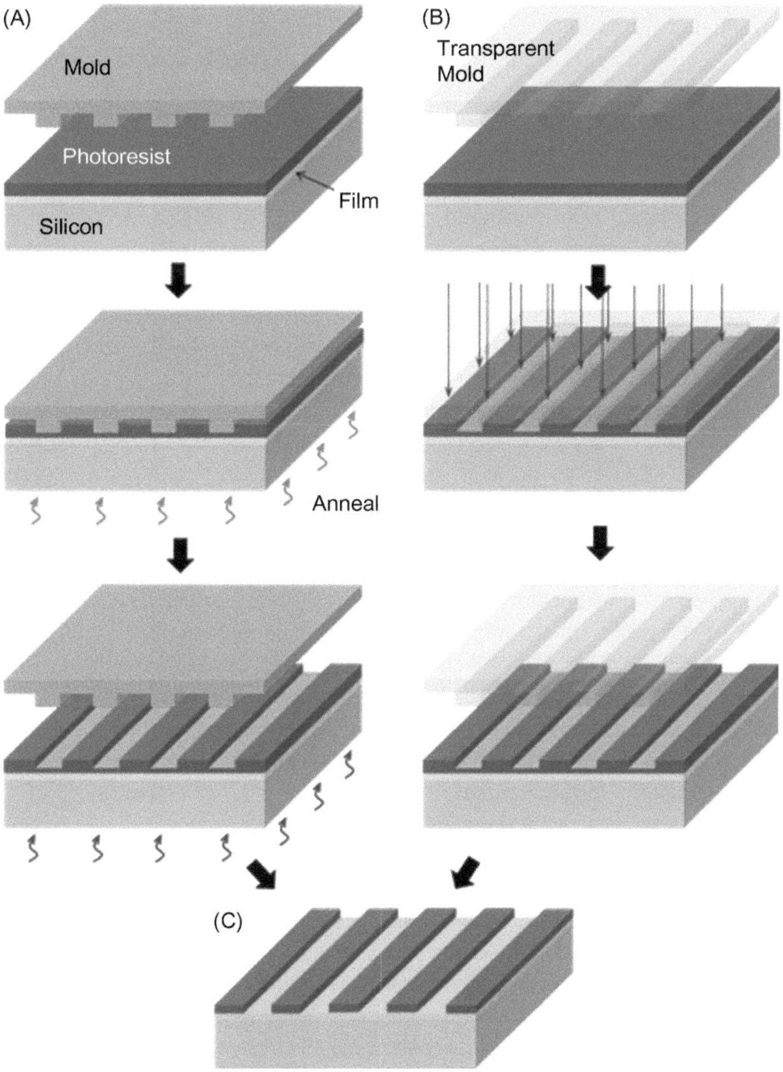

FIGURE 4.23

A schematic of (A) NIL and (B) SFIL processes and (C) the resulting pattern on the wafer.

contrast on the PR is a reversed image of the original mold. The residual layer after embossing is removed via a dry etching process, completing the pattern transfer into the PR layer (Figure 4.23C). Patterning of arrays of pillars and holes with the feature size of 10 nm has been demonstrated with NIL [66]. However, the high viscosity of the polymers constrains the range of the pattern density due to the limited distance that polymer reflows during imprinting. A few modified processes have been subsequently developed to expand the scope and applicability of the NIL technique. Step and flash imprint lithography (SFIL) utilizes an UV-curable resist and an UV-transparent mold and enables patterning of less than 100-μm features at room temperature and higher throughput [67,68] (Figure 4.23B). SFIL also allows low-viscosity fluids to be used and helps to avoid incomplete mold filling issue and increase the pattern density. More recently, laser-assisted direct imprint has been developed—by rapidly heating a thin layer of the substrate material using excimer laser pulses and embossing a mold into the resulting liquid layer, various nanostructures of polymeric and other materials were fabricated with resolution better than 10 nm [69].

Unlike the beam- or probe-based techniques, NIL is a high-throughput process due to its parallel nature of printing and etching, bearing resemblance to optical lithography. For this reason, NIL has been considered a serious contender to replace optical lithography in the microelectronic industry, but the fabrication of silicon microprocessors requires much tighter tolerance for defects thus making NIL an unlikely fit at the current stage. More detailed information on the working principle and the requirements on materials for the mold and resists can be found elsewhere [70]. NIL has also been frequently employed to build nanochannels of various sizes and materials. The Guo group at University of Michigan developed a technique to form nanofluidic channels by directly imprinting hydrophilic hydrogen silsequioxane (HSQ) resist and sealing with another HSQ thin film [71]. The same group also utilized the modified NIL to create an array of nanofluidic channels of sub-100-nm in size[72]. When a very thin polymer layer is used during imprinting, the trenches in the mold are partially filled, generating enclosed nanochannels (see Figure 4.24). The Pang group at University of Michigan developed the 3D nanochannel network in width from 200 nm to 2 μm using the sequentially stacked thermal NIL on planarized self-supporting sealing materials over polymer sacrificial layers (see Figure 4.25) [73,74]. The development of these nanochannels can result in practical tools for the analysis of the biomolecules.

4.2.5 Soft lithography

Soft lithography was so named because soft, elastomeric elements are used for pattern formation and transfer [75,76]. While the conventional radiation-based lithography (photolithography, e-beam lithography, etc.) techniques are extremely successful on ultra-flat surfaces, it is difficult, or impractical, to adapt these lithographic methods for emerging applications that require the patterning of unusual materials (e.g., biomaterials) and/or on the nonconventional and nonplanar (e.g., etched, curved, or porous) substrates (e.g., plastics). Another distinct benefit of soft lithography is that it does not involve high capital and operation costs. Some of the oldest and conceptually simple machining ideas—embossing, molding, stamping, printing, writing, etc.—are borrowed in soft lithography to create nanometer as well as large size patterns. Numerous variations in soft lithography have been proposed and demonstrated in the literature, but soft lithography is generally divided into three main categories: (1) microcontact printing (μCP), (2) micromolding, and (3) microtransfer molding (μTM) [77].

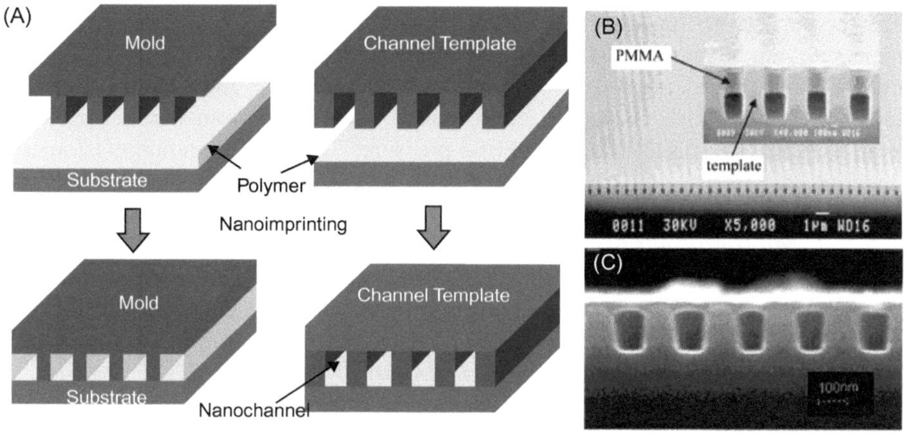

FIGURE 4.24

Schematics of (a) (LEFT) the conventional NIL process of using a mold with surface protrusion patterns to imprint into a polymer resist and (RIGHT) the nanofluidic channel fabrication by using a template mold to imprint into a thin polymer layer to leave unfilled and self-enclosed channels, and SEM micrographs of imprinted nanofluidic channels of the cross sections (b) 300 × 500 nm and (c) 75 × 120 nm channels.

Source: *[72] Guo, L.J.,X. Cheng ,C.-F. Chou, (2004). Fabrication of size-controllable nanofluidic channels by nanoimprinting and its application for DNA stretching.*

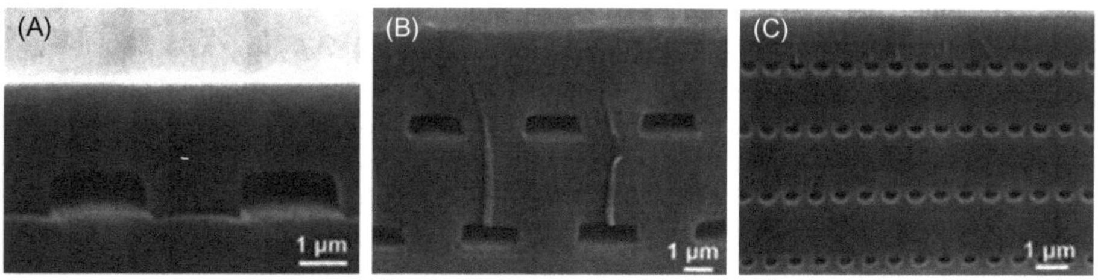

FIGURE 4.25

Nanochannels fabricated by bonding a reverse UV-imprinted SU-8 layer with a thin SU-8 adhesion layer. SEM images of: (A) one-level 2 μm wide channels, (B) two-level 2 μm wide channels, and (C) four-level 400 nm wide channels.

Source: *[74] Rogers JA, Nuzzo RG. Recent progress in soft lithography.*

Common to soft lithography, regardless of its variations, are a master (or mold) and a stamp. Figure 4.26 shows the formation of these two key elements. First, create a master mold with the patterns desired for fabrication. Next, pour an elastomer polymer into the mold to make a stamp. Finally, the stamp is removed from the mold and can be used to create a replica or other printing applications. The masters used for soft lithography can be formed using the micro- or nanolithography techniques that are already established. The key to the success in soft lithography is its ability to form a conformal contact between the elastomer stamp and the surface of the substrate. When the substrate is rough, uneven, or curved, the stamp can adapt itself to the overall shape of the substrate, leading to an intimate contact without voids. Van der Waals-based adhesion forces allow the elastomer to spontaneously create a conformal contact with these nonflat surfaces, even without the application of external pressure. The most common elastomeric material used in soft lithography is polydimethyl siloxane (PDMS).

4.2.5.1 Microcontact printing

Microcontract printing (μCP) uses a high-resolution elastomeric stamp with a chemical ink capable of forming a pattern of SAM on a target substrate. The concept of SAM, the key element in μCP, is discussed in Chapter 3. Although μCP was originally developed for printing SAMs of alkanethiols, $CH_3(CH_2)_nS-$, on gold surfaces [78], it has been adapted to create patterns in a wide range of materials, including metals, organic molecules, polymers, nanomaterials, and biological entities [57,79]. The basic process sequence of μCP on the model system (thiols on gold) is shown in Figure 4.27 and is as follows: (1) Soak a PDMS stamp into a solution of thiol-forming ink by placing a drop of ink onto the stamp or selectively contact inking the stamp. (2) Place the stamp with the SAM-coated surface downward on a gold-coated substrate (forming a conformal contact between the stamp and the substrate) and transfer the alkanethiol layer from only the surface protruding from the PDMS stamp onto the substrate. (3) The alkanethiol layer will form a chemical bond with the Au substrate, and when the stamp is removed from the substrate, the "ink" pattern remains on the substrate surface. (4) The alkanethiol layer can act as an intermediate resist against a gold etching, and therefore the pattern of a gold film can be generated. In some cases, a patterned thiol layer was utilized as a mask for subsequent plating process to form various 3D structures [80]. The same thiol chemistry and μCP can also work for other coinage metals such as silver and copper. Another popular material system is the silane-based SAM on the metal oxide surface. For example, octadecyltrichlorosilane, $CH_3(CH_2)_{17}SiCl_3$, can be patterned and printed onto a SiO_2 or glass substrate with the μCP technique.

Features smaller than 100 nm have been patterned using μCP [81]. The resolution limit for patterning the SAM with μCP is partly due to the intrinsic nature of the stamp and its fabrication and partly due to the mechanism of transfer. While PDMS offers flexibility and conformability, its mechanical instability including large deformation, thermal expansion, and chemical-induced swelling/shrinkage can make it unsuitable for large-scale industrial processes. Accurate reproduction

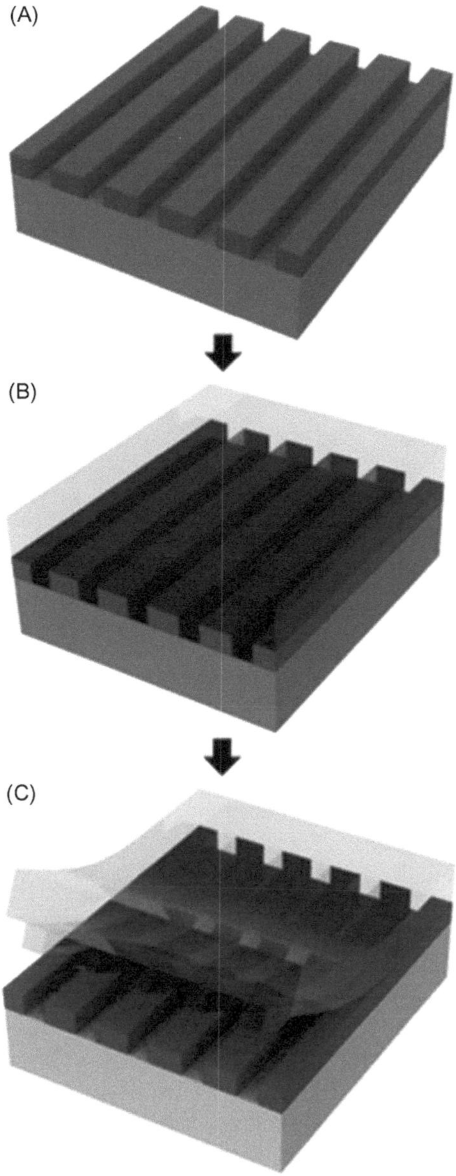

FIGURE 4.26

A process sequence for fabricating a mold and a stamp for soft lithography.

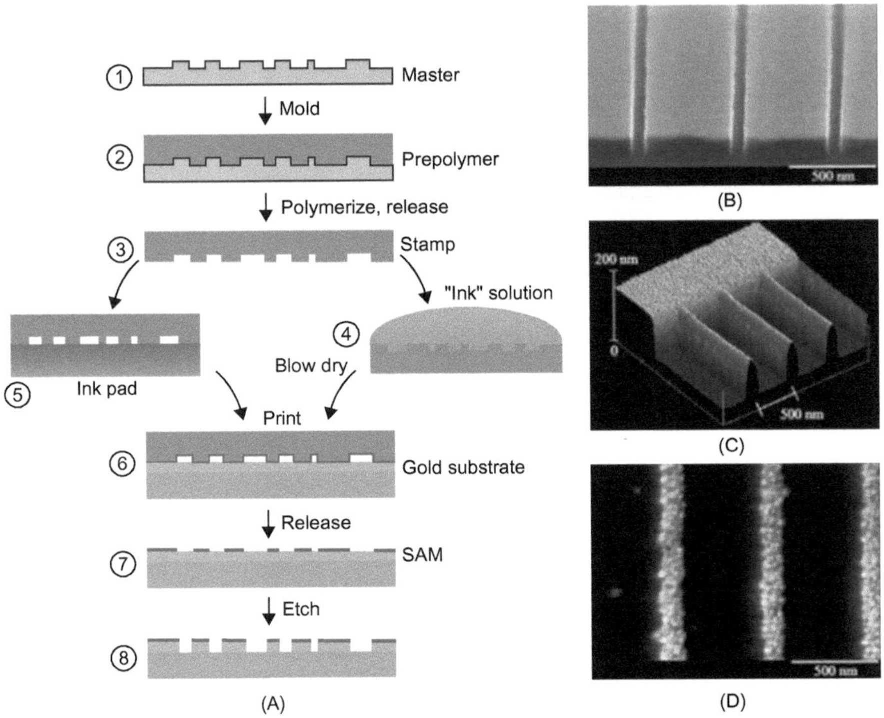

FIGURE 4.27

(A) A process sequence of the μCP and subsequent pattern transfer: A prepolymer (2) covering the master (1) is cured by heat or light, and demolded to form an elastomeric stamp (3). The stamp is inked by immersion (4) or contacted with an ink pad (5), and printed onto the substrate (6), forming a SAM. The ink pattern (7) is then transferred into the substrate by a selective etch (8). SEM images of (B) the master, (C) the stamp, and (D) the printed and etched pattern.

Source: *[76] Michel B, Bernard A, Bietsch A, Delamarche E, Geissler M, Juncker D, et al. Printing meets lithography: soft approaches to high-resolution patterning.*

of small patterns can also be compromised by the diffusion of ink molecules away from the contact areas [76]. Longer contact time and sufficient reactants are required to form a defect-free monolayer but unfortunately can allow the thiol molecules to diffuse and widen the width of the pattern. It was shown that the alkanethiol with the higher molecular weight decreased the surface diffusion of the ink during printing and thus helped to form 100-nm Au lines [76].

4.2.5.2 Micromolding and replica

The principle of various micromolding techniques is very similar to μCP, but in micromolding a PDMS stamp is used in a different way that allows polymers and materials other than SAMs to be patterned. A variety of micromolding techniques have been developed in the literature including replica molding (RM), microtransfer molding (μTM), micromolding in capillaries (MIMIC), and solvent-assisted micromolding.

4.2.5.2.1 Replica molding

RM is a type of casting process capable of replicating complex nano- and microstructures in a single step. Duplicating the morphology information on the surface of the mold, the replicated features possess almost the same dimensions and topologies as the mold. The fidelity of the process depends on van der Waals interactions, wetting, and filling factors [77]. RM first starts with casting an elastomer mold (either PDMS or polyurethane) against a rigid master. The elastomer mold bears the negative structures with respect to the rigid master. Then UV or thermally curable prepolymer is poured and replicated against this negative elastomer mold. Therefore, the structures on the resulting replica are complementary to those on the mold and very similar to those on the original rigid master. Polyurethane or epoxy-based materials have been replicated using RM. Hydrophilic micro- to nanochannels composed of a poly(ethylene glycol) (PEG)-based polymer were fabricated using RM and UV-assisted irreversible sealing (Figure 4.28) [82]. The PEG surfaces of the channels are nonbiofouling, permitting to study a capillary flow of biological fluids. Senn et al. [83] created the multilevel micro-/nanostructures using the RM process to develop microfluidic device of fuel delivery system for micro fuel cell applications.

4.2.5.2.2 Micromolding in capillaries

Bringing a mold into contact with a flat substrate forms voids or microchannels, and one way to create patterns is to fill these voids with liquid materials and solidify them (Figure 4.29). Liquid prepolymers, liquid metals, or solutions containing nanomaterials can be filled into the microchannels by capillary force or external pressure. The former technique, filling the mold by capillary force, is often referred to as MIMIC [85]. Once the liquid materials are solidified through evaporation or curing, the mold is removed from the surface. Due to its compliance and formation of reversible bonding, PDMS and other elastomeric materials are frequently used as a mold. MIMIC is a perfect example of the microfluidic components assisting the fabrication of micro- and nanostructures, and therefore, understanding a fluid behavior in micro- or nanochannels would provide a better design rule and defined fabrication limitation. For example, a capillary flow in these channels is very low (especially true for polymers with high viscosity), limiting the cross-section dimension and channel length of the microchannels to be filled. In addition, a continuous network of microchannels is needed to introduce the liquid in the channels, and therefore isolated features and contour patterns cannot be fabricated with MIMIC.

4.2.5.2.3 Microtransfer molding

In μTM, a small amount of liquid prepolymer is first placed on microchannels and microcavities on the surface of an elastomeric mold. After removing the excessive liquid, the mold is brought into contact with a target substrate, on which the liquid prepolymer becomes solidified thermally or photochemically. When the mold is peeled away, the polymeric microstructures are left behind on the substrate, completing the transfer process [86]. Compared to MIMIC, the μTM

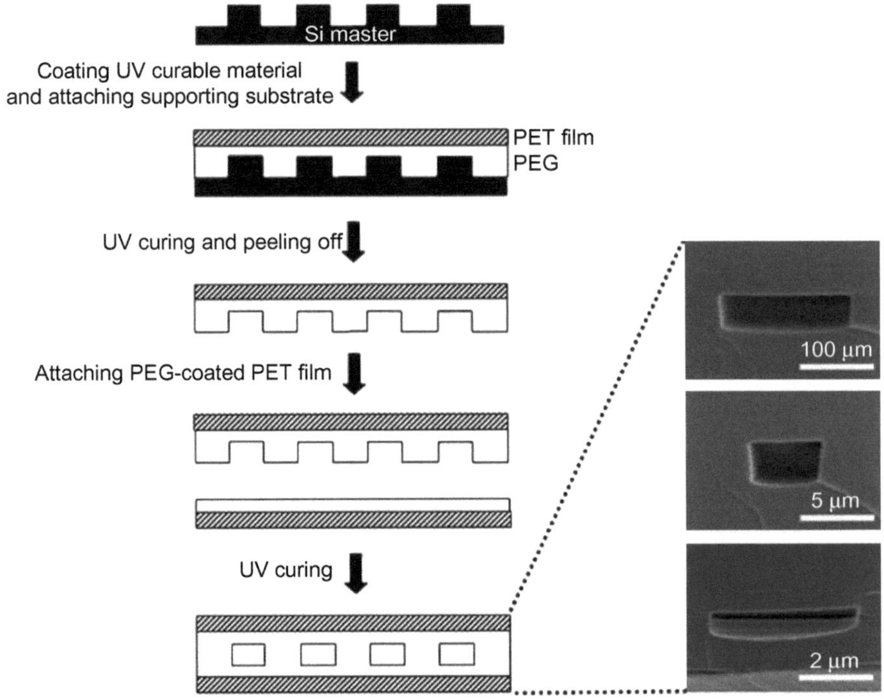

FIGURE 4.28

A schematic diagram of the fabrication procedure for the PEG micro-/nanochannels (patterned by RM) and SEM images of the three PEG channels. The geometries of the channels were: 145 × 46, 4 × 3, and 4 × 0.2 μm^2 (width × height) from top to bottom. PET: polyethylene terephthalate.

Source: *[82] Jeong HE, Ki P, Kwak MK, Seo CH, Suh KY. Capillary kinetics of water in homogeneous, hydrophilic polymeric micro- to nanochannels.*

process is fast and scalable and also has more geometric freedom for the types of the patterns to be generated: isolated features, ring-like enclosed patterns, and so on. Multilayer microstructures, freestanding microstructures, and patterns on the curved/nonplanar surfaces can also be fabricated based on μTM.

4.2.6 Self-assembly and self-organization

Self-assembly is defined in a broad term as the spontaneous formation of well-defined structures from "disordered" components [87]. Self-assembly is ubiquitous in nature and universe from atomic to astronomical scale, including atoms forming molecules, molecules crystallizing and aggregating into a larger cluster, surfactants forming micrometer-size micelles, lipids forming the cell membranes, and even planets and stars forming galaxies. Self-assembly is considered to be the most important and practical bottom-up approach to nanotechnology, and the examples of nanoscale building block so far investigated for self-assembly include

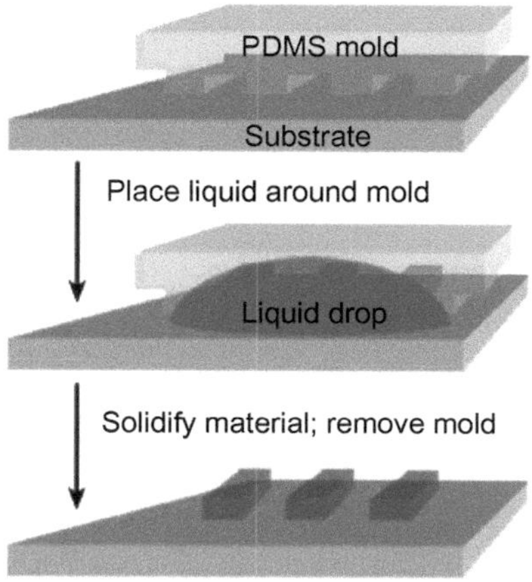

FIGURE 4.29

Schematic illustration of MIMIC——a molding process to pattern liquid materials in microchannels formed by the contact of a mold with a substrate.

Source: *[84] Menard E, Meitl MA, Sun Y, Park J-U, Shir DJ-L, Nam Y-S, et al.Micro-and nano-patterning techniques for organic electronic and optoelectronic systems.*

nanoparticles and colloids [88,89], block copolymers (BCPs) [90,91], liquid crystals [92], and viruses and biological entities [93,94]. In order for self-assembly to be successful, a few criteria in terms of the materials and environments must be met. First, it is a spontaneous process in which the associated free energy change should be negative. Second, the materials involved should interact with each other in a reversible manner, allowing the components to adjust their positions to be ordered. Therefore, the weak interaction like the van der Waals force is appropriate for self-assembly. Lastly, the environment should allow the components to be mobile for self-assembly, i.e., most of the self-assembly takes place in liquid environment. Here, we focus our attention on the patterning capability of self-assembly and limit our discussion to the 2D colloidal assembly and BCP.

4.2.6.1 Nanosphere lithography

Highly monodispersed colloidal nanospheres can self-assemble into 2D colloidal crystals under appropriate conditions. A technique to create a monolayer array of colloidal nanospheres (or microspheres) on a substrate via self-assembly is called nanosphere lithography [95,96]. When nanospheres with diameters ranging from 200 to 1000 nm are deposited onto the surface, they are self-assembled into an ordered nanostructure mostly with the hexagonal close-packed (hcp) array [97]. The hcp structure is preferred because it is the thermodynamically stable position with minimum free energy. Solvent evaporation and spin coating are two dominant techniques in assembling these nanospheres on the substrate surface. Attractive capillary forces, convective transport of the nanospheres, concentration of the colloid suspension, and centrifugal force are all playing an important role in determining the ordering and quality of the obtained arrays. Nanosphere

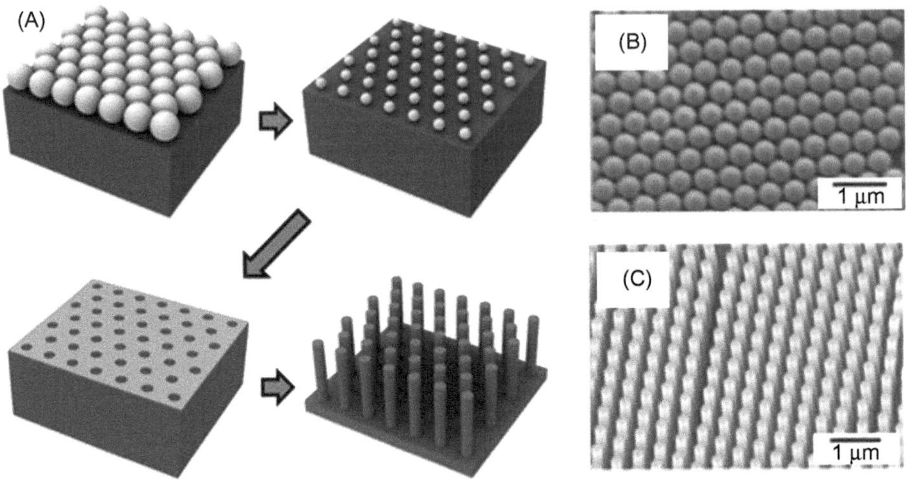

FIGURE 4.30

(A) A process sequence to create an array of vertically aligned silicon nanowires from nanosphere lithography and SEM images of (B) hcp array of 500 nm polystyrene nanospheres and (C) the resulting silicon nanowire arrays.

Source: *[42] In HJ, Field CR, Pehrsson PE. Periodically porous top electrodes on vertical nano-wire arrays for highly sensitive gas detection.*

lithography is recognized as a facile and inexpensive nanofabrication technique that does not rely on expensive tools, and these arrays of colloids have been used as masks or templates for deposition, etching, imprinting, etc. One example of nanosphere lithography is shown in Figure 4.30. The polystyrene nanosphere is first spin-coated onto the silicon surface to form a 2D hcp array. The diameter of the spheres can be reduced by RIE etching using O_2 plasma. Then, a thin layer of noble metal (e.g., Au) is evaporated onto the surface and lift-off in the acetone to remove the polystyrene beads, resulting in a perforated Au film (nanomesh) on the silicon surface. With MACE wet etching, an ordered array of vertically aligned silicon nanowires is created on the surface [42,98].

4.2.6.2 Block copolymers

A block copolymer (BCP) is a polymer made of two or more monomer subunits that are tethered covalently. Diblock and triblock copolymers are BCPs with two or three repeating, distinct blocks. The advent of BCP in the scene of nanotechnology came about in the early 1990s—the ability of each polymer unit to be self-assembled in nanometer length scale on the technologically relevant surfaces [99,100]. Patterning in BCP takes place by thermodynamic instability between two blocks, in which a collection of diblock molecules is self-assembled through microphase separation—the contact between similar and dissimilar blocks are maximized and minimized, respectively [101]. The free energy cost of contact between dissimilar monomers can be described by the temperature-dependent Flory–Huggins interaction parameter, which along with the degree of polymerization and the relative composition fractions enables to predict the morphologies of the diblock molecules after processing [102]. The resulting morphologies of BCP thin films are also affected by the other processing parameters or substrate treatment, including the film thickness, geometric confinement on surface, surface chemistry, presence of field, type of solvents for annealing, and annealing temperature and time, and so on [103]. For example, the importance of the film thickness as a boundary condition is shown in Figure 4.31, and the transition of the morphologies is evident from thinner to thicker films [104].

The patterning by the self-assembly of BCPs is advantageous because it is parallel, does not rely on expensive facilities, and has a tunability of the size, shape, and composition. Numerous applications have been proposed including nanolithography [105–107], nanotemplating [108,109], and nanoporous membranes [110,111], and the lithography technique based on the self-assembly of BCP is recognized as a serious candidate for the next generation of lithography in the International Technology Roadmap for the Semiconductor Industry [112]. Figure 4.32A–C illustrates the BCP-based nanolithography—transferring periodic microdomains to the substrates using the patterned BCP as a resist—and several

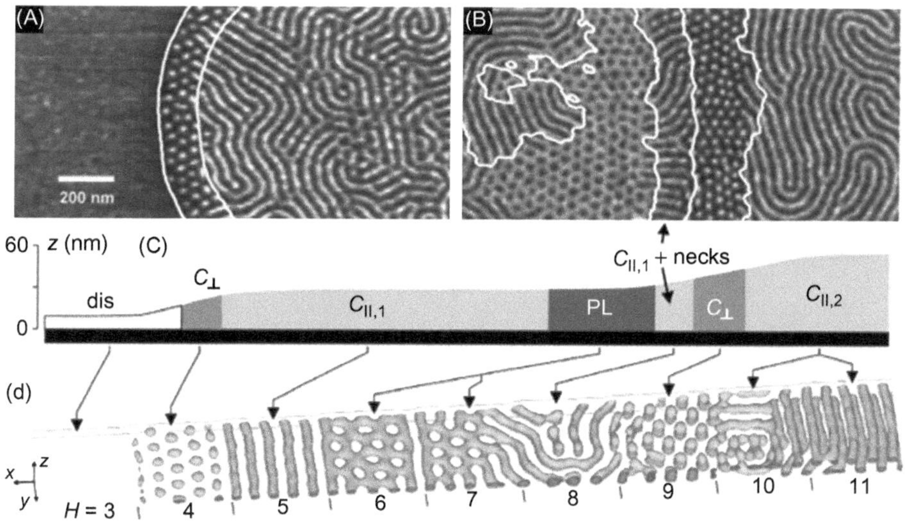

FIGURE 4.31

Effect of the film thickness on the self-assembled morphology: (A) and (B) AFM phase images of a thin poly(styrene-*b*-butadiene-*b*-styrene) triblock copolymer film on Si substrates as a function of film thickness, (C) a schematic height profile of the phase images shown in (A) and (B), and (D) a simulation of the same block copolymer.

Source: *[104] Knoll A, Horvat A, Lyakhova KS, Krausch G, Sevink GJA, Zvelindovsky AV, et al. Phase behavior in thin films of cylinder-forming block copolymers.*

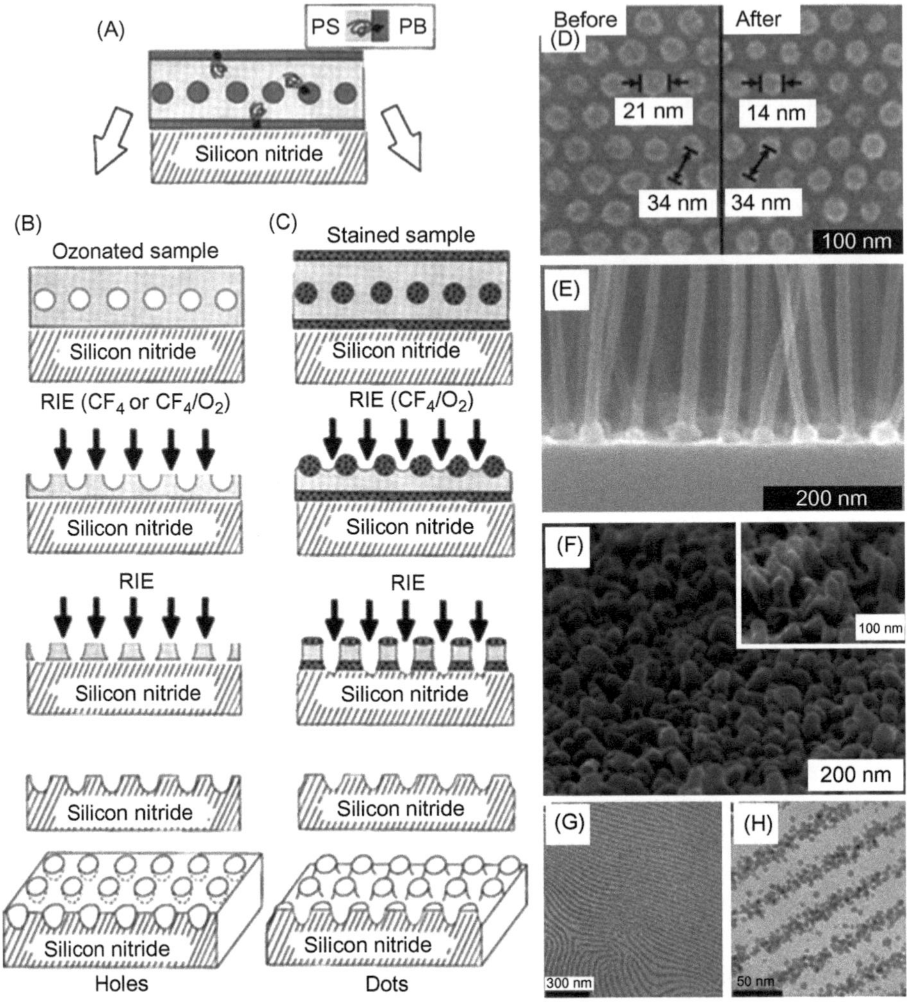

FIGURE 4.32

Schematic of a BCP nanolithography process: (A) cross-sectional view of a nanolithography template consisting of a uniform monolayer of polybutadiene (PB) spherical microdomains on Si_3N_4 (reproduced with permission from the American Association for the Advancement of Science from [100]), (B) process flow when an ozonated copolymer film is used as a positive resist, which produces holes in Si_3N_4 and (C) process flow when an Os-stained copolymer film is used as a negative resist, which produces dots in Si_3N_4 [100]. SEM images of (D) nanopatterned iron catalyst particles patterned by the PS-*b*-PMMA template and (E) CNTs grown from the catalyst nanoparticles [113]. (F) Polypyrrole nanorods grown by electropolymerization from the PS-*b*-PMMA template [114]. (G) and (H) TEM micrographs of the Pd nanoparticles assembled in PS-*b*-PMMA.

Source: *[115] Horiuchi S, Fujita T, Hayakawa T, Nakao Y. Three-dimensional nanoscale alignment of metal nanoparticles using block copolymer films as nanoreactors.*

nanostructures created from BCP-based nanotemplates. More detailed discussion on BCP applications are summarized in the literature [91,116–119].

4.2.6.3 Porous alumina template

A use of the membrane with well-defined holes of nanometer scale has gained tremendous attention in the past two decades. The holes can be filled up with desired materials, leading to the formation of the nanostructure arrays. A wide range of materials has been synthesized into nanostructures from this templated growth. Porous aluminum oxide membrane is particularly interesting because (i) it forms self-ordered hexagonal arrays of cells containing cylindrical pores that are intrinsically in nanometer dimensions, (ii) the pore size can be controlled over large areas, and (iii) the pores are straight and have HAR (>1000). Porous alumina template is fabricated via anodization in a strong acid—a competing reaction of oxide growth and partial dissolution of aluminum oxide generating pores [120]. The geometry of the porous alumina resembles a honeycomb structure with each columnar cell containing a cylindrical pore at the center. Achievable pore diameters range from 20 to 500 nm and the pore densities from 10^9 to 10^{11} cm^{-2}. The pore regularity was remarkably improved by two-step anodization as the first long anodization step removes the thick oxide layer and highly ordered pores are obtained by a subsequent anodization [121]. As shown in Figure 4.33, the porous membrane has long-range-ordered channel array, potentially used as a vertically aligned nanochannel-array architecture for data storage system, nanofluidic devices, and photonic crystal applications [122]. Examples of nanomaterials fabricated by porous alumina templates were reviewed elsewhere [123].

4.2.6.4 DNA origami

The self-assembly of the biomolecules relies on the complementary nature of the molecular recognition. Biological components such as peptide nucleic acid, deoxyribonucleic acid (DNA), ribonucleic acid, and viruses can be self-assembled to create devices and nanostructures [112]. Among those, DNA has drawn most attention since its theoretical inception of DNA nanostructuring [124] and experimental demonstration of a topologically equivalent DNA cube [125]. DNA is highly attractive in terms of structural building blocks for nanotechnology because of its (i) well-defined nanometer geometry (a 3.4 nm helical repeat and ~2 nm diameter), (ii) high specificity of the interaction between complementary base pairs that are predictable and programmable, (iii) combined stiffness and flexibility, stiffness with a persistence length of 50 nm as a building block and flexibility from a bend in any branched DNA structure, and (iv) well-established chemistry and methodology for isolation, purification, manipulation, labeling, etc. [126]. Recently, the DNA self-assembly of arbitrary nanostructures was reported using the process called DNA origami [127]. A single long-stranded DNA molecule was folded in one step with the aid of a number of short DNA strands (namely staples) to form 2D structures of arbitrary shape. The multiple parts of the long scaffold DNA are complementary to short staple DNAs, and when hybridized, the scaffold DNA was folded into a designed shape. The computer program was used to design individual staple strands and calculate their

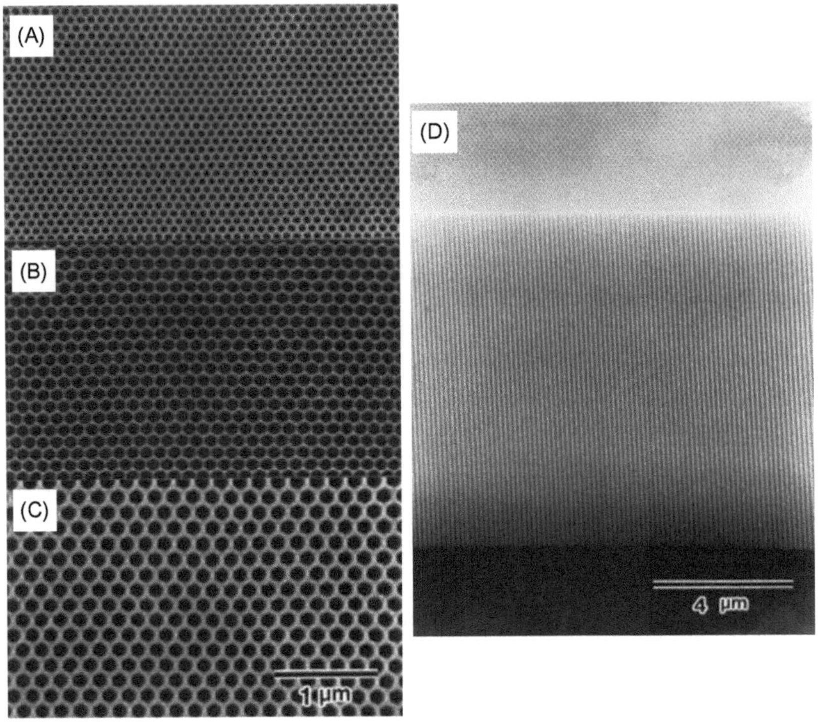

FIGURE 4.33

SEM micrographs of porous alumina membrane after two-step anodization with different intervals of (A) 100 nm, (B) 150 nm, and (C) 200 nm, and (D) cross-sectional view of channel array.

Source: *[122] Masuda H, Yamada H, Satoh M, Asoh H, Nakao M, Tamamura T. Highly ordered nanochannel-array architecture in anodic alumina.*

placement relative to the template DNA. Figure 4.34 illustrates the various nanostructures obtained by DNA origami and the corresponding AFM images [127]. Squares, rectangles, disks, and five-point stars of roughly 100 nm in length were demonstrated with a spatial resolution of 6 nm. An example of 3D nanostructures—a DNA box—was created by folding six DNA origami sheets [128]. These DNA self-assembled structures are used as scaffolds for templated self-assembly of proteins [129], gold nanoparticles [130], and other nanomaterials [126].

4.2.7 Other emerging techniques

4.2.7.1 Interference lithography

Patterns in interference lithography (IL) are formed by the interference of a small number (usually two) of coherent optical beams and are periodic in nature. For example, interferometers with two beams can produce a pattern which is periodic in one dimension, and this interference pattern consists of a periodic series of fringes representing the maxima and minima of light intensity. When this pattern is exposed on the PR, a set of parallel lines (i.e., a grating) is created. The actual intensity pattern formed from IL is sinusoidal—thus the formation of the discrete lines is aided by the nonlinear nature of PRs. Two criteria should be satisfied for IL to work successfully: a light source has to be (i) spatially coherent with normal incidence and (ii) monochromatic [131].

The obvious limitation in IL is that arbitrary patterns cannot be generated. In fact, periodicity in more than 1D requires either more than two interfering beams or more than one exposure. However, IL is arguably the best and often only lithographic method to create periodic devices because other modes cannot attain the

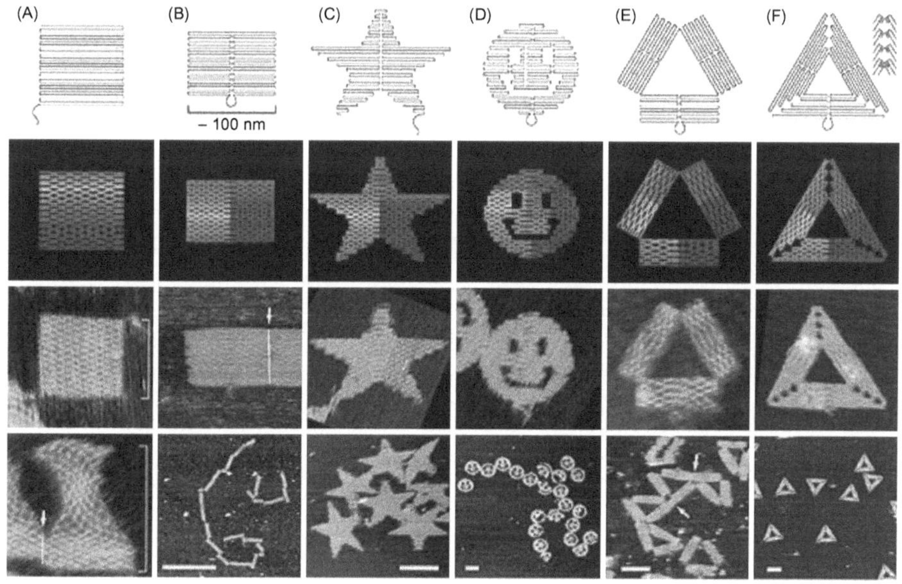

FIGURE 4.34

Examples of DNA origami for arbitrary geometry: Top row—folding paths. (A), square; (B), rectangle; (C), star; (D), disk with three holes; (E), triangle with rectangular domains; (F), sharp triangle with trapezoidal domains and bridges between them (red lines in inset). Second row from top—diagrams showing the bend of helices at crossovers (where helices touch) and away from crossovers (where helices bend apart). Color indicates the base-pair index along the folding path; red is the 1st base, purple the 7000th. Bottom two rows, AFM images. White lines and arrows indicate blunt-end stacking. White brackets mark the height of an unstretched square and that of a square stretched vertically (by a factor >1.5) into an hourglass. White features in f are hairpins. All images and panels without scale bars are of the same size, 165 nm × 165 nm. Scale bars for lower AFM images: (B), 1 μm; (C–F), 100 nm [127]. (For interpretation of the references to color in this figure legend, the reader is referred to the web version of this book.)

spatial-phase coherence of a truly periodic structure. Densely packed features over a wide area (the entire wafer surface) can be quickly generated with decent uniformity. Another merit of IL is its ability to make smaller features than any other optical scheme at a given wavelength, rivaling even high-resolution e-beam lithography. Recall that the minimum linewidth, d_{min}, in optical lithography is determined by the wavelength, NA, and k_1. The advanced industrial tools in optical lithography exhibit $k_1 = 0.4$ and NA = 0.75, resulting in $d_{min} \sim 0.53\lambda$, with the requirement of multimillion dollar lithography equipment. If we consider more affordable lithographic systems for the research community, d_{min} is around 2.5λ, limiting the best resolution to around 1 μm. For IL, it has been demonstrated that with the affordable lithographic system, $k_1 = 0.25$ and NA $\sim$ 0.95 can be achieved, allowing $d_m \sim 0.26\lambda$, half of the state-of-the-art industrial system.

4.2.7.2 Grayscale lithography

Grayscale lithography utilizes a photomask that has 2D binary patterns of subresolution. When a thin PR film is exposed to UV light through the mask, the intensity of the light is locally modulated by diffraction limit and the PR layer is exposed to varying depths resulting in a gradient height profile in PR [132,133]. 3D structures in PR can be fabricated upon the development process. The intensity modulation is given by the use of different sized subresolution patterns (gray-levels) on the photomask. Due to high-resolution requirement (subwavelength pattern size), the photomask for grayscale lithography is fabricated through e-beam lithography. Grayscale lithography is primarily used to economically fabricate 3D MEMS structures [134], microoptical components including microlenses [135], and diffractive elements or hybrids. The optical structures in the PR can also be transferred into the substrate by RIE. The capability of creating 3D structures in a single development step is useful in microfluidic devices, and tapered microfluidic channels were successfully fabricated using this technique (Figure 4.35) [136].

4.3 Materials in micro- and nanofluidics

The selection of materials in engineering systems is driven by their functionality, geometric (design) constraint, and manufacturability. For example, the material properties such as mechanical strength, optical transparency, surface chemistry, biocompatibility, chemical resistance, and thermal characteristics are not only critical for the function and performance of micro- and nanofluidic devices but also affect the decision of the best fabrication processes to make them. Traditionally, silicon and glass have been the dominant substrate materials in many lab-on-a-chip applications due to the fact that the fabrication methods were well developed by the semiconductor and MEMS industries, and the surface functionalization and derivatization were characterized by the chromatography and other industries [137]. Recently, more micro- and nanofluidic devices have been developed in polymeric or plastic platforms. The benefits of using polymers include (i) their extremely diverse set of material characteristics in polymers

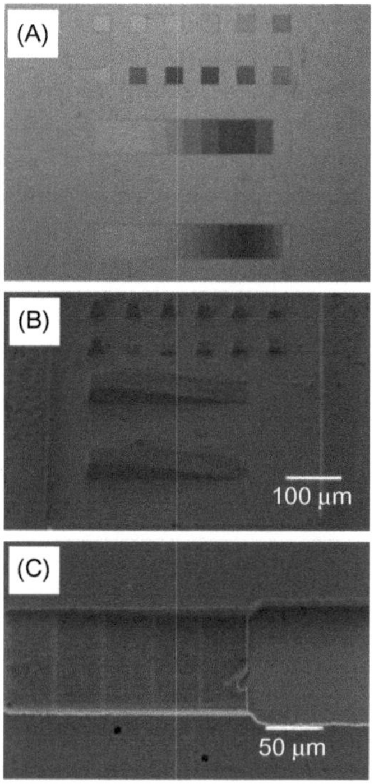

FIGURE 4.35

(A) An optical micrograph of the calibration mask, (B) an SEM image of the corresponding micro tapered structures in thick positive PR AZ4562, and (C) an SEM image of the tapered microchannel in PR AZ4562.

Source *[134]: Nock V, Blaikie RJ. Fabrication of optical grayscale masks for tapered microfluidic devices*

enable them to be synthesized or engineered to possess the properties required for targeted functionality [138] and (ii) inexpensive manufacturing procedures are available and continuously developed in academia and industries.

Substrates are not the only materials that interest us. Coating materials on these substrates and substances that flow in the micro-/nanofuidic devices are also a great interest to us. Especially, nanomaterials and biomaterials are extremely important because they are often the subject matters of investigation or the target materials for syntheses and manipulation. A complete review of these materials is beyond the scope of the book and therefore only a brief account on some most relevant materials is given.

4.3.1 Silicon

A large number of microfabrication techniques employed in MEMS and micro- and nanofluidic devices are inherited from integrated circuit fabrication. Silicon is undoubtedly the most prevalent substrate material in microelectronics, driving the information technology revolution of the current era. The dominance of silicon in the semiconductor industries can be attributed to its superior electrical and mechanical properties as well as its abundance and manufacturability—silicon is one of the very few materials that can be economically manufactured in single-crystalline form. Though there are other semiconductor elements/compounds exhibiting higher electrical conductivity than silicon, the electrical conductivity of silicon can be precisely modulated by impurity doping, which is very important in diodes and transistors. Due to the technological and economical significance of the material, silicon and its compounds have been studied extensively and exhaustively, and the remarkable wealth of information has been accumulated over the last few decades enabling innovative and new areas of Si-based applications beyond integrated circuits. For example, silicon-based MEMS transducers such as accelerometers, pressure sensors, gyroscopes, and micromirrors have been commercially successful, and their market continues to grow. Silicon is also a core element of numerous other applications such as solar cells and photodetectors. Riding on silicon-based MEMS technology, early development of microfluidics was built upon Si substrates—applications bridging between MEMS and microfluidics appeared including microvalves, micropumps, and micromixers [139].

Si is an excellent material in mechanical standpoint. It is an ideal elastic material with little hysteresis up to the yield point. Its Young's modulus can be as high as 190 GPa (for $<111>$ orientation, 160 GPa for $<110>$ orientation, 130 GPa for $<100>$ orientation) similar to that of steel. The yield strength of silicon is 7 GPa at room temperature (compared to 2–4 GPa of different steel family) and its Poisson ratio is 0.27, one of the lowest among all solids. Silicon is brittle and will fracture beyond the yield point with a large fracture strain of 4%. The thermal behavior and dimensional stability of Si is also superior. Si has a melting temperature of 1410°C and a glass transition temperature of Si is around 700°C above which silicon starts to soften and suffers from plastic deformation under load. Moreover, the coefficient of thermal expansion of Si is 2.6×10^{-6} per °C, much lower than most metals. Therefore, under all the operating and most of processing conditions of micro-/nanofluidics devices, Si is a reliable substrate. A high thermal conductivity of Si (~150 W/m K) helps to remove heat away from the substrate used as an efficient heat sink. However, Si is not optically transparent; the attenuation depth of light in Si (the distance light travels through the material before the intensity drops to 36% of its initial value) is 2.7 μm at 633 nm and 0.2 μm at 436 nm. Silicon is transparent only at infrared wavelengths above 1.1 μm. Because most detection and analysis in micro-/nanofluidic devices rely on the optical methods with UV = visible wavelength (250–700 μm), different window materials (glass or polymer) are often integrated with silicon substrates.

In regard to different microstructures, Si exists in three forms: crystalline, polycrystalline, or amorphous. The Czochralski growth of single-crystalline silicon starts with purifying silica sand into metallurgical grade silicon (MGS), and a large cylindrical ingot is then drawn/solidified from the molten MGS with a single-crystalline seed, which determines the resulting crystal orientation. The single crystal ingot is sliced into wafers along the specific crystal planes and is available in various sizes (2–12 in. in diameter). The cut plane, i.e., the top surface of the wafer, is denoted as the wafer orientation. Figure 4.36A shows a (100) wafer, (100) being the most common orientation with a (110) crystal plane as a major flat cut. Currently, 4-in. wafers are primarily used for the research community because of the prevalence of 4-in. compatible processing

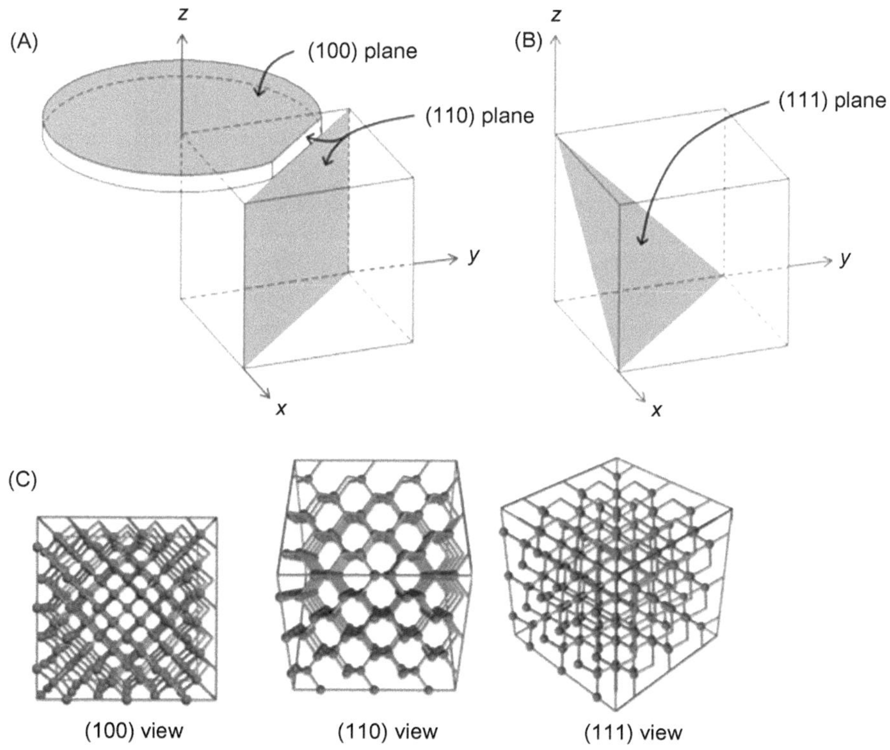

FIGURE 4.36

(A) Illustrating a (100) Si wafer that is cut to reveal one of the (100) planes on the wafer surface and the major flat is on the (110) plane, (B) (111) Si crystal plane, (C) silicon crystal viewed from different angles: face view (100), edge (110), and vertex view (111) (Figure C © 2010, reproduced with permission from John Wiley and Sons from [140]).

tools. Silicon has a diamond-cubic lattice structure; the packing density of atoms varies with the crystal plane orientation, accounting for its anisotropic nature in the material properties. A higher packing density on the (111) plane not only leads to the higher mechanical strength in that direction but also makes silicon more resistant to some types of chemical etchants (Figure 4.36B,C). Anisotropic wet etching of Si producing trapezoidal structures is discussed in Section 4.1.4.1.

Polycrystalline and amorphous silicon are also important materials in the IC, MEMS, and solar cell industries. Both types of silicon are typically deposited as a thin film using CVD. Many MEMS devices such as actuators and sensors have been built based on surface micromachining of polysilicon or amorphous silicon. However, these two materials are not frequently encountered in micro- and nanofluidic devices.

4.3.2 Glass and quartz

Glass is known to many of us as amorphous silica (SiO_2), but varieties of glasses are produced with the inclusion of other compounds to help possess different thermal, mechanical, optical, and chemical properties. The most common type of glass is soda-lime glass widely used for windows, bottle containers, and other everyday commodities. The foreign inclusion (e.g., Na_2O, CaO, Al_2O_3, K_2O, MgO, TiO_2) in soda-lime glass lowers the process temperature, making it into desirable shapes at affordable costs. However, most of the laboratories and technical glassware are made of borosilicate glass (BSG). The main inclusion other than silica is boric oxide (B_2O_3), which requires high temperature for manufacturing, but in turn BSG is more resistant to chemical attack and thermal shock than soda-lime glass. BSG are sold under the brand names Pyrex (Corning), Duran (Schott), Bomex (KBG), etc.

The reasons that glass substrates are very popular as microfluidic substrates are not only, like Si, rigid, dimensionally stable, and available in a flat wafer form but also optically transparent allowing for the optical methods to image and interrogate fluids and substances under study. Cover slips and microscope slides, usually made of BSG, are routinely used as platforms for many microfluidic chips, and glass microchannel devices such as capillary electrophoresis are commercially available.

Though BSG has a relatively small coefficient of thermal expansion (3.5×10^{-6} per °C), some applications entail even smaller CTEs and more chemical inertness, which can be offered by quartz or fused silica. Quartz is a naturally occurring crystalline form of silica—a silicon atom surrounded by four oxygen atoms in tetrahedral coordination. Fused (or synthetic) quartz can be synthesized by melting naturally occurring quartz at around 2000°C. Fused silica is a synthetic glass consisting of silicon dioxide with no or little amount of other components and amorphous in structure. Both quartz and fused silica have more than 99% of SiO_2, and their purity, chemical inertness, extremely low CTE ($\sim 0.5 \times 10^{-6}$ per °C), high resistance to thermal shock, high melting temperature ($\sim$1610°C), and superior UV transmission (cutoff frequency of $\sim$170 nm) make them suitable for (i) high-temperature applications including furnace tubes, crucibles, and refractory materials and (ii) high-precision optical elements (e.g., optical fibers, lens, photomasks). The piezoelectric property of quartz enables the application, such as the quartz crystal microbalance, to be useful for the accurate measurements of small mass changes.

The primary way of micromachining glass or quartz substrates is wet etching using concentrated hydrofluoric acid (HF) or buffered oxide etchant (BOE) [141,142]. Due to its isotropic nature of etching, the sidewall profiles of etched micro- and nanochannels are typically rounded, therefore limiting the aspect ratio to around 1. When more directional etching of glass is desirable, plasma etching using SF_6 or Freon discharges has been implemented [143]. Etch rates of glass and quartz are rather slow ($<$0.1 μm/min) due to high binding energy of SiO_2, and thru-etching of glass substrates would be impractical with wet or dry etching. Therefore, holes and channels in glass/quartz substrates are more commonly created using laser ablation [144,145], ultrasonic drilling [146], and micropowder blasting [147].

4.3.3 Other ceramics

4.3.3.1 Alumina (Al_2O_3)

Alumina is one of the most important ceramic materials [148]. It has a high melting temperature of 2054°C and is chemically stable and unreactive, leading to applications including high-temperature components/housing, catalyst supports, and biomedical implants. The hardness, strength, and wear resistance of alumina are among the highest for oxides, making it suitable for abrasive materials and bearings and as a component in cutting tools. Alumina is most popular in the form of a thin film, which can be deposited by CVD, ALD, or reactive sputtering, and it is used as an electrical insulator, thanks to its high electrical resistance. Its crystalline form, known as sapphire, has a superior optical transparency—high transmission in UV = visible frequencies. Porous (anodized) alumina is important in nanotechnology and micro-/nanofluidics because of its utility as a filter paper, nanocapillary arrays, and a template for nanomaterial syntheses. By adjusting anodizing conditions such as applied voltage, electrolyte concentration, and temperature, one can create a self-ordered (typically in a hexagonal lattice) porous alumina structure with an interpore distance of

50–500 nm. Template growth of nanomaterials from anodized nanoporous alumina has been covered previously.

4.3.4 Polymers

4.3.4.1 Polydimethyl siloxane

PDMS is a type of organosilicon compound and an elastomeric polymer, and its chemical formula is $CH_3[Si(CH_3)_2O]_nSi(CH_3)_3$. Apart from its use in soft lithography and micro-/nanofluidics, PDMS can be found in our everyday products: contact lenses and medical devices, food additives, anti-foaming agents, etc. The popularity of PDMS for micro-/nanofluidic device fabrication can be explained in many folds: (i) its optical transparency (from 250 to 1100 nm) enabling the observation of the contents in micro-/nanochannels by optical microscope, (ii) its widely accepted biocompatibility, (iii) its ability to form a temporary bond with most clean surfaces and its ability to rapidly form a permanent bond on the oxide surface with plasma activation, (iv) its gas permeability allowing cell culture, and (v) its abundance and low cost. However, PDMS has its negative aspects with respect to a use of microfluidics materials: (i) dimensional instability—it swells in many solvents and sometimes bonding interface can be degenerated, (ii) low modulus and poor mechanical strength make it extremely difficult to deposit a thin film (metallic or dielectric) on the surface and thus hard to integrate electrodes or other electrical/mechanical components, and (iii) unstable surface—highly mobile moiety at the surface keeps surface functionalization from lasting for an extended period.

4.3.4.2 SU-8

SU-8, first developed by IBM, is an epoxy-based, negative PR. A thickness ranging from 0.5 μm to a few hundred microns can be prepared using spin coating. The structures as thick as 2 mm and with an aspect ratio higher than 20 have been demonstrated with contact lithography. This feat of achieving HAR structures in a simple lithographic step is mainly due to the high transparency and low absorption of SU-8 in the UV range (365 nm). The ability to create the HAR structures as well as 3D patterns without expensive equipment (e.g., X-ray lithography in LIGA or deep reactive ion etcher) is the key reason for the initial success of SU-8. A myriad of applications for SU-8 HAR and 3D structures include high-luminosity lighting panels, photonic crystals, and biosensors (refer to the review paper by Del Campo [149] for details). Its mechanical stability and low Young's modulus as well as excellent chemical resistance enable SU-8 to be directly used as a structural material for MEMS and microfluidics [150,151]. SU-8 is also known as one of the best biocompatible materials, gaining popularities in bioMEMS and lab-on-a-chip applications [152].

4.3.5 Nanomaterials

The origin of nanotechnology mainly stems from the remarkable physical properties of nanomaterials that may be significantly different from those of bulk counterparts. The unique properties of nanomaterials include (1) a significantly lower melting point or phase transition temperature due to the increased fraction of the surface atoms with respect to the total amount of atoms, (2) a decrease in electrical conductivity due to increased surface scattering, (3) high (sometimes near-theoretical) mechanical strength, thanks to reduced imperfections in nanocrystalline structures, and (4) novel optical/electrical/magnetic properties owing to quantum confinements. Depending on their shape and morphology, nanomaterials are classified into three main categories: nanotubes, nanowires, and nanoparticles. Recently, other exotic shapes of nanomaterials are synthesized and utilized as nanoplasmonic sensors, catalysts, supports, etc. [153]. In micro-/nanofluidics, nanomaterials are often a subject of manipulation and/or interrogation [154–156].

4.4 Fabrication case study

4.4.1 Case 1. Fabrication of gas microvalve and micropump

A group at University of Illinois has developed an electrostatically actuated micropump device with two passive membrane-based check microvalves [157]. As shown in Figure 4.37, the pump device consists of four layers: (from bottom to top) silicon valve die, polyimide valve membrane, silicon pump die, and pump diaphragm. This example is chosen because it showcases the complex process flow of a multilayer microfluidic device involving the many fabrication processes we covered in this chapter.

First, standard double-sided photolithography is implemented to pattern PRs on both sides of the silicon wafer. Gas channels and inlet/outlet holes are

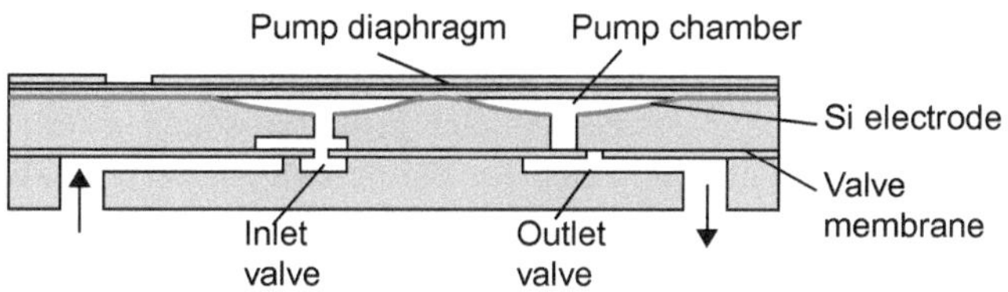

FIGURE 4.37

A cross-sectional view of the electrostatic microgas pump integrated with two passive membrane valves.

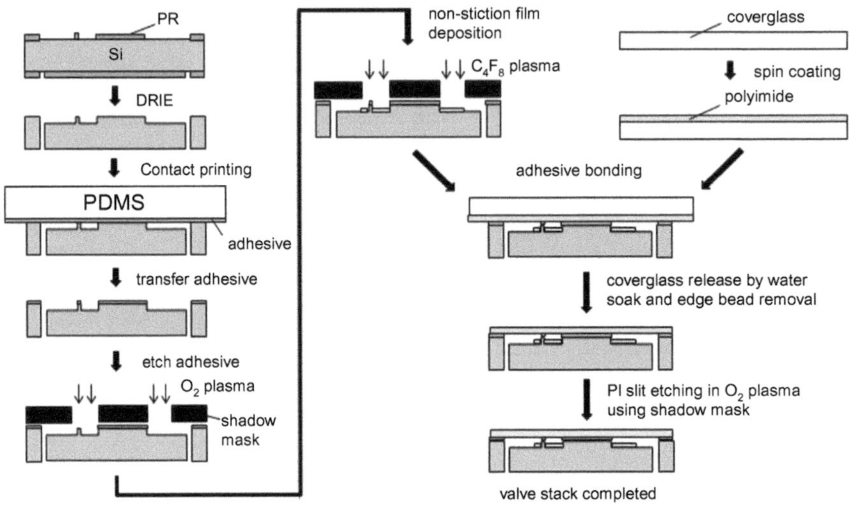

FIGURE 4.38

Fabrication sequence of the Si valve die and the polyimide valve membrane.

etched using DRIE to create a valve die. μCP is used to pattern and transfer an adhesive layer, and a plasma deposition is performed through a shadow mask to selectively deposit a non-stiction coating on the valve seat. Second, a polyimide (PI) membrane is spin-coated and cured on a coverglass, and likewise, an adhesive layer and a non-stiction coating are selectively transferred to and deposited on the desired areas of the membrane. Adhesive bonding technique is used to bond the silicon die and PI valve membrane. A hole and slit in the PI membrane as fluidic interconnect is created by RIE dry etching (see Figure 4.38).

Next, the silicon pump die is fabricated with double-sided photolithography and DRIE etching. Thermal oxide is grown to protect the surface from getting damaged in the subsequent chemical mechanical polishing, which creates curved pump cavities. Spin-on dopant is then spin-coated and diffused in high-temperature furnace, followed by thermal oxidation for electrical isolation. Another lithographic process defines contact pads, on which a thin gold film is deposited via evaporation and annealed for ohmic contact. A non-stiction coating is selectively deposited through a shadow mask. A multilayer of the conducting diaphragm, PI/Cr/Au/Cr/PI, is prepared on a coverglass using spin coating and metal evaporation (Figure 4.39).

Lastly, each layer is now bonded to form a micropump stack using the adhesive bonding technique. The adhesive layer is patterned and transferred via μCP. Due to the low-temperature operation for adhesive bonding, polyimide

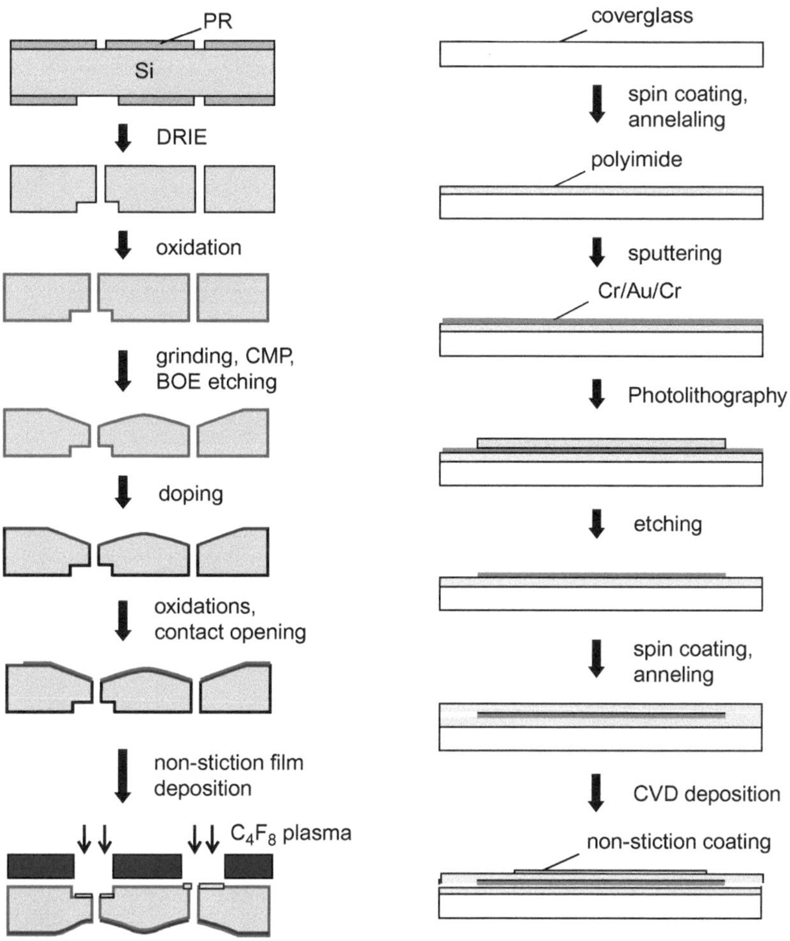

FIGURE 4.39

Fabrication sequence of (A) the Si pump die and (B) the conducting diaphragm.

membranes, metal electrodes, and non-stiction coating, materials are not damaged during the process (see Figure 4.40).

4.4.2 Case 2. Fabrication of single nanopore

Synthetic single nanopores are the ultimate 1D nanofluidic element and have garnered significant attention over the last couple of decades. A nanopore whose

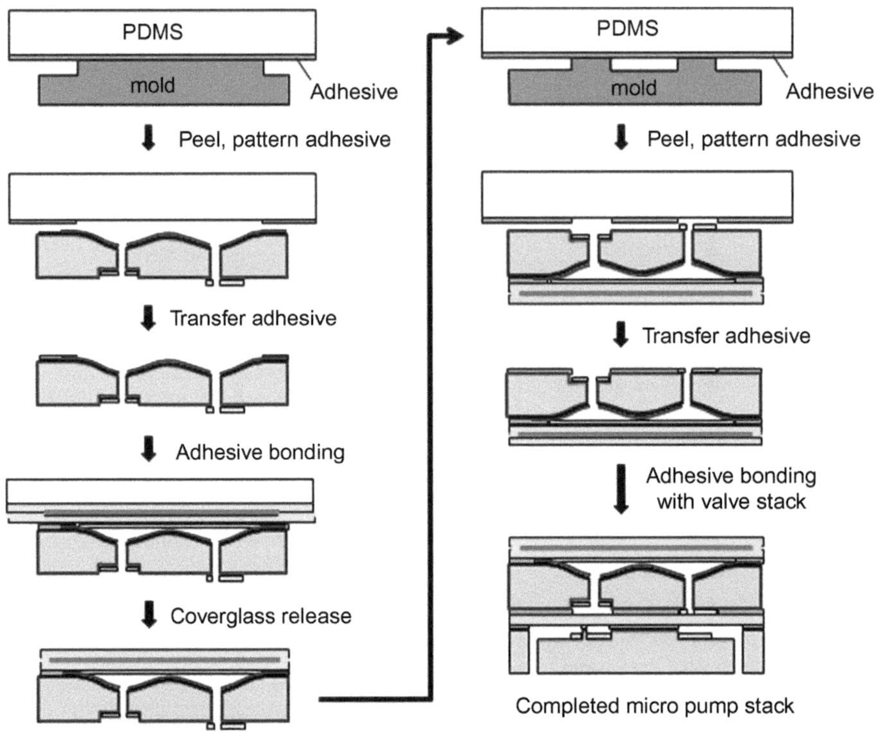

FIGURE 4.40

The final bonding sequence to form a complete microgas pump stack.

size is equivalent to biological entities of interest is created in an electrically insulating membrane; it can act as a biological protein channel and be used as a single-molecule detector. The solid-state nanopores in DNA sequencing and other biological applications will be discussed in Chapter 5. The most common way to create a single nanopore in a controllable fashion is to use electron or ion-beam etching. Depending on the target size of the nanopore, electrons or ions can be focused on the surface and drill a hole through the thin membrane. The top example in Figure 4.41 shows a single nanopore of 5 nm in diameter [158]. A nanopore of a larger diameter (20 nm) was first opened on the silicon nitride membrane using e-beam lithography, and high-energy irradiation of e-beam in transmission electron microscope reduced the pore size to a single nanometer scale. In another example (bottom in Figure 4.41), a 50-nm hole is drilled through the PMMA–metal–PMMA composite structure using focused ion beam [158].

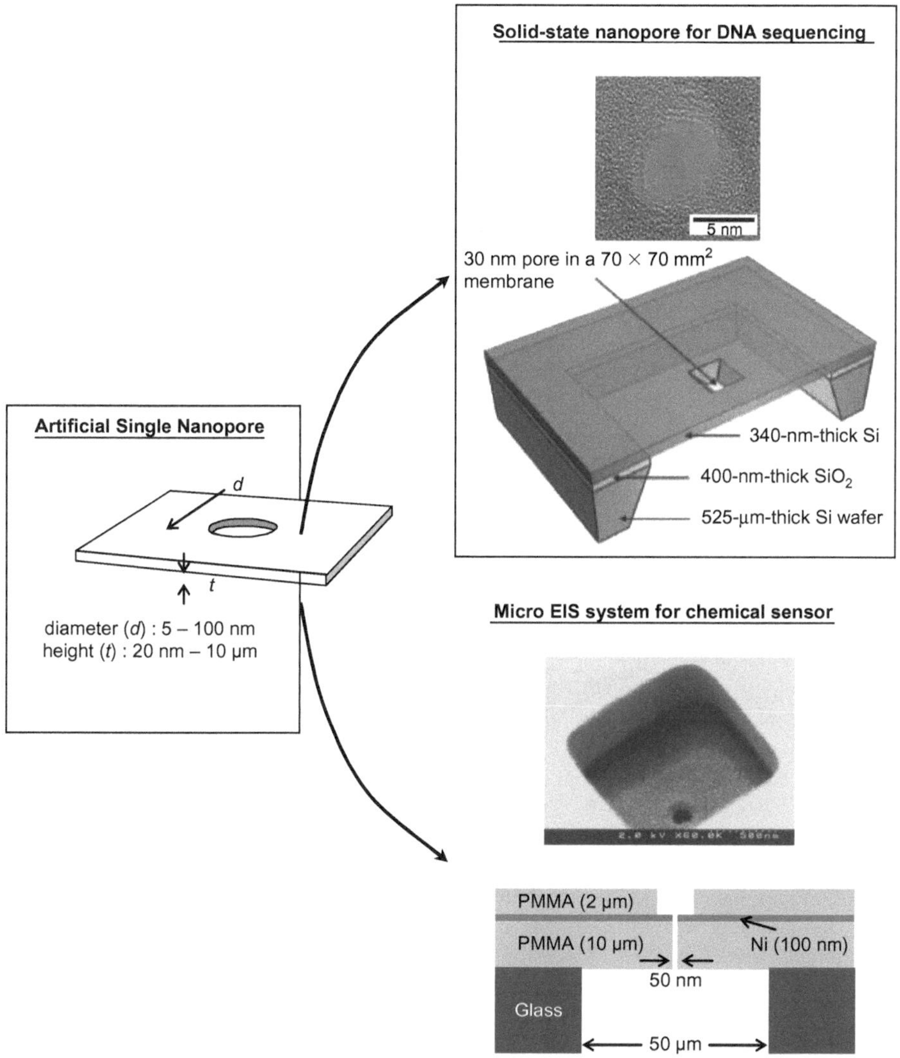

FIGURE 4.41

Fabrication example of a single nanopore created (top) by transmission electron microscope and (bottom) by focused ion beam.

Source: *[158] Storm AJ, Chen JH, Ling XS, Zandbergen HW, Dekker C. Fabrication of solid-state nanopores with single-nanometre precision and [157] Prakash S, Yeom J, Jin N, Adesida I, Shannon MA. Characterization of ionic transport at the nanoscale.*

References

[1] Duffy DC, McDonald JC, Schueller OJA, Whitesides GM. Rapid prototyping of microfluidic systems in poly(dimethylsiloxane). Anal Chem 1998;70:4974–84.

[2] Tritchkov A, Jeong S, Kenyon C. Lithography enabling for the 65 nm node gate layer patterning with alternating PSM—optical microlithography XVIII. Proc SPIE 2005;5754:215–25.

[3] Perlitz S, Buttgereit U, Scherübl T, Seidel D, Km Lee, Tavassoli M. Novel solution for in-die phase control under scanner equivalent optical settings for 45-nm node and below. Proc SPIE 2007;6607.

[4] Randall J, Tritchkov AV, Jonckheere RM, Jaenen P, Ronse KG. Reduction of mask induced CD errors by optical proximity correction. Proc SPIE Int Soc Opt Eng 1998;3334:124–30.

[5] Sheppard CJR, Hegedus Z. Resolution for off-axis illumination. J Opt Soc Am A 1998;15(3):622–4.

[6] 193 nm immersion lithography: Status and challenges [Internet]. SPIE; 2007.

[7] Lin BJ. The k_3 coefficient in nonparaxial λ/NA scaling equations for resolution, depth of focus, and immersion lithography. J Micro/Nanolithogr MEMS MOEMS 2002;1:7–12.

[8] Wang Y, Miyamatsu T, Furukawa T, Yamada K, Tominaga T, Makita Y, et al. High-refractive-index fluids for the next-generation ArF immersion lithography. Proc SPIE 2006;6153:61530A.

[9] Wei Y, Back D. 193nm immersion lithography: Status and challenges. SPIE 2007;22 (March):10.1117/2.1200703.0001.

[10] Maluf N, Williams K. An introduction to microelectromechanical systems engineering. Norwood, MA: Artech House; 2004.

[11] Deal BE, Grove AS. General relationship for the thermal oxdiation of silicon. J Appl Phys 1965;36(12):3770–8.

[12] Cao H, Yu Z, Wang J, Tegenfeldt JO, Austin RH, Chen E, et al. Fabrication of 10 nm enclosed nanofluidic channels. Appl Phys Lett 2002;81(1):174–6.

[13] Carter G, Colligon JS. Ion bombardment of solids. New York, NY: Elsevier; 1968.

[14] Sigmund P. Theory of sputtering I: sputtering yield of amorphous and polycrystalline targets. Phys Rev 1969;184(2):383–416.

[15] Stuart RV, Wehner GK. Sputtering yields at very low bombarding energies. J Appl Phys 1962;33:2345–52.

[16] Bodhansky J, Roth J, Bay HL. An analytical formula and important parameters of low-energy ion sputtering. J Appl Phys 1980;51:2861–5.

[17] SPECS GmbH. n.d. Useful information and facts about the practice of sputtering. SPECS GmbH, Germany. Available from: <http://www.specs.de/cms/upload/PDFs/IQE11-35/sputter-info.pdf>.

[18] Choy KL. Chemical vapor deposition of coatings. Prog Mater Sci 2003; 48:57–170.

[19] Suntola T, Hyvarinen J. Atomic layer epitaxy. Ann Rev Mater Sci 1985;15:177–95.

[20] Leskela M, Rital M. Atomic layer deposition (ALD): from precursors to thin film structures. Thin Solid Films 2002;409(1):138–46.

[21] Intel. Introducing the 45nm Next-Generation Intel® Core™ Microarchitecture. White Paper. Available from: <http://www.intel.com/content/dam/doc/white-paper/45nm-next-generation-core-microarchitecture-white-paper.pdf>; 2007.

[22] Döpper J, Clemens M, Jung WES, Kämper K-P, Lehr H. Micro gear pumps for dosing of viscous fluids. J Micromech Microeng 1997:7.

[23] Ho C-H, Chin K-P, Yang C-R, Wu H-M, Chen S-L. Ultrathick SU-8 mold formation and removal, and its application to the fabrication of LIGA-like micromotors with embedded roots. Sens Actuat Phys 2002;102:130–8.

[24] Williams KR, Muller RS. Etch rates for micromachining processing. J Microelectromech Syst 1996;5:256–69.

[25] Williams KR, Gupta K, Wasilik M. Etch rates for micromachining processing—part II. J Microelectromech Syst 2003;12:761–78.

[26] Durand NFY, Renaud P. Label-free determination of protein–surface interaction kinetics by ionic conductance inside a nanochannel. Lab Chip 2009;9:319–24.

[27] Gervais L, Hitzbleck M, Delamarche E. Capillary-driven multiparametric microfluidic chips for one-step immunoassays. Biosens Bioelectron 2011;27:64–70.

[28] Zhu H, Holl M, Ray T, Bhushan S, Meldrum DR. Characterization of deep wet etching of fused silica glass for single cell and optical sensor deposition. J Micromech Microeng 2009;19:065013.

[29] Prakash S, Pinti M, Bhushan B. Theory, fabrication, and applications of microfluidics and anofluidic biosensors. Philos Trans Royal Soc A 2012;370:2269–303.

[30] Tsujino K, Matsumura M. Morphology of nanoholes formed in silicon by wet etching in solutions containing HF and H_2O_2 at different concentrations using silver nanoparticles as catalysts. Electrochim Acta 2007;53(1):28–34.

[31] Lee C-L, Tsujino K, Kanda Y, Ikeda S, Matsumura M. Pore formation in silicon by wet etching using micrometre-sized metal particles as catalysts. J Mater Chem 2008;18:1015–20.

[32] Li X, Bohn PW. Metal-assisted chemical etching in HF/H_2O_2 produces porous silicon. Appl Phys Lett 2000;77:2572–4.

[33] Tsujino K, Matsumura M. Helical nanoholes bored in silicon by wet chemical etching using platinum nanoparticles as catalyst. Electrochem Solid-State Lett 2005;8 (12):C193–5.

[34] Hidreth OJ, Lin W, Wong CP. Effect of catalyst shape and etchant composition on etching direction in metal-assisted chemical etching of silicon to fabricate 3D nanostructures. Nano Lett 2009;3(12):4033–42.

[35] Huang Z, Shimizu T, Senz S, Zhang Z, Zhang X, Lee W, et al. Ordered arrays of vertically aligned [110] silicon nanowires by suppressing the crystallographically preferred $<100>$ etching directions. Nano Lett 2009;9:2519–25.

[36] Chen H, Wang H, Zhang X-H, Lee C-S, Lee S-T. Wafer-scale synthesis of single-crystal zigzag silicon nanowire arrays with controlled turning angles. Nano Lett 2010;10:864–8.

[37] Zhang ML, Peng KQ, Fan X, Jie JS, Zhang RQ, Lee ST, et al. Preparation of large-area uniform silicon nanowires arrays through metal-assisted chemical etching. J Phys Chem C 2008;112:4444–50.

[38] Peng K, Xu Y, Wu Y, Yan Y, Lee S-T, Zhu J. Aligned single-crystalline Si nanowire arrays for photovoltaic applications. Small 2005;1:1062–7.

[39] Garnett EC, Yang P. Silicon nanowire radial p–n junction solar cells. J Am Chem Soc 2008;130:9224–5.

[40] Peng K, Jie J, Zhang W, Lee S-T. Silicon nanowires for rechargeable lithium-ion battery anodes. Appl Phys Lett 2008;93:033105.

[41] Peng K-Q, Wang X, Lee S-T. Gas sensing properties of single crystalline porous silicon nanowires. Appl Phys Lett 2009;95:243112.

[42] In HJ, Field CR, Pehrsson PE. Periodically porous top electrodes on vertical nanowire arrays for highly sensitive gas detection. Nanotechnology 2011;22:355501.

[43] Huang Z, Geyer N, Werner P, de Boor J, Gösele U. Metal-assisted chemical etching of silicon; a review. Adv Mater 2011;23:285–308.

[44] Bollinger LD. Ion beam etching with reactive gases. Solid State Technol 1983;26 (1):99–108.

[45] Bang ZL, MacDonald NC, editors. An RIE process for submicron, silicon electromechanical structures. International conference on solid-state sensors and actuators. San Francisco, CA; 1991.

[46] Yeom J, Wu Y, Selby JC, Mark A, Shannon J. Maximum achievable aspect ratio in deep reactive ion etching of silicon due to aspect ratio dependent transport and the microloading effect. Vac Sci Technol B 2005;23:2319. doi:10.1116/1.2101678.

[47] Clerc P-A, Dellmann L, Grétillat F, Grétillat M-A, Indermühle P-F, Jeanneret S, et al. Advanced deep reactive ion etching: a versatile tool for microelectromechanical systems. J Micromech Microeng 1998;8:272–8.

[48] Jansen H, Boer MD, Burger J, Legtenberg R, Elwenspoek M. The black silicon method: a universal method for determining the parameter setting of a fluorine-based reactive ion etcher in deep silicon trench etching with profile control. J Micromech Microeng 1995;5:115–20.

[49] Niklaus F, Stemme G, Lu J-Q, Gutmann RJ. Adhesive wafer bonding. J Appl Phys 2006;99:031101.

[50] Mao P, Han J. Fabrication and characterization of 20 nm planar nanofluidic channels by glass–glass and glass–silicon bonding. Lab Chip 2005;5:837–44.

[51] Delft KMV, Eijkel JCT, Mijatovic D, Druzhinina TS, Rathgen H, Tas NR, et al. Micromachined Fabry–Perot interferometer with embedded nanochannels for nanoscale fluid dynamics. Nano Lett 2007;7(2):345–50.

[52] Durand NFY, Renaud P. Label-free determination of protein–surface interaction kinetics by ionic conductance inside a nanochannel. Lab Chip 2009;9:319–24.

[53] Radadia AD, Salehi-Khojin A, Masel RI, Shannon MA. The fabrication of all-silicon micro gas chromatography columns using gold diffusion eutectic bonding. J Micromech Microeng 2010;20:015002.

[54] Blanco FJ, Agirregabiria M, Garcia J, Berganzo J, Tijero M, Arroyo MT, et al. Novel three-dimensional embedded SU-8 microchannels fabricated using a low temperature full wafer adhesive bonding. J Micromech Microeng 2004;14:1047–56.

[55] Agirregabiria M, Blanco FJ, Berganzo J, Arroyo MT, Fullaondo A, Mayora K, et al. Fabrication of SU-8 multilayer microstructures based on successive CMOS compatible adhesive bonding and releasing steps. Lab Chip 2005;5:545–52.

[56] Cui Z. Nanofabrication: principles. Capabilities and limits. 1st ed. Springer; 2008.

[57] Xia Y, Rogers JA, Paul KE, Whitesides GM. Unconventional methods for fabricating and patterning nanostructures. Chem Rev 1999;99(7):1823–48.

[58] Yasin S, Hasko DG, Ahmed H. Fabrication of <5 nm width lines in poly(methylmethacrylate) resist using a water:isopropyl alcohol developer and ultrasonically-assisted development. Appl Phys Lett 2001;78:2760–2.

[59] Perry JM, Zhou K, Harms ZD, Jacobson SC. Ion transport in nanofluidic funnels. ACS Nano 2010;4:3897–902.

[60] Devlin NR, Brown DK, Kohl PA. Patterning decomposable polynorbornene with electron beam lithography to create nanochannels. J Vac Sci Technol B 2009;27:2508–11.
[61] Volkert CA, Minor AM. Focused ion beam microscopy and micromachining. MRS Bull 2007;32:389–95.
[62] Giannuzzi LA, Stevie FA. Introduction to focused ion beams: instrumentation, theory, techniques, and practice. New York: Springer; 2005.
[63] Chou SY, Krauss PR, Renstrom PJ. Imprint lithography with 25-nanometer resolution. Science 1996;272:85–7.
[64] Schift H. Nanoimprint lithography: an old story in modern times? a review. J Vac Sci Technol B 2008;26:458–80.
[65] Guo LJ, Cheng X, Chou C-F. Fabrication of size-controllable nanofluidic channels by nanoimprinting and its application for DNA stretching. Nano Lett 2004;4(1):69–73.
[66] Krauss PR, Chou SY. Nano-compact disks with 400 Gbit/in^2 storage density fabricated using nanoimprint lithography and read with proximal probe. Appl Phys Lett 1997;71(21):3174–6.
[67] Bailey T, Choi BJ, Colburn M, Meissl M, Shaya S, Ekerdt JG, et al. Step and flash imprint lithography: template surface treatment and defect analysis. J Vac Sci Technol B 2000;18(6):3572–7.
[68] Stewart MD, Willson CG. Imprint materials for nanoscale devices. MRS Bull 2005;30:947–52.
[69] Chou SY, Keimel C, Gu J. Ultrafast and direct imprint of nanostructures in silicon. Nature 2002;417:835–7.
[70] Guo LJ. Nanoimprint lithography; methods and materials requirements. Adv Mater 2007;19:495–513.
[71] Cheng L-J, Chang S-T, Guo LJ. Nanoimprint of nanofluidic channels by using hydrophilic hydrogen silsequioxane (HSQ). Proceedings of MicroTAS. Boston, MA, USA; 2005 p. 518–20.
[72] Guo LJ, Cheng X, Chou C-F. Fabrication of size-controllable nanofluidic channels by nanoimprinting and its application for DNA stretching. Nano Letters 2004;4(1):69–73.
[73] Reano RM, Pang SW. Sealed three-dimensional nanochannels. J Vac Sci Technol B 2005;23:2995–9.
[74] Rogers JA, Nuzzo RG. Recent progress in soft lithography. Mater Today 2005;8(2):50–6.
[75] Yang B, Pang SW. Multiple level nanochannels fabricated using reversal UV nanoimprint. J Vac Sci Technol B 2006;24:2984–7.
[76] Michel B, Bernard A, Bietsch A, Delamarche E, Geissler M, Juncker D, et al. Printing meets lithography: soft approaches to high-resolution patterning. IBM J Res Develop 2001;45(5):697–719.
[77] Xia Y, Whitesides GM. Soft lithography. Ann Rev Mater Sci 1998;28:153–84.
[78] Kumar A, Whitesides GM. Features of gold having micrometer to centimeter dimensions can be formed through a combination of stamping with an elastomeric stamp and an alkanethiol "ink" followed by chemical etching. Appl Phys Lett 1993;63(14):2002–4.
[79] Whitesides GM, Ostuni E, Takayama S, Jiang X, Ingber DE. Soft lithography in biology and biochemistry. Ann Rev Biomed Eng 2001;3:335–73.

[80] Jackman RJ, Brittain ST, Adams A, Prentiss MG, Whitesides GM. Design and fabrication of topologically complex, three-dimensional microstructures. Science 1998;280:2089–91.
[81] Kumar A, Biebuyck HA, Whitesides GM. Patterning self-assembled monolayers: applications in material science. Langmuir 1994;10(5):1498–511.
[82] Jeong HE, Ki P, Kwak MK, Seo CH, Suh KY. Capillary kinetics of water in homogeneous, hydrophilic polymeric micro- to nanochannels. Small 2007;3(5):778–82.
[83] Senn T, Esquivel JP, Lörgen M, Sabaté N, Löchel B. Replica molding for multilevel micro-/nanostructure replication. J Micromech Microeng 2010;20:115012.
[84] Menard E, Meitl MA, Sun Y, Park J-U, Shir DJ-L, Nam Y-S, et al. Micro- and nanopatterning techniques for organic electronic and optoelectronic systems. Chem Rev 2007;107:1117–60.
[85] Kim E, Xia Y, Whitesides GM. Polymer microstructures formed by moulding in capillaries. Nature 1995;376:581–4.
[86] Zhao X-M, Xia Y, Whitesides GM. Fabrication of three-dimensional micro-structures: microtransfer molding. Adv Mater 1996;8:837–40.
[87] Blaaderen AV, Ruel R, Wiltzius P. Template-directed colloidal crystallization. Nature 1997;385:321–4.
[88] Whitesides GM, Grzybowski B. Self-assembly at all scales. Science 2002;295:2418–21.
[89] Rabani E, Reichman DR, Geissler PL, Brus LE. Drying-mediated self-assembly of nanoparticles. Nature 2003;426:271–4.
[90] Lazzari M, López-Quintela MA. Block copolymers as a tool for nanomaterial fabrication. Adv Mater 2003;15:1583–94.
[91] Lopes WA, Jaeger HM. Hierarchical self-assembly of metal nanostructures on diblock copolymer scaffolds. Nature 2001;414:735–8.
[92] Saez M, Goodby JW. Supermolecular liquid crystals. J Mater Chem 2005;15:26–40.
[93] Douglas T, Young M. Host–guest encapsulation of materials by assembled virus protein cages. Nature 1998;393:152–5.
[94] Douglas T, Young M. Virus particles as templates for materials synthesis. Advanced Materials 1999;11(8):679–81.
[95] Denkov ND, Velev OD, Kralchevsky PA, Ivanov IB, Yoshimura H, Nagayama K. Mechanism of formation of two-dimensional crystals from latex particles on substrates. Langmuir 1992;8:3183–90.
[96] Micheletto R, Fukuda H, Ohtsu M. A simple method for the production of a two-dimensional, ordered array of small latex particles. Langmuir 1995;11:3333–6.
[97] Zhang J, Li Y, Zhang X, Yang B. Colloidal self-assembly meets nanofabrication: from two-dimensional colloidal crystals to nanostructure arrays. Adv Mater 2010;22:4249–69.
[98] Huang Z, Fang H, Zhu J. Fabrication of silicon nanowire arrays with controlled diameter, length, and density. Adv Mater 19;744–48.
[99] Mansky P, Chaikin PM, Thomas EL. Monolayer films of diblock copolymer microdomains for nanolithographic applications. J Mater Chem 1995;30:1987–92.

[100] Park M, Harrison C, Chaikin PM, Register RA, Adamson DH. Block copolymer lithography: periodic arrays of $\sim 10^{11}$ holes in 1 square centimeter. Science 1997;276:1401–4.
[101] Darling SB. Directing the self-assembly of block copolymers. Prog Polym Sci 2007;32(10):1152–204.
[102] Segalman RA. Patterning with block copolymer thin films. Mater Sci Eng R Rep 2005;48(6):191–226.
[103] Albert III JNL, Epps TH. Self-assembly of block copolymer thin films. Mater Today 2010;13:24–33.
[104] Knoll A, Horvat A, Lyakhova KS, Krausch G, Sevink GJA, Zvelindovsky AV, et al. Phase behavior in thin films of cylinder-forming block copolymers. Phys Rev Lett 2002;89(3):035501.
[105] Ruiz R, Kang H, Detcheverry FA, Dobisz E, Kercher DS, Albrecht TR, et al. Density multiplication and improved lithography by directed block copolymer assembly. Science 2008;321:936–9.
[106] Stoykovich MP, Nealey PF. Block copolymers and conventional lithography. Mater Today 2006;9(9):20–9.
[107] Rodwogin MD, Spanjers CS, Leighton C, Hillmyer MA. Polylactide–poly(dimethylsiloxane)–polylactide triblock copolymers as multifunctional materials for nanolithographic applications. ACS Nano 2010;4(2):725–32.
[108] Park S, Wang J-Y, Kim B, Xu J, Russell TP. A simple route to highly oriented and ordered nanoporous block copolymer templates. ACS Nano 2008;2(4):766–72.
[109] Pai RA, Humayun R, Schulberg MT, Sengupta A, Sun J-N, Watkins JJ. Mesoporous silicates prepared using preorganized templates in supercritical fluids. Science 2004;303:507–10.
[110] Thurn-Albrecht T, Steiner R, DeRouchey J, Stafford CM, Huang E, Bal M, et al. Nanoscopic templates from oriented block copolymer films. Adv Mater 2000;12(11):787–91.
[111] Yang SY, Ryu I, Kim HY, Kim JK, Jang SK, Russell TP. Nanoporous membranes with ultrahigh selectivity and flux for the filtration of viruses. Adv Mater 2006;18(6):709–12.
[112] Galatsis K, Wang KL, Ozkan M, Ozkan CS, Huang Y, Chang JP, et al. Patterning and templating for nanoelectronics. Adv Mater 2010;22:769–78.
[113] Lee DH, Shin DO, Lee WJ, Kim SO. Hierarchically organized carbon nanotube arrays from self-assembled block copolymer nanotemplates. Adv Mater 2008;20(13):2480–5.
[114] Lee JI, Cho SH, Park S-M, Kim JK, Kim JK, Yu J-W, et al. Highly aligned ultrahigh density arrays of conducting polymer nanorods using block copolymer templates. Nano Lett 2008;8(8):2315–20.
[115] Horiuchi S, Fujita T, Hayakawa T, Nakao Y. Three-dimensional nanoscale alignment of metal nanoparticles using block copolymer films as nanoreactors. Langmuir 2003;19(7):2963–73.
[116] Hamley IW. Nanostructure formation using block copolymers. Nanotechnology 2003;14:R39–54.
[117] Park C, Yoon J, Thomas EL. Enabling nanotechnology with self-assembled block copolymer patterns. Polymer 2003;44:6725–7032.

[118] Kim JK, Yang SY, Lee Y, Kim Y. Functional nanomaterials based on block copolymer self-assembly. Prog Polym Sci 2010;35:1325–49.
[119] Li M, Ober CK. Block copolymer patterns and templates. Mater Today 2006;9:9.
[120] Menon L. Porous alumina templates for nanofabrication. In: Dekker C, editor. Encyclopedia of nanoscience and nanotechnology. 2nd ed. Taylor & Francis; 2009.
[121] Masuda H, Fukuda K. Ordered metal nanohole arrays made by a two-step replication of honeycomb structures of anodic alumina. Science 1995;268:1466–8.
[122] Masuda H, Yamada H, Satoh M, Asoh H, Nakao M, Tamamura T. Highly ordered nanochannel-array architecture in anodic alumina. Appl Phys Lett 1997;71(19):2770–2.
[123] Shingubara S. Fabrication of nanomaterials using porous alumina templates. J Nanopart Res 2003;5:17–30.
[124] Seeman NC. Nucleic acid junctions and lattices. J Theor Biol 1982;99:237–47.
[125] Chen J, Seeman NC. Synthesis from DNA of a molecule with the connectivity of a cube. Nature 1991;350:631–3.
[126] Lin C, Liu Y, Rinker S, Yan H. DNA tile based self-assembly: building complex nanoarchitectures. Chem Phys Chem 2006;7:1641–7.
[127] Rothemund PWK. Folding D.N.A. to create nanoscale shapes and patterns. Nature 2006;440:297–302.
[128] Andersen ES, Dong M, Nielsen MM, Jahn K, Subramani R, Mamdouh W, et al. Self-assembly of a nanoscale DNA box with a controllable lid. Nature 2009;459:73–6.
[129] Yan H, Park SH, Finkelstein G, Reif JH, LaBean TH. DNA-templated self-assembly of protein arrays and highly conductive nanowires. Science 2003;301:1882–4.
[130] Le JD, Pinto Y, Seeman NC, Musier-Forsyth K, Taton TA, Kiehl RA. DNA-templated self-assembly of metallic nanocomponent arrays on a surface. Nano Lett 2004;4(12):2343–7.
[131] Walsh ME. On the design of lithographic interferometers and their application. Cambridge, MA. MIT; 2004.
[132] Henke W., Hoppe W., Quenzer H.J., Staudt-Fischbach P., Wagner B., editors. Simulation and experimental study of gray-tone lithography for the fabrication of arbitrarily shaped surfaces. Proc IEEE Micro Electro Mechanical Systems (MEMS), Oiso, Japan; 1994.
[133] Waits CM, Ghodssi R, Ervin MH, Dubey M, editors. MEMS-based Gray-Scale Lithography. International Semiconductor Device Research Symposium (ISDRS). Washington, DC; December 5–7, 2001.
[134] Waits CM, Modafe A, Ghodssi R. Investigation of gray-scale technology for large area 3D silicon MEMS structures. J Micromech Microeng 2003;13(2):170–7.
[135] Hung K-Y, Tseng F-G, Chou H-P. Application of 3D gray mask for the fabrication of curved SU-8 structures. Microsyst Technol 2005;11:365–9.
[136] Nock V, Blaikie RJ. Fabrication of optical grayscale masks for tapered microfluidic devices. Microelectron Eng 2008;85:1077–82.
[137] Becker H, Locascio LE. Review: Polymer microfluidic devices. Talanta 2002;56:267–87.
[138] Lee S-J, Sundararajan N. Microfabrication for microfluidics. Norwood, MA: Artech House; 2010.

[139] Ho C-M, Tai Y-C. Review: MEMS and its applications for flow control. J Fluids Eng Trans ASME 1996;118(3):437–46.

[140] Franssila S. Introduction to microfabrication. 2nd edition John Wiley & Sons; 2010.

[141] Goto M, Sato K, Murakami A, Tokeshi M, Kitamori T. Development of a microchip-based bioassay system using cultured cells. Anal Chem 2005;77: 2125–31.

[142] Sun K, Yamaguchi A, Ishida Y, Matsuo S, Misawa H. A heater-integrated transparent microchannel chip for continuous-flow PCR. Sens Actuat B Chem 2002;84:283–9.

[143] Park JH, Lee N-E, Lee J, Park JS, Park HD. Deep dry etching of borosilicate glass using SF6 and SF6/Ar inductively coupled plasmas. Microelectron Eng 2005;82: 119–28.

[144] Hnatovsky C, Taylor RS, Simova E, Rajeev PP, Rayner DM, Bhardwaj VR, et al. Fabrication of microchannels in glass using focused femtosecond laser radiation and selective chemical etching. Appl Phys Mater Sci Process 2006;84:47–61.

[145] Matsuo S, Tabuchi Y, Okada T, Juodkazis S, Misawa H. Femtosecond laser assisted etching of quartz: microstructuring from inside. Appl Phys Mater Sci Process 2006;84:99–102.

[146] Diepold T, Obermeier E. Smoothing of ultrasonically drilled holes in borosilicate glass by wet chemical etching. J Micromech Microeng 1996;6:29–32.

[147] Sung SH, Hwang IC, Park S-J, Eden JG. Interchannel optical coupling within arrays of linear microplasmas generated in 25–200 μm wide glass channels. Appl Phys Lett 2010;97:231502.

[148] Shackelford JF, Doremus RH. Ceramic and glass materials; structure. Properties and processing. New York, NY: Springer; 2008.

[149] Del Campo A, Greiner C. SU-8: a photoresist for high-aspect-ratio and 3D submicron lithography. J Micromech Microeng 2007;17:R81–95.

[150] Abgrall P, Conedera V, Camon H, Gue A-M, Nguyen N-T. SU-8 as a structural material for labs-on-chips and microelectromechanical systems. Electrophoresis 2007;28:4539–51.

[151] Yoon Y-K, Park J-H, Allen MG. Multidirectional UV lithography for complex 3-D MEMS structures. J Microelectromech Syst 2006;15(5):1121–30.

[152] Rajaraman S, Choi S-O, Shafer RH, Ross JD, Vukasinovic J, Choi Y, et al. Microfabricated technologies for a coupled three-dimensional microelectrode, microfluidic array. J Micromech Microeng 2007;17:163–71.

[153] Yanik AA, Huang M, Artar A, Chang T-Y, Altug H. Integrated nanoplasmonic–nanofluidic biosensors with targeted delivery of analytes. Appl Phys Lett 2010;96:021101.

[154] Singh A, Limaye M, Singh S, Lalla NP, Malek CK, Kulkarni S. A facile and fast approach for the synthesis of doped nanoparticles using a microfluidic device. Nanotechnology 2008;19(24):245613.

[155] Cottam BF, Krishnadasan S, deMello AJ, deMello JC, Shaffer MSP. Accelerated synthesis of titanium oxide nanostructures using microfluidic chips. Lab Chip 2007;7:167–9.

[156] Dendukuri D, Doyle PS. The synthesis and assembly of polymeric microparticles using microfluidics. Adv Mater 2009;21:4071–86.

[157] Han J, Yeom J, Mensing G, Flachsbart B, Shannon MA. Characteristics of electrostatic gas micro-pump with integrated polyimide passive valves. J Micromech Microeng 2012;22:095007.
[158] Storm AJ, Chen JH, Ling XS, Zandbergen HW, Dekker C. Fabrication of solid-state nanopores with single-nanometre precision. Nat Mater 2003;2:537−40.
[159] Prakash S, Yeom J, Jin N, Adesida I, Shannon MA. Characterization of ionic transport at the nanoscale. Proc IMechE J Nanoeng Nanosyst 2006;220:45−52.
[160] Guo, L. J., X. Cheng, C.-F. Chou, (2004). Fabrication of size-controllable nanofluidic channels by nanoimprinting and its application for DNA stretching. Nano Letters 4(1): 69−73.
[161] Douglas, T. and M. Young (1999). Virus particles as templates for materials synthesis. Advanced Materials 11(8): 679−681

Select bibliography

Abgrall P, Nguyen N-T. Nanofluidics. Norwood, MA: Artech House; 2009.
Beckel D, Bieberle-Hutter A, Harvey A, Infortuna A, Muecke UP, Prestat M, et al. Thin films for micro solid oxide fuel cells. J Power Sources 2007;173:325−45.
Campbell SA. The science and engineering of microelectronic fabrication. 2nd ed. New York: Oxford University Press; 2001.
Chakraborty S. Microfluidics and microfabrication. Springer; 2010.
Edel JB, de Mello AJ. Nanofluidics; nanoscience and nanotechnology. RSC Publishing; 2009.
Holleman AF, Wiberg E. Inorganic chemistry. San Diego, CA: Academic Press; 2001.
Jaeger RC. Film deposition introduction to microelectronic fabrication. 2nd ed. Upper Saddle River, NJ: Prentice-Hall; 20020-201-44494-1.
Lee S-JJ, Sundararajan N. Microfabrication for microfluidics. Norwood, MA: Artech House; 2010.
Levinson H. Principles of lithography. 2nd ed. SPIE—The International Society for Optical Engineering; 2004.
Madou M. Fundamentals of microfabrication. 2nd ed. CRC Press; 2002.
Maluf N, Williams K. An introduction to microelectromechanical systems engineering. 2nd ed. Norwood, MA: Artech House; 2004.
Prakash S, Pinti M, Bhushan B. Theory, fabrication, and applications of microfluidics and nanofluidic biosensors. Philos Trans Royal Soc A 2012;370:2269−303.
Tabeling P. Introduction to microfluidics. New York: Oxford; 2005.
Xia Y, Rogers JA, Paul KE, Whitesides GM. Unconventional methods for fabricating and patterning nanostructures. Chem Rev 1999;99:1823−48.

CHAPTER

Lab-on-a-Chip and Fluid Manipulation Applications*

5

CHAPTER OUTLINE

5.1 Introduction

Microfluidic and nanofluidic systems have been the subject of intense investigations for a variety of "on-chip" operations. In all such applications the goal is to

*Sections in this chapter have been reproduced, with kind permission of:
IEEE © 2008, for text from Ref [44]; Royal Society Publishing © 2012, for text from Ref [46]; Sage Publishing © 2007, for text from Ref [204]; Springer Science and Business Media © 2012, for text from Ref [205].

S. Prakash & J. Yeom: Nanofluidics and Microfluidics. DOI: http://dx.doi.org/10.1016/B978-1-4377-4469-9.00005-6

develop a series of unit operations starting with fluid and sample manipulation to complete diagnosis of the fluid. For example, the introduction of chemical processes on-chip in the mid-1990s led to development of lab-on-a-chip (LoC) or micro-total analysis systems (μ-TAS) with the aim of performing common laboratory scale techniques for separations and detections on chip. This idea has spawned an entire generation of applications such as "factory-on-chip," which refers to performing synthesis and assembly operations on chip to build micro- or nanoscale components, and "body-on-chip," which refers to performing several functions of fluid handling and processing commonly performed in the human body. For instance, micro- and nanopumps can be used to imitate the behavior of the heart to move a mass of fluid, organs like liver are imitated through a series of on-chip embedded proteins for fat and lipid separations, and salt from water is filtered with membranes to mimic kidney operation. Overall, the goal is to use miniaturized systems that can manipulate fluids to perform a variety of unit operations on a single platform thereby making the platform a multi-function unit cell.

In this chapter we will present a review of existing scientific literature that covers the main categories of unit operations needed for most microfluidic and nanofluidic systems.

5.2 Fluid manipulation

Fluid manipulation falls in two major categories. First is movement of a fluid or species within the fluid from one location to another on a chip. Typically, such manipulation will involve convection with momentum imparted to the fluid. Therefore, such actions require micro- and nanopumps. The second category involves selective transport of fluid and species, i.e., valving of fluidic behavior. Table 5.1 presents a brief summary of common fluid manipulation techniques for sample transfer, the common principles involved, and some representative examples for some of the techniques listed here.

Table 5.1 A Brief Summary of Common Fluid Manipulation Techniques in μ-Na*fl* Devices

Sample Transfer	Principle	References
Pumps	Electrokinetic, magnetohydrodynamic, mechanical pumping	[3]
Valves	Surface modification	[4,5]
Droplets	Electrically controlled surface tension drives liquid droplets	[6]
Arrays/mixing	Arrays of nanoholes used for depletion zone mixing	[7]

Reprinted with permission from [47].

Fluid manipulation can take many forms. One of the emerging areas of fluid and species manipulation borrows concepts to create "fluidtronics" by creating fluid analog devices of electronic systems. Nanofluidic channels and nanopores allow for ion transport manipulation that parallels electronic devices such as field effect transistors (FETs) [1] and diodes [2]. The ability to control charge movement is due to the fact that the characteristic length of the channel is on the same order as the size of the EDL. An applied potential across the length of a channel drives an electroosmotic flow. A perpendicular electric field is created by controlling the surface charge through a third electrode on the channel wall. The magnitude of this transverse electric field serves to 'gate' the flow in a manner similar to solid-state electronic FETs and such devices are called flow FETs. [8-13] Such devices have been used for selective protein transport allowing for manipulation of biomolecules in picoliter size samples [13]. A nanofluidic channel that has a surface charge of equivalent magnitude but opposite polarity on the left and right halves of the channel can serve as a fluidic diode, allowing electrolyte to pass under forward bias but blocking flow during reverse bias as long as the double layers do not overlap [14, 15]. Recently, tunable transport characteristics and diode behavior have been demonstrated in NCAMs through induced charge inversion [16, 17].

5.3 Microvalves and nanovalves

We begin our discussions by reviewing the current state of literature on microscale and nanoscale valves typically employed in μ-Na*fl* systems. Figure 5.1 below presents one overview of valve classification. In the discussion to follow, we hope to show that micro- and nanovalves have a huge variety based on the application of interest. Over the years, several review articles on microvalves and books on MEMS and microfluidics also include detailed sections and chapters including design and operational principle discussions. Many entries are listed in the Select bibliography to direct the more interested reader to several of these sources. The main aim of this section is to demonstrate the wide range of options available to the valve user and provide a basic understanding of how a valve can be chosen for different applications.

Micro- and nanovalves need to accurately meter flows with increasingly smaller devices, different ranges of operating pressures (increasing with decreasing critical dimensions) and electric fields, need for implantable medical devices requiring biocompatibility, and faster response time. Therefore, it is important to note the main characteristics needed in such valves. The main criteria for defining specifications on valves are:

a. Leak rate: an ideal valve will have no leak when close, and permit flow only when open. Therefore, a leak ratio, LR, is defined as:

$$\mathrm{LR} = \frac{Q_{i,\mathrm{c}}}{Q_{i,\mathrm{o}}} \tag{5.1}$$

where Q_i denotes the flow rate in the closed or open state, respectively. The inherent assumption here is that the valve operation occurs at constant pressure.

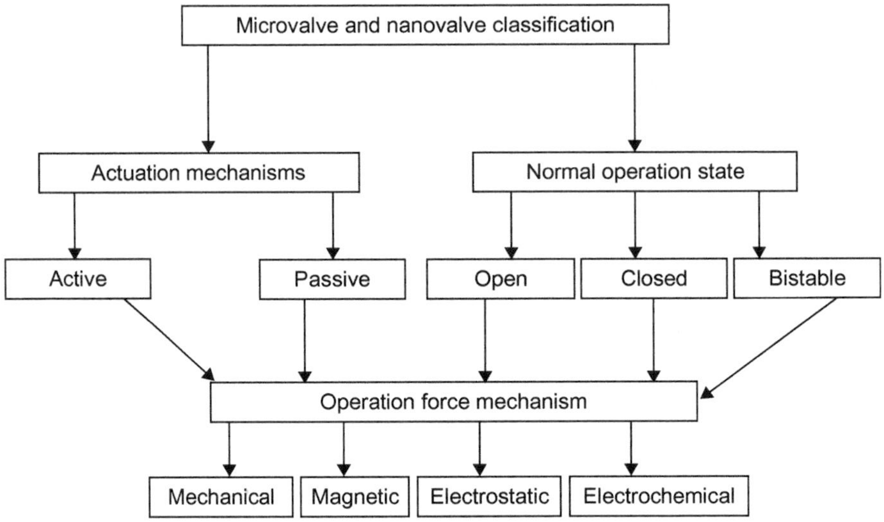

FIGURE 5.1

A flow diagram showing microvalve and nanovalve classification and the main forces that can be used to operate the valves.

b. Valve conductance: is the maximum flow rate permissible through a valve. Note that the valve conductance will be a function of the applied pressure. Therefore, for given operating conditions, the same valve may record different valve conductances and is consequently a design parameter.

c. Response time: For active valves, the response time refers to the time for the valve to control/meter the flow after the actuator is engaged. The time can range from 100 μs (or faster) for electrostatically actuated valves to 1 s (or slower) for pneumatically actuated valves.

d. Operational conditions: Given that valves are usually made of some physical material and need to be part of devices, conditions for operational temperature, chemical environment, power consumption for active valves and compatibility with the device are often design considerations.

It can be seen from Table 5.1 and Figure 5.1 that several technology combinations can be generated that can meet the valve criteria listed above. Consequently, several different types of microvalves have been developed and demonstrated. Nanovalving is a relatively new area and only a handful of valves designed specifically for nanofluidic systems have been demonstrated. Next, we summarize and discuss a variety of previously reported micro- and nanovalves.

One of the first valves developed for microfluidic flows was a mechanical solenoid plunger valve. The valve follows a simple operation mechanism with a diaphragm sealing the flow in a channel, schematically depicted in Figure 5.2. The valve seat was made of silicon with a flexible nickel diaphragm actuated by a solenoid for use in gas chromatography. However, these valves were typically located external to the device with piping or hose connections providing large dead volumes and consequently integrated micro-gas chromatographs have been developed with on-chip valving as discussed in Section 5.9.

Another category of active valves uses external magnetic fields to actuate the valves. These valves can operate at low opening pressures in the range of 2 kPa and can work at higher pressures as well reaching nearly 200 kPa, providing a versatile operation range. The work by Oh et al. [18] developed an in-line micro ball valve with a polymer tubing, a nickel ball with a diameter of 760 μm, and a Teflon™ tube for the housing. The device showed a leakage flow rate of 30 μl/min at 2.1 kPa. Fu et al. [19] developed a normally open microvalve which consists of a micro ball valve actuated by a magnetic field. The micro ball valve consists of three hot-embossed polysulfone layers and three FeNiCr metal layers. The valve used iron balls with a diameter of 3 mm as moving parts driven by a commercially available coil. As an on–off switching valve, the valve can operate with an inlet pressure ranging from 50 to 200 kPa. The LR for this valve was nearly 0.3.

Another type of active valve is the pinch-type valve where the valve mainly consists of a height-adjustable solenoid, a spring-loaded plunger and a silicone tube. The valve is usually in the normally closed operation mode and kept in place by the compressive force of the spring causing the plunger to pinch down the silicone tube. When the solenoid is actuated, it lifts the plunger from the tube causing the valve to open. These valves have been used by Oh et al. [20] in an integrated microfluidic biochemical detection system for performing the immunoassays. Ahn et al. developed microvalves consisting of an integrated inductor and a silicon membrane with an Ni/Fe thin film and a stationary inlet/outlet valve seat (cited in [21]). The microvalve consists of three layers, with the bottom two layers making up the normally closed valve. The top layer is the actuator which contains the flux generator on its top surface combined with Ni/Fe plated through

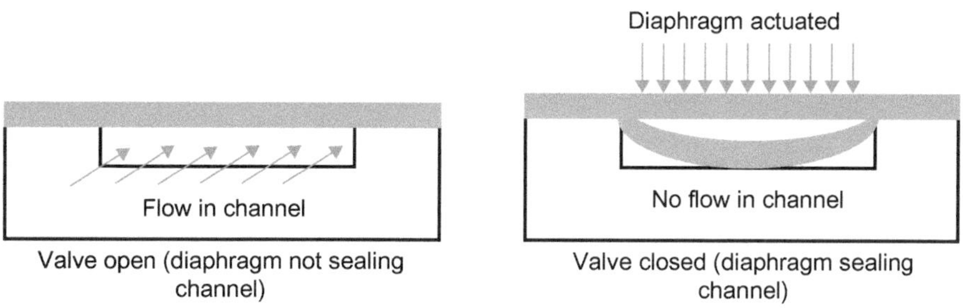

FIGURE 5.2

A schematic depicting diaphragm-based valve operation and showing the open state that permits flow through a microchannel and the closed state that does not permit flow through a microchannel.

holes for guiding the flux to the valve. The actuator and valve components were separately fabricated and then mounted to a glass "motherboard" which contains both fluid flow channels and patterned gold traces for making electrical connections. In addition to providing an easy means for testing, the "motherboard" allowed for the later attachment of other microfluidic components to create a complete microfluidic total analysis system (μ-TAS) on a single substrate. An inductor acted as a flux generator and produced sufficient forces to pull the silicon membrane by magnetically coupling with the Ni/Fe permalloy electroplated on the silicon membrane. The measured LR was 20.5 μl/min at an input pressure of 12.4 kPa, 9.4 μl/min at 7.6 kPa, and 5.6 μl/min at 4.8 kPa. For the case of DI water the LR was 10.5 μl/min at 8.3 kPa and 3.9 μl/min at 4.1 kPa.

A major class of valves utilizes electrostatics for operation. Among these, microvalves that control both gas flows and liquid flows have been extensively studied. For example, in one application, the actuator consisted of a pair of planar electrodes sandwiching a conductive film that has an S-shaped bend in the middle, the movement of which is controlled as the voltage is alternately applied between each of the electrodes and the film. The propagation speed of the S-bend was 4.0 m/s at an applied voltage of 150 V. The valve allowed a flow rate of 10 sccm at a pressure of 60 Pa for air as the working fluid. In another example, Goll et al. [22, 23] developed a microvalve by actuating a moving electrode membrane. The movable membrane acting as an electrode consists of two insulating layers with a conductive gold layer in between. The 3 μm thick membrane was attracted to the upper or lower valve chamber leading to either opening or closing of the inlet to the microvalve as a function of the excitation voltage magnitude.

For a more complete illustration of this class of valves, let us consider the work done at the University of Illinois by the Shannon group in developing polyimide-based active valves for gas flow control. In an effort to develop a micro-gas chromatography system (see discussion below), Bae et al. developed a bi-directional electrostatic microvalve that can switch a gas flow in less than 100 μs under high-pressure operation [24]. As shown in Figure 5.3, the valve is in a normally closed position and flow is present between Port A and C. The closing action is assisted by the voltage applied between the valve closing electrode and membrane electrode. When a voltage is applied between the valve opening electrode and membrane electrode (while removing it between the other two), the movable membrane is deflected toward the valve seat, opening the flow path from A to C. High-pressure operation can be achieved as the pneumatic pressure is applied to the other side of the membrane electrode to either enhance the hold up pressure or increase the closing speed. Another key component is a hydrophobic layer deposited on every touching interface to prevent any stiction from occurring and to reduce the required valve opening pressure [25, 26]. The microvalves can open or close in 50 μs or less for applied pressure up to 126 kPa with 47 million cycles of lifetime.

Electrokinetic-based valves have also been developed. Kirby [27] et al. developed a microvalve which is opened and closed by electrokinetic pressures. Laser

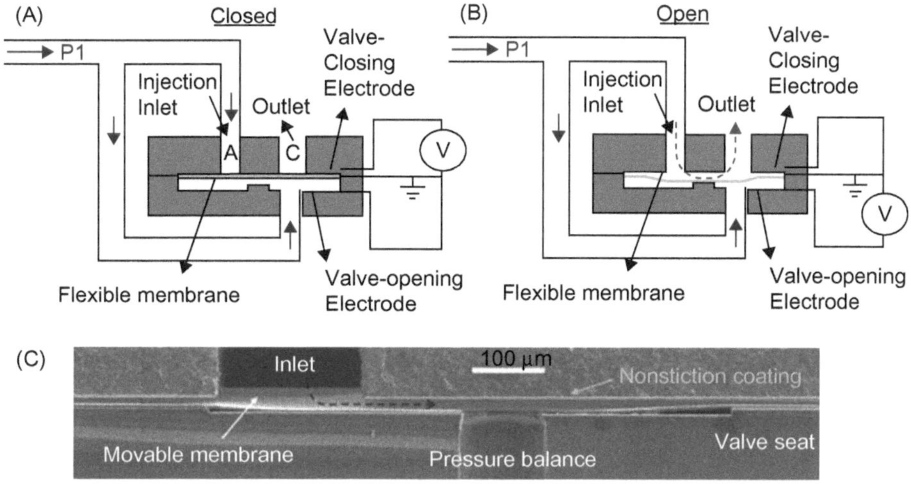

FIGURE 5.3

Diagrams of the valve when (A) closed and (B) opened shown with fluid flow paths. (C) An SEM image of the cross-sectional view of the fabricated microvalve consisting of a flexible membrane, electrodes, and silicon channels. © 2007 IEEE. Reprinted with permission from [24].

photopatterning of polymer monoliths inside glass microchannels is used to fabricate mobile fluid control elements. The glass substrates and cross-linked polymer monoliths operate in water—acetonitrile mixtures and can hold pressures as high as 350 bar. Open/closed flow ratios of 10^4 to 10^6 have been demonstrated over a pressure range of 1.5–70 bar. The pressure–leak relationship showed potential uses for valving of flow through packed or monolithic chromatography columns (see Figure 5.4).

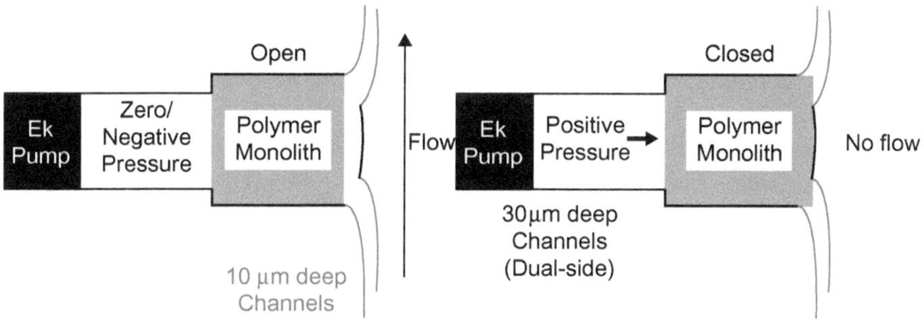

FIGURE 5.4

A monolithic valve based on electrokinetic principles to operate at relatively high pressures (up to 350 bar). Reprinted from [27], Copyright 2002, with permission from Elsevier.

From the discussions above, it should be evident that microvalves span a very broad range of ideas that have been implemented. Other active valve actuation mechanisms include piezoelectric materials that can generate forces leading to pressures of several MPa, thermal methods that often use bi-metallic alloys with differential coefficients of thermal expansion generating variable internal and external forces when heated, thermopneumatic microvalves which are operated by the volumetric thermal expansion of a sealed or enclosed liquid or solid coupled to membrane deflection arising from the volumetric change of the enclosed liquid or solid. Such valves have been shown to generate a 16 Pa pressure rise with 500 mW power consumption, while closing of the membrane was maintained with a 30 mW steady input power. The microvalve generated a flow rate of 400 ml/min under 11.3 Pa differential pressure and leakage flow rates as low as 1 μl/min.

In the work reported by Ahn et al., and summarized in the review by Oh et al., [20, 21] microvalves with an intregrated inductor on a silicon membrane with an nickel-iron thin film was developed. The valve had a stationary inlet/outlet valve seat. In this microvalve, the inductor acted to provide electrical flux and generate force to actuate the silicon membrane through electromagnetic coupling with the nickel-iron thin film. The leakage ratio (LR) was measured to be 20.5 μl/min. at an input pressure of 12.5 kPa. As pressure was lowered to 7.6 kPa and then to 4.8 kPa, LRs of 9.4 μl/min. and 5.6 μl/min. respectively were reported. De-ionized (DI) water was also tested as a working fluid and the LR as function of pressure was again reported to be at 10.5 μl/min. at 8.3 kPa and 3.9 μl/min. at 4.1 kPa.

Microfabricated shape memory alloys (SMA) have been used to develop thin-film based actuators for microvalves. In work by Kohl et al. [28–30] a three-metal Ni/Ti/Pd thin film SMA was used to control pressure differences below 250 kPa. Under these conditions, a gas flow of 360 ml/min. was reported with a valve stroke of 20 μm. Given that SMAs provide several interesting material combinations to work with, a pinch-type SMA microvalve was developed for application in an implantable drug delivery system. In work by Reynaerts et al. [31], a Ni/Ti wire acted like a hose clamp to "pinch down" a silicone rubber tube and was found to be effective as a microvalve to pressures reaching 200 kPa. The Ni/Ti SMA microvalve wire had a diameter of 120 μm and it was reported to work at a power consumption of 120 mW required for actuation with an opening time-scale reaching over 650 ms. Similar to SMA valves, polymeric materials such as hydrogels also provide actuation mechanisms. For example, stimuli responsive hydrogels (e.g., temperature, pH, and pressure) have also been used for local flow control. Other researchers have also used Ni/Ti SMA wires to create other pinch type valves for PDMS. In the work by Pemble et al. [32], the Ni/Ti SMA wires were able to withstand a pressure of 207 kPa but in contrast consumed nearly 400 mW of power to generate a net flow rate of 16.8 ml/min. Similar to SMA valves, polymeric materials such as hydrogels also provide actuation mechanisms. For example, stimuli responsive hydrogels (e.g., temperature, pH, and pressure) have also been used for local flow control. It is evident from these examples that even with similar materials microvalves can generate very different outputs based on specific use and design of the microvalves.

Microvalves based on properties other than mechanically actuated parts have also been developed. Specifically, methods that use rheological properties of fluids, phase change, and electrochemical reactions have also been used to develop microvalves. For example, in the work by Neagu et al. [33, 34] electrolysis of water was used to generate bubbles (presumably oxygen or hydrogen) to create a gas pressure for actuating the membrane for valving action.In yet another example for electrochemical valving, Zhu et al. [36] developed a self-regulating hydrogen generator based on a combination of a dissolvable metallic layer for the metal hydride reactions followed by a virtual or passive gas valve. When a hydrogen consumption rate is reduced from the micro fuel cell, the hydrogen pressure in the microchannel increases, pushing the liquid (reactants) out of the reaction zone and decreasing (or stopping) the hydrogen generation rate (Figure 5.5).

In the work by Yang and Maeda [35, 36] the focus was on developing devices with microfluidic connections to enable device integration. In the passive valve developed by Yang et al. [38], a pair of silicon microvalves was used. They reported generating water flow rates reaching 1.6 ml/min. at a forward pressure of 4 kPa. However, there was a much lower flow rate of 50 μl/min. that was reported at a backward pressure of 4 kPa. The examples for device integration relate to passive valving. A generic socket with a mounted microfluidic device with a potential for multiple connections is shown in Figure 5.6 to illustrate the need to perform multiple unit operations on-chip as opposed to off-chip.

As discussed in Chapters 2 and 3, capillary forces can be very useful in microfluidics and nanofluidics. These principles have also been implemented for

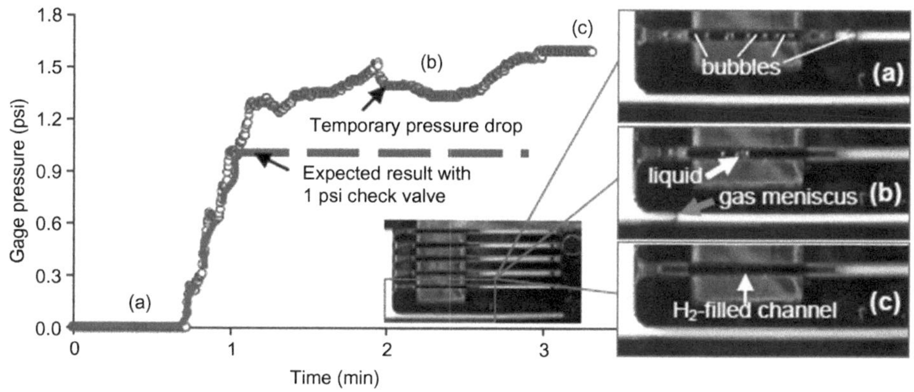

FIGURE 5.5

Hydrogen pressure at outlet during stop-process in on-board self-regulating and self-circulating hydrogen generator. Inset (A) shows the hydrogen bubbles pumping liquid without power consumption during normal operation. Inset (B) shows gas entering return loop and pushing liquid into the already cleared reaction channel. Inset (C) shows a reaction microchannel completely filled with hydrogen, stopping the further hydrogen generation © 2009 IEEE. Reprinted with permission from [37].

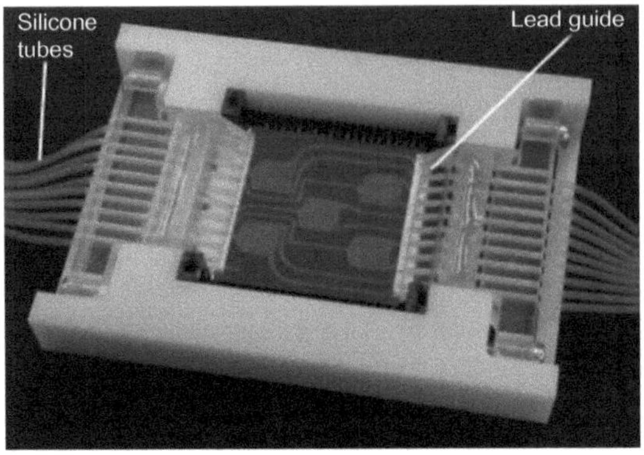

FIGURE 5.6

Socket mounted with a 5-micromixer array chip with modular valves.

developing a class of valves driven by surface forces or "surface-mediated valving." One of the earliest demonstrations of this work include using electrocapillary, thermocapillary, or surface energy effects. The electrocapillary effect was used to develop electrowetting and is now extensively used for droplet-based microfluidics. The work at University of California – Los Angeles and Duke University has shown significant promise towards developing droplet microfluidics. For example, in Figure 5.7, we see an electrowetting array in which the droplets are moved from one electrode pad to another to perform a series of combinatorial activities that could be useful in many different chemical or biological applications.

Use of surface energy and more recently use of surface charge to control flows has also been developed. Patterning surfaces with hydrophilic and hydrophobic regions inside microchannels and nanochannels have been used to create valves for system where air–water or air–liquid interfaces may exist. For example, in Figure 5.8, the Prakash group at Ohio State has shown selective flow control in 100 μm wide and 20 μm deep channels made in glass chips to direct flow by functionalizing surfaces with a fluorinated molecule to render channels hydrophobic. Figure 5.8 shows an optical microscope image of the flow in channels as a consequence of part of the device with a Y-shaped channel geometry showing flow control.

In a recent advance in surface-directed flows, the Prakash group has also shown use of surface charge to develop valves for microchannels with 15 μm depth and 100 μm width as well as for nanochannels with 30–100 μm width channel depth varying from 250–500 nm. In this case, the glass channels also in the Y-shaped design were chemically modified with an aminated surface in one of the Y-channel legs to generate a surface that in principle can be positively charged when exposed to a neutral or acidic pH electrolyte due to protonation of the amine groups. However, AFM measurements indicated that even aminated groups remain negatively charged in a broad pH range, and acquire a positive charge at pH approaching 4 (Figure 5.9).

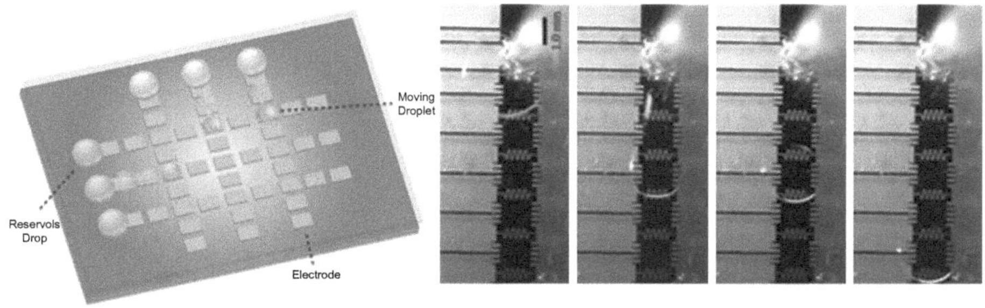

FIGURE 5.7

The left panel shows an array being developed at UCLA for droplet manipulation (courtesy of Professor C.J. Kim, UCLA Mechanical and Aerospace Engineering Department) and the right panel shows a sequence of images from Duke University, where the droplet moves from top to bottom of the picture in the 4-frame array as shown (Reprinted with permission from [40]).

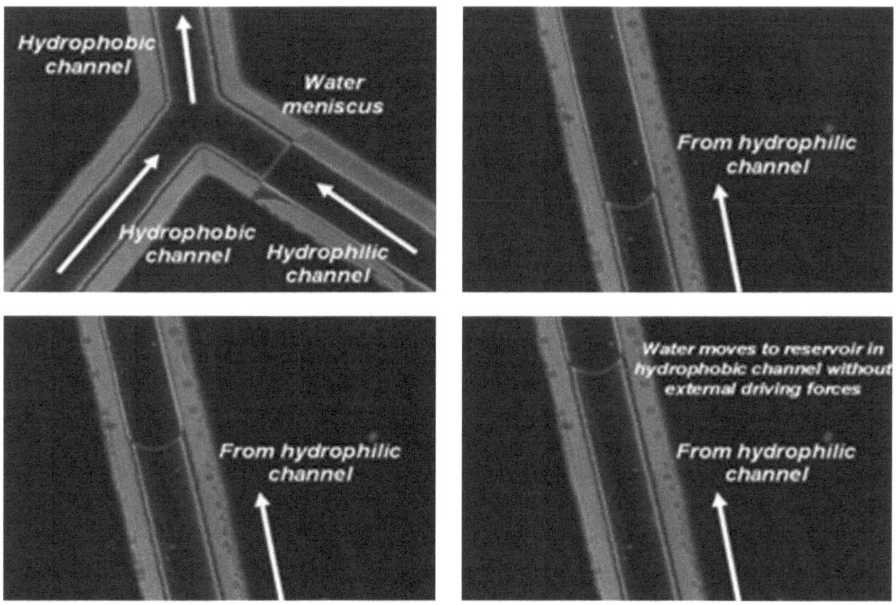

FIGURE 5.8

Selective transport of DI water in a channel network functionalized with a fluorinated molecule, FDTS or heptadecafluoro-1,1,2,2,hydrodecal trichlorosilane.

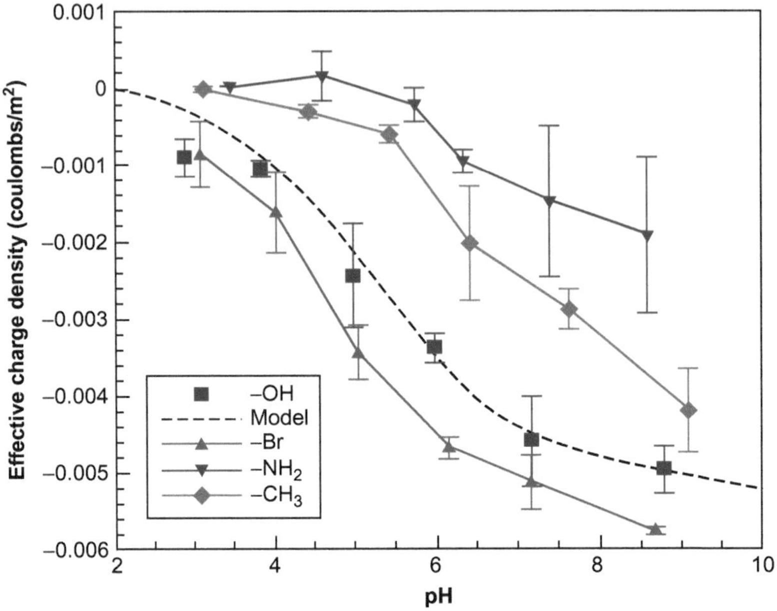

FIGURE 5.9

Charge density from an AFM measurement for the tip-force interaction showing measured effective charge density on a variety of functionalized silica surfaces as a function of electrolyte pH. Reprinted with permission from [39].

However, the aminated coating was still implemented and showed that successful valving can be achieved in a short duration of device operation (~100 s or less) but at long durations (~approaching 400 s) the valve "leaks" and can cause some flow in the opposite polarity channels. This proof-of-concept for a surface valve is presented in Figure 5.10.

5.4 Micropumps and nanopumps

In any μ-Na*fl* device, apart from selectively directing a flow from one part of the device to another or metering flows, a fluid needs to be convected from one location to another. This is usually done by imparting momentum to the fluid. On the macroscale most fluid transfer occurs by adding a mechanical pressure to the fluid and imparting momentum. The mechanical pressure is applied by using pumps. Just like our discussion with microvalves and nanovalves, the variety of operating principles for pumps is enormous. Over the years, several review articles and sections or chapters in books have discussed micropumps in detail. With the advent of nanofluidics, scaling of micropumps and development of new methods to create nanoscale pumping are being actively investigated. We direct the interested

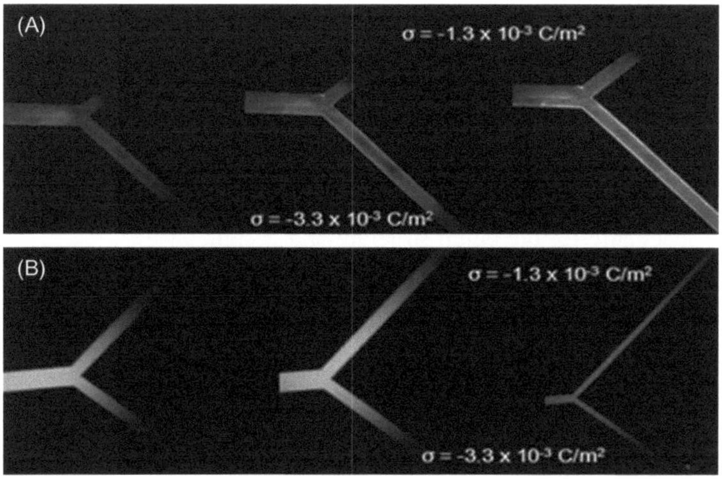

FIGURE 5.10

Image from an optical microscope showing proof-of-concept testing for a surface-mediated valve. Note figure A shows that the red dye (positively charged, Rhodamine B) prefers the more negative channel, and the negatively charged green dye (fluorescien, figure B) prefers the more positive channel. However, both dyes eventually penetrate the channels as the time elapsed images reach 400 s (left to right).

reader to several of these in-depth discussions through the wide-ranging bibliography presented at the end of this chapter.

Using the commonly expressed classification of pumps in literature, pumps impart momentum through two primary mechanisms: (i) displacement pumps that rely on a moving boundary to exert physical pressure through the boundary on the working fluid, and (ii) dynamic pumps that add energy to the fluid thereby imparting momentum to the fluid or generating a pressure. The first micropumps reported were miniaturized versions of macroscale pumps, the displacement pumps (and some dynamic pumps like the centrifugal pumps) which are also classified as mechanical pumps. Pumps that use electrokinetic, electrohydrodynamic, or magnetohydrodynamic methods for pumping the fluid (usually dynamic pumps) were also reported for fluid pumping. These types of pumps that use a body force to drive the fluid are also classified as non-mechanical pumps.

Table 5.2 summarizes reported pressures and maximum achieved flow rates for a variety of pump types including the working fluid.

The most common types of micropumps that have been reported include pumps with a large range of operating principles like reciprocating pumps that use a diaphragm to create a differential force for pumping, peristaltic pumps, valve-less rectification pumps, rotary pumps like gear pumps and centrifugal pumps, ultrasonic pumps, electroosmotic, electrohydrodynamic, and magnetohydrodynamic pumps. The main design and use considerations include estimating the dead volume, required flow rate and pressure, energy consumption for pump operation, actuation mechanisms, and overall pump efficiency.

One of the ways to illustrate the key ideas about micropumps is to look at a case study example for a gas micropump developed by the Shannon group at the University of Illinois. In this example, we will discuss the development of a gas pump with a flow rate objective of 10 sccm with a passive valve with a stroke volume of 3–6 μl and a chamber pressure rise of 10 kPa. Figure 5.11 shows below the schematic for the pump structure and design.

The membrane diaphragm controlling flows were actuated electrostatically with two passive valves to direct the flow. At a finite design pressure, the valve opens allowing the membrane to deflect and permit gas flow. For pumping action, the valves open and close along with the actuator to create a "zipping" action. Consequently, as the actuator diaphragm "zips-in" the outlet valve has to open with the inlet valve closed and as the diaphragm pulls out, the inlet valve has to open with the outlet valve closed. Figure 5.12 shows digital photographs of the microfabricated gas pump as part of a test set-up and the key components that go in the gas gump.

As an example for the micro-gas pump operation, the actuation voltage for the polyimide membrane acting as the diaphragm was 250 V with the maximum stroke volume of 2.6 μl (96% of the designed stroke volume) permitting a maximum flow rate of 20 μl/min with an operational frequency of 0.3 Hz. As with any of the micropumps that operate on electrostatic actuation of a diaphragm, the flow rate is a function

Table 5.2 Summary for Several Micropumps with Key Reported Parameters

Pump Type	Reported Operating Pressures (kPa)	Reported Flow Rates (μl/min)	Working Fluid for Maximum Flow Rate
Reciprocating (diaphragm-based)	0.1–200	0.1–35,000	Air (35,000 μl/min); Water (16,000 μl/min)
Gear pumps	2.4	5.5	Water
	100	180–190	Glycerine-water mixture
Peristaltic	2.4–5.8	3–30,000	Air (30,000 μl/min); Water (8000 μl/min)
Electrohydrodynamic	~0.5	0.1–15,000	Water-based (typically with salt)
	~0.4	0.1–14,000	Ethanol or other alcohol-based mixtures
Electroosmotic	5–2000	0.005–10	Water-based (typically with salt)
	4000–20,000	0.02–0.2	Organic solvent mix
Magnetohydrodynamic	~0.2	>60	Water-based (typically with salt)

of the stroke volume and frequency; therefore, design analysis is usually used to determine the optimal conditions for pump operation. The detailed fabrication procedures to create this micro-gas pump can be found in the Case Study at the end of Chapter 4.

Unlike nanovalves, development of nanopumps relying on mechanical parts has been difficult due to challenges in fabrication. Therefore, the most common "nanopumps" rely on "injections," i.e., delivery of small packets of fluids through nanostructures such as nanochannels and nanopores during electrokinetic flow. Therefore, nanofluidics can add functionality for sample manipulations in analytical chemistry, such as sample injections, separation, purifications, and preconcentration for quantitative and qualitative identification. For example, NCAMs functioning as controllable molecular gates can mediate digital transfer [41] of fluid voxels from one microfluidic channel to another. In addition, NCAMs presenting different pore sizes [11] can transfer analytes with disparate mass characteristics at different rates to achieve intelligent fluidic control. This capability has consequently found use in a miniaturized lead sensor that uses DNAzyme [42] as a probe and for preparative post-separation processing for mass limited samples [43]. Figure 5.13 shows a figure for the molecular gate concept that was used for rapid injections from a nanocapillary array membrane or NCAM to a microfluidic channel as discussed in the review by Prakash et al. [44].

As an aside, NCAMs refer to the general category of nanopores fabricated in membranes that can be used as vertical interconnects to form 3D hybrid microfluidic–nanofluidic structures enabling vast functionality. While the nanopores can be fabricated by a variety of means, one method (see Chapter 4 for further

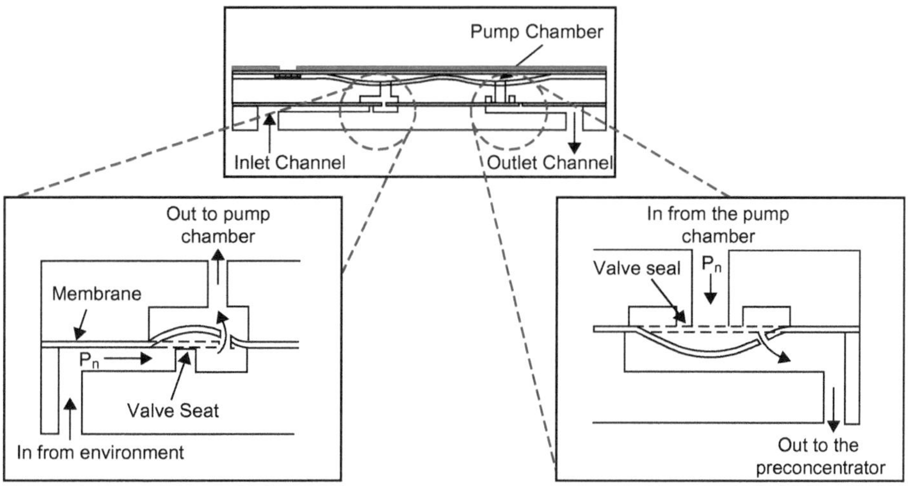

FIGURE 5.11

A schematic showing the design for electrostatic pump actuation for the case study example of a micro-gas pump developed at the University of Illinois. Reproduced with permission from [26], image courtesy M. A. Shannon, University of Illinois.

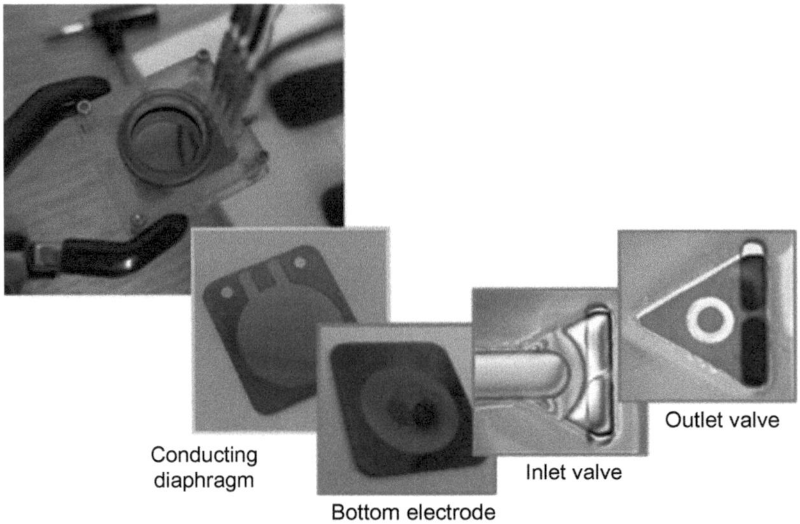

FIGURE 5.12

Digital photographs of a micro-gas pump developed at the University of Illinois. Reproduced with permission from [26], images courtesy J. Han and M. A. Shannon, University of Illinois.

details) uses a TEM to fabricate nanopores. In fact, as of 2012, the Prakash group at Ohio State has used the TEM to fabricate the world's smallest elliptical and triangular cross-section nanopores in 30 nm thick silicon nitride membranes. An example of the elliptical nanopores is seen in Figure 5.14.

5.5 On-chip separations and combinations

Chemical and biological separations form an essential component in analytical, synthetic, and combinatorial chemistry and biology. The ability to take a fluid comprising several components and separate them into individual components is the main task. However, combinatorial processes for drug testing and evaluations or on-chip reactions require the opposite problem to be solved, i.e., several components must be systematically combined to yield the desired product. Consequently, a series of unit operations may need to be performed on a single chip including sample injections, manipulations (as discussed in Sections 5.1.3 and 5.2–5.4 above), separations, reactions, and/or sensing and detection of final products (Section 5.7 discusses biosensors as an example). Let us begin our discussions in this part by looking at each of these individual unit operations. Figure 5.15 shows an intellectual flow diagram for processing samples requiring multiple unit operations.

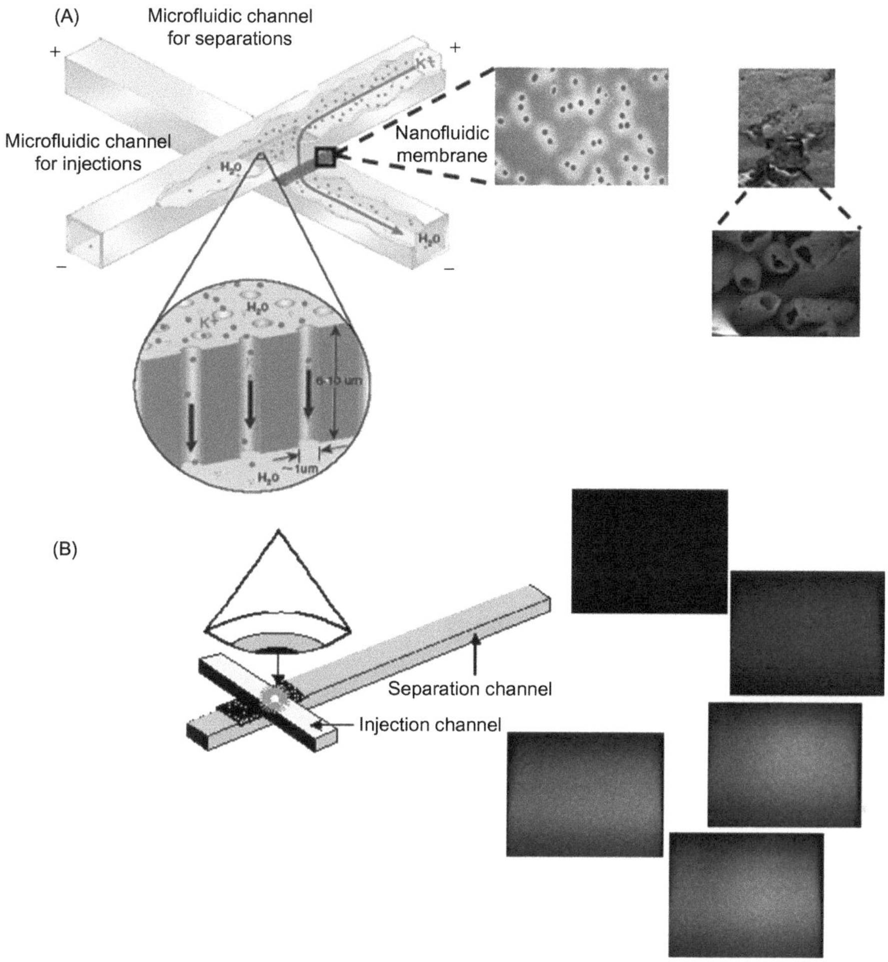

FIGURE 5.13

Conceptual schematic in (A) and example of operation in (B) for rapid injection between microchannels using molecular gates. Reproduced with permission from [45].

5.5.1 Sample injection and separation

Using the example of the molecular gate from Figure 5.13 to illustrate sample injections for nanofluidic devices, we see that an NCAM sandwiched between two microfluidic channels can be used to electrokinetically inject samples from a source microchannel to a separation microchannel, where electrophoretic separations are performed. Initially a potential is applied across the separation channel with the waste reservoir of the separation channel grounded. During this step the source channel is allowed to float. In this example, a picoliter (10^{-12} l) volume from the source side is

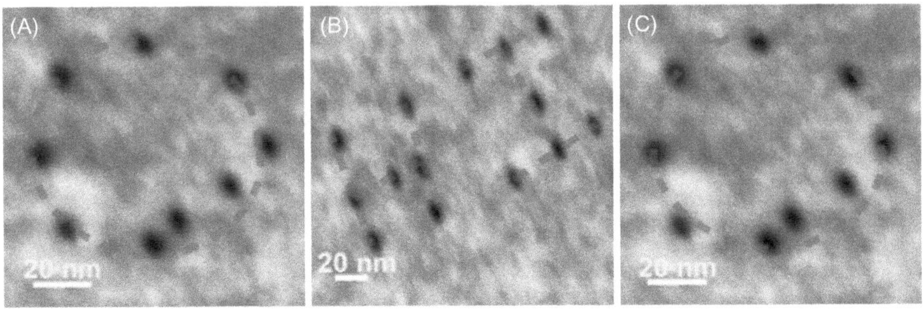

FIGURE 5.14

Fabrication of nanopore arrays using a transmission electron microscope (TEM). Ordered arrangements were fabricated to illustrate the word "OHIO" with elliptical cross-section nanopores. The red-dotted line is meant to serve as an eye-guide. The pores in A are $4 \pm 0.4 \times 2.6 \pm 0.6$ nm The nanopores that spell out "H" and "I" are $5.8 \pm 0.7 \times 3.0 \pm 0.5$ nm as shown in B. © 2008 IEEE. Reprinted with permission from [44].

injected through the NCAM by applying an equal potential to either side of the source channel and grounding the waste reservoir of the separation channel. This results in the injection of a well-defined sample plug into the separation channel. Injection times are short, typically a few hundred milliseconds. Electrophoretic separation is then performed on the sample plug before another sample injection [41]. This design can also be adapted for sample collection. If a third channel is added and separated by an NCAM from the separation channel, the sample plug can be transferred down the separation channel by electroosmosis and then collected in the third channel by applying an equal bias to both reservoirs of the separation channel and grounding the collection channel [10, 11]. These studies illustrate the ability to manipulate flows of mass limited samples [43, 46]. The transport of target molecules through NCAMs can be controlled by ionic strength, magnitude, and polarity of the applied potential, pore size, and surface chemistry (i.e., surface charge density and surface energy). NCAMs allow for electrically controlled valving and rapid sample injections [48–52] with high sample plug reproducibility, separation resolution, and reproducibility approaching 1% for some studies [41]. Following the success of these initial demonstrations of molecular gates, new planar microchannel–nanochannel hybrid devices have been developed that allow for controlled dosing of a variety of transfection agents to single cells through electroporation at the microchannel–nanochannel interface [53] with potential applications in high-throughput drug delivery or transfection of materials to individual cells (see Figure 5.16).

Similar injection/collection strategies have been implemented to perform two-stage sample separation. The first stage is electrophoretic separation based on the electrophoretic mobilities of varying species, while the second stage involving chiral amino acid mixture separation is achieved through micellar electrokinetic chromatography with a chiral selector. Separation depends critically on injection

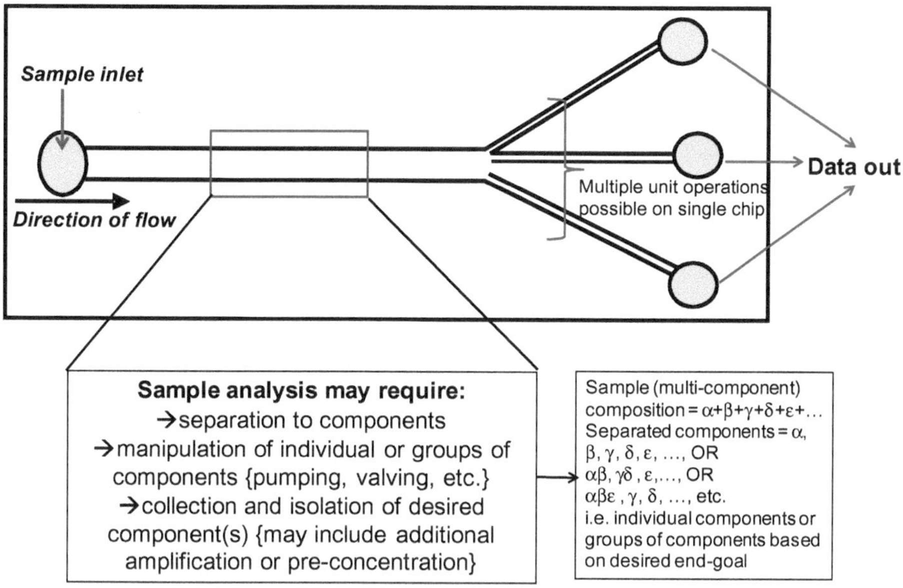

FIGURE 5.15

The schematic shows a conceptual framework for a fluid sample analysis on a microfluidic or nanofluidic chip. It should be noted that there are several steps requiring adequate sample processing for analysis including separations. In addition, most microfluidic or nanofluidic systems will provide the ability to perform several unit operations on a single chip or platform including the ability for bio- or chemical assays, measurements and transduction of signals based on mechanical, electrical, optical, or electrical properties of the sample leading to data output for further analysis or interpretation.

Discussions with Prof. A.T. Conlisk and Prof. D. Hansford are acknowledged for the concepts presented in this figure.

and collection capabilities in the device. Such separations are important to biochemical studies that require the separation of complex mixtures of analytes with orthogonal separation principles [54].

Apart from hybrid microfluidic–nanofluidic devices, separation literature just focusing on LoC applications is vast. In bio-related applications, chip-scale separations for molecular scale species such as DNA and proteins have been demonstrated. In addition, cell sorting forms another class of chip-scale separations. On the chemical side, a variety of "phoretic" techniques have been shown. In the next few paragraphs, we present an overview of several separation methods and direct the interest reader to the Select bibliography for more details on specific separation processes.

The key to any separation process is to identify a fundamental property for the species of interest that can help isolate single components or groups of similar components displaying the chosen fundamental property. Table 5.3 identifies some commonly used properties that have been used for separation processes.

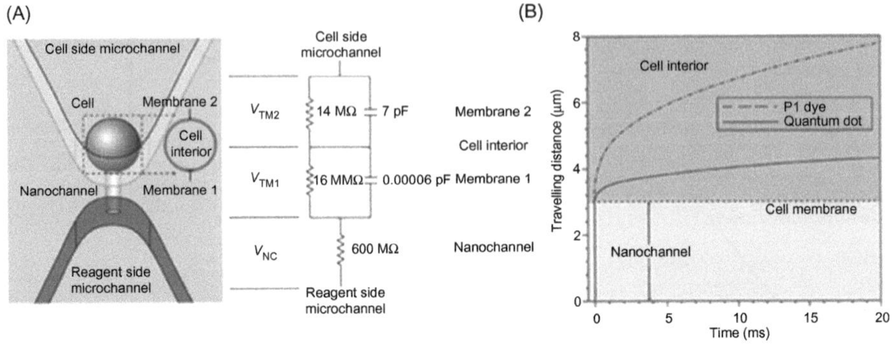

FIGURE 5.16

Use of a planar micro-nanochannel hybrid device for the electroporation of an intact cell. The electroporation is achieved by the high electric field that can be concentrated at the micro-nanochannel interface. © 2011. Reprinted by permission from Macmillan Publishers Ltd from [53].

Table 5.3 Summary of Common Separation Parameters Considered in a Variety of Applications

Fundamental property	Example of separation technique
Size	Filtration or sieving, dialysis
Charge on species or device walls	Electrophoresis, isoelectric focusing, ion-exchange chromatography, electrodialysis
Species polarity	Adsorption chromatography, hydrophobic interaction chromatography
Solubility	Electrocoagulation, chemical coagulation, selective precipitate formation

5.5.1.1 Sieving or filtration

One of the simplest forms of separations relies on selecting different sizes of species. Nanoscale structures such as nanopores and nanochannels allow for the fabrication of molecular filters of sieves since the size of many biomolecules is on the same order as the characteristic size of these structures themselves. Transport through pores is affected by steric hindrance or exclusion due to molecular entropy, hydrodynamic hindrance, caused by viscous forces associated with the walls of the pores, and charge interactions between the molecule and the surface of the nanopore or nanochannel. Molecules that are larger than the size of the pores themselves can be electrically driven through the pore forcing the molecules to change their shape and conformation (e.g., DNA, see discussion on DNA later). Pore/channel size, surface chemistry (i.e., charge), applied potentials and multi-step filtrations with different membranes can be used to perform highly efficient separations [2, 12]. Furthermore, methods for active manipulation of the pore geometry using stimulus-responsive polymers, through temperature-mediated shape changes in polymer brushes [55] and pressure-induced insertion of ion-channels in lipid bilayers [56], have been demonstrated that permit actively controlled size-selective transport across NCAMs. Three-dimensional systems using nanofluidic interconnects, which provide fluidic communication among microfluidic channels in vertically separated layers, have been of interest for the purpose of fluid handling in mass-transported limited samples [57, 58]. In these micro–nano hybrid systems size-based selectivity for various dextrans has been demonstrated [10, 41]. Size-based selectivity can be understood based on the Stokes–Einstein relation. For a chemical species in bulk solution the diffusion coefficient is governed by

$$D_i = kT/6\pi\eta r_i \tag{5.2}$$

As the size of a nanochannel, nanopore, or nanotube decreases, or as the size of the molecules increase, one expects to observe the physical effects of hindered diffusion. For a molecule of radius r_i diffusing within a nanoscale channel of comparable radius r_p, the centre of the molecule cannot approach the wall within a distance r_i. The extent to which the diffusion coefficient D_i of a molecule in the confined nanoscale space is reduced in contrast to its value in the bulk solution D_{sol} is given by the Renkin equation

$$D_i/D_{sol} = 1 - 2.104\lambda + 2.09\lambda^3 - 0.95\lambda^5, \tag{5.3}$$

where $\lambda = r_i/r_p$. A similar analysis has been presented by Bayley and Martin in their review article on resistive-pulse sensing [59].

As discussed extensively in Chapters 2 and 3, all microfluidic and nanofluidic systems are characterized by high *SA/V* ratios. Consequently, in confined flows with particles the rigid walls of the device can be used to generate wall–particle interactions for size-based separation. For example, engineered separation of microfluidic devices where the Stokes number is used as a controlled variable leads to denser particles by trapping them at the walls but allowing lower density to go through with the bulk flow. In many cases with the same material particles but different sizes, the larger particles will stay at the walls while the smaller particles will travel through with the bulk flow.

$$St = \frac{\mathrm{RT_P}}{\mathrm{CT_F}} = \frac{\rho_\mathrm{P} a^2 U}{18\eta\rho_\mathrm{F} D_\mathrm{h}} \tag{5.4}$$

where $\mathrm{RT_P}$ is the relaxation time for the particle and $\mathrm{CT_F}$ is the characteristic time scale for the flow. ρ_P and ρ_F denote particle and fluid densities, while other terms have the same definitions as defined previously in Chapter 2. As Stokes number (*St*) increases the tendency for the particle to follow particle inertia and velocity increases as opposed to following the trajectory due to bulk fluid motion. Consequently, local regions of acceleration (or deceleration) can design parameters for the microchannels through pre-selected or engineered changes in channel geometry to cause particle separation. In addition, channels with curvature also rely on the Dean number, *De*, to generate separations by causing a differential centrifugal force with changing radius of curvature of the microchannel (denoted by R for a hydraulic diameter, D_h) on particles as they negotiate the bend or curved part of the channels. De is defined as:

$$De = Re\sqrt{\frac{D_h}{R}} \tag{5.5}$$

St and De have been used to generate a variety of design for sorting cells and particles. For example, Seo et al. [60] developed a spiral channel separator for microfiltration applications to selectively remove particles at a flow rate of 92 mm/s with an efficiency of ~99%. Other sieving techniques developed include use of deterministic lateral displacement with particles interacting with arrays of posts developed by the Austin group at Princeton University for size-based separation of microbeads and blood cells. In another approach, Seki et al. [61] developed "pinched flow fractionation" for particle separation by using St as a design parameter. The size-based filtration has been extended to the nanoscale as well. The Shannon group at the University of Illinois has developed series of triangular posts in a microchannel to separate colloidal size particles (100 nm to ~1–2 μm) in an aqueous solution. The key was to include an acoustic wave to generate engineered standing waves with nodes at the posts to collect the particles within the post structures.

5.5.1.2 Rapid reagent mixing

Incorporating NCAMs into microfluidic structures affords a highly tunable, nanoscale pathway for the transport of molecules from one chemical environment to another [62]. Typically 5–50 μm thick, NCAMs support rapid electroosmotic flow with relatively low electrical potentials. Under such conditions linear velocities of order ~1 mm/s can be achieved across the nanopore. In addition, permitting the digital switching of fluid packets across the nanopores and coupling of rapid flow in the nanopores to the microfluidic channels results in rapid mixing. Generally, electrokinetic transport in microfluidic channels occurs at low Reynolds number (see Chapter 2 for further details), and as a result, mixing of molecular entities across streamlines relies on diffusion. However, nanochannel-mediated electroosmotic flow encourages convective mixing at the micro/nanochannel junction, as demonstrated by effective mixing within 3D structures comprised of orthogonal microfluidic channels surrounding a polymeric NCAM [63]. This strategy has been successfully used for injecting Ca^{2+} into a channel containing a fluorogenic Ca^{2+} probe, or molecular beacon action by introducing Pb^{2+} into a channel containing a DNA aptazyme, which is quenched in its native state but unquenched after analyte-mediated dehybridization [42]. Similarly, large biomolecules, such as enzymes suspended in a microfluidic channel can be electrokinetically injected across an NCAM into an orthogonal microchannel containing the substrate, where the two entities convectively mix and react at the nanopore [49, 50]. These operations, which exploit the rapid response to control signal for electrokinetic transport and rapid mixing, mimic the behavior of stopped-flow reactors, but on volumes that are orders of magnitude smaller than bench-top stopped-flow instruments.

Microfluidic and nanofluidic systems also present several challenges to mixing fluid streams. Fundamental studies have evaluated the role of nondimensional parameters such as the Peclet number to determine the effects of diffusive broadening as flow proceeds along a microchannel. The broadening can be a detriment to separation processes; however, control over such phenomena can also lead to the development of novel applications such as the laminar flow fuel cells as discussed in Chapter 6. Other mixers have included serpentine channels, 3D ridges and a variety of structural designs to generate circulation and vortex-like regions in the flow field to permit better mixing of incoming fluid streams.

5.5.1.3 Analyte preconcentration

Nanofluidic components integrated into microfluidic devices can be used to preconcentrate the sample before injecting it to the microchannel. In one demonstration, fluorescein, a negatively charged dye, was combined with a buffer solution. A positive potential was applied to the source side of the device while the receiving side (other side of NCAM) was grounded in a 3D interconnect device formed like a sandwich structure with the nanocapillaries between two vertically separated microchannels. The fluorescein dye was repelled away from the positive electrode and collected on the source side of the NCAM. The surface of the NCAM was positively charged corresponding to negative mobile counter ions in the pores. This prevented fluorescein from simply passing through the pore for the initial voltage configuration. The dye collected near the NCAM after 40 min was 300 times the initial concentration of the fluorescein for both 1 μM and 0.1 μM initial concentrations, indicating that the concentration was independent of the repulsive forces between the fluorescein molecules [64]. When the bias was reversed, the preconcentrated sample was transported through the NCAM to the receiving side. Preconcentration has applications in biochemical studies where small amounts of biomolecules can be preconcentrated before detection, allowing lower detection limits. Studies have used techniques based on preconcentration in a T-shaped device for preconcentration of DNA [58]. Iannacone et al. [65] demonstrated that NCAMs can be used to enhance the intensity of an insulin peak introduced into a mass spectrometer from a microfluidic channel using electrospray ionization; this occurs by reducing the salt adducts inherent in the sample preparation process. In addition, various receptors can be incorporated into NCAMs. This strategy could be beneficial for sample preconcentration and purification of biomolecules in microfluidic devices. Both antibodies [66] and DNA probes [67] have been successfully incorporated in NCAMs for applications in protein purification and preconcentration. Preconcentration factors of up to 300 have been achieved for single polarity species [64] in NCAMs and up to 10^6 for proteins and peptides [68] in simple nanofluidic structures.

One interesting phenomenon that takes place at the micro-nanochannel interface is concentration polarization, with one end being enriched and the other depleted for a given species [69]. Gradients in ionic concentration cause added impedance to solute rejection, and the resulting concentration polarization serves as a limiting factor for water flux through reverse osmosis membranes. Bio-inspired systems are envisaged for energetically favorable pumping of hydrated ions through membranes by modulating pore potentials [70]. The phenomenon of concentration polarization, in combination with pressure-driven flow, has also been used to create ion and particle depletion zones around membranes to develop devices for water desalination [71] (see also Chapter 6 for additional discussion on water desalination).

With pore dimensions as small as several nanometers, the fluid transport across NCAMs differs considerably from those in microfluidic structures. In particular, lateral (radial) diffusion can efficiently deliver molecules from the center of the channel to the walls at the nanoscale and, thus, can effectively be combined with electrokinetically driven axial flow. Applications employing nanochannel-immobilized enzymes are particularly attractive. With relatively short distances for substrates to traverse, it is expected that they will encounter surface-bound enzymes many times during transit through the nanochannel, thereby rendering enzyme kinetics within confined geometries superior to those in bulk solution [72]. Figure 5.17 from [72] illustrates the immobilization of the enzyme, invertase, onto a nanoporous membrane for the conversion of sucrose into glucose and fructose. The flow cell device used in this study not only exhibits a higher rate of enzyme/substrate interaction than in bulk [73], but also requires a lower substrate concentration for product detection. In another study, reaction-diffusion modeling was combined with measurements of enzymatic reaction velocities during injection-relaxation cycles across an NCAM to show that the enzymatic activity of immobilized horseradish peroxidase in NCAMs is up to 100 times higher than in free solution [51, 52]. Most notably, electrokinetic transport in NCAMs affords rapid convective delivery of substrates to active sites in order to improve turnover of mass transport limited enzymatic reactions. Thus, enzyme-functionalized NCAMs utilized within microfluidic devices, as described above, show a great promise for in situ generation of biological reagents for downstream reactions.

5.6 Transport and analysis systems for DNA

DNA stands for deoxyribonucleic acid and is a member of a family of biopolymers. Nucleic acids were discovered by Friedrich Miescher in 1869. This class of biopolymers is considered to be the building block for several life forms as nucleic acids exists in all life forms, including within bacteria, mitochondria, chloroplasts, and viruses. The basic component of biological nucleic acids is a nucleotide which contains a sugar molecule, a phosphate group, and a nucleobase. The sugar molecule forms the backbone for a DNA molecule with the hydrophilic OH groups. The nucleobases exist in a sequence of nitrogenous bases, adenine (A), guanine (G), cytosine (C), and thymine (T). Hydrogen bonding is known to occur between A and T bases and between C and G bases leading to the formation of the double helix structure. It is now generally accepted that for double-stranded DNA (dsDNA) the physical properties of the linear chain polymer are a function of the length and nearly independent of the base sequence. DNA characteristic lengths can run from ~2 nm for the molecule diameter to ~10–100 nm for the persistence length, P_l. P_l is defined as the mean radius of curvature of a large polymeric molecule at some average temperature, T, due to thermal fluctuations. Given the importance of DNA to biological systems and the fact that the main scales are accessible to microfluidics and nanofluidics, numerous devices and approaches have been developed to study and manipulate DNA structure, function, and properties. Commonly, DNA exists in two states: as a dsDNA with a P_l at physiological salt conditions of about 50 nm, or as a single-stranded molecule (ssDNA) lacking the complementary strand with a much smaller persistence length of about 3 nm. A typical DNA molecule sits in a coiled state, akin to a plate of noodles or spaghetti pasta with the base sequences and "data" codes all along the length of the DNA. Consequently, a "tape-reader" is needed to linearize the DNA molecule and "read" the codes along the length to provide key biological information. This reading process is termed as DNA sequencing. However, because DNA is located within a cell, a long process is needed to sequence DNA as depicted schematically in Figure 5.18.

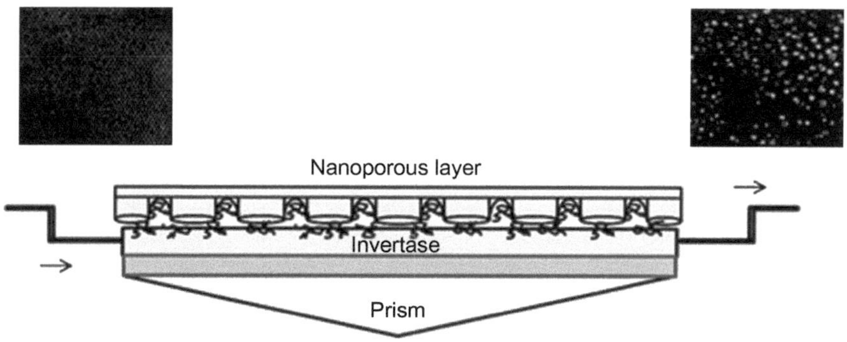

FIGURE 5.17

Flow cell to monitor real-time production of glucose and fructose. Substrate (sucrose) is introduced convectively to nanoporous anodic alumina featuring immobilized invertase from left to right. Sucrose/invertase interaction is monitored by surface plasmon resonance (SPR) measurement. © 2011 American Chemical Society, reproduced with permission from [72].

Several microfluidic systems have been built to carry out each operation, but an integrated system to carry out the complete process is still elusive. In 2008, commercial DNA sequencing was estimated to cost \$100,000 with the process taking approximately 2 months to complete. The human genome sequencing project cost ~\$3 billion and led to several technological innovations; however, the goal of cheaper and faster DNA sequencing (\$1000 with time on the order of 1 h or less) is still being pursued. One of the promising technologies for DNA sequencing is the development of nanopores for DNA sequencing. The nanopore sequencing method is based on the Coulter concept (Figure 5.19), where a flow of liquid through a small orifice made in a thin wall is maintained by a pressure difference. The flow can carry micrometer-sized particles that produce pulses in the capillary conductance, g. The number of pulses per unit measurement time (usually seconds) indicates the particle concentration; the particle size can be inferred from the pulse amplitude. In a conventional Coulter counter, small particles are forced by pressure difference through a small aperture (on the micrometer scale) made in an insulating wall such as glass. When the particles pass through the aperture they displace the electrolyte in the solution, causing blockades in an ionic current flowing through the aperture, observed as momentary change in the conductance of the aperture (see Figure 5.19). The magnitude of these blockades is roughly proportional to the volume of each particle, and depends somewhat on the particle shapes. The dynamics of DNA translocation can be directly measured at the single molecule level. When the DNA enters the pore, it gives rise to an abrupt drop in the ionic current flowing through the pore. This signal is used to detect the residence time of the biopolymers in the pore.

With progress in fabrication (for an example of nanopore fabrication see Figure 5.14) the development of nanopore and nanochannel sequencers has been a major research topic in the past few years. As an example of the ion current vs time measured for a nanopore, see Figure 5.20(a). When 11 kbp dsDNA is added, downward spikes appear. Each spike indicates an event where a single DNA molecule traverses the 10 nm nanopore. The bottom panel zooms in on one event.

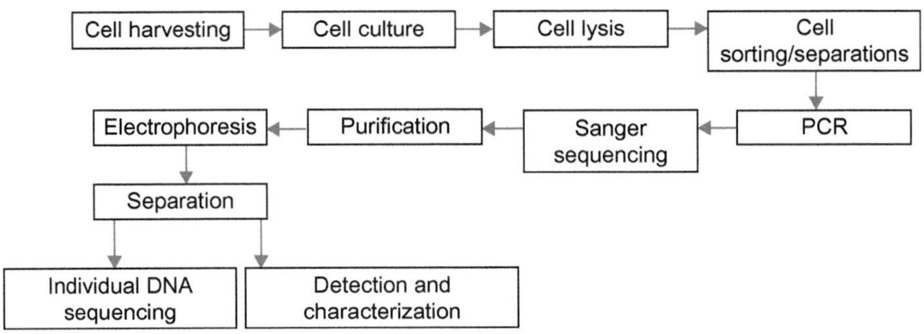

FIGURE 5.18

Sequence of steps needed for DNA analysis, sequencing, and other studies.

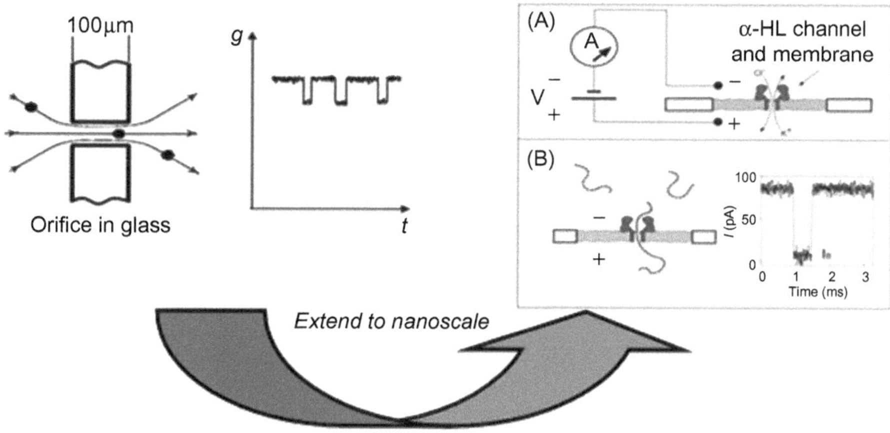

FIGURE 5.19

Conceptual depiction of the idea behind the Coulter counter for particles and how it can be used for DNA sequencing through a measurement of blockage current.

Figure 5.20(b) shows a scatter plot where the amplitude of the current change is plotted vs. the time of the DNA translocation. Each point is a separate translocation event. This particular example is the data for 11 kbp circular nicked dsDNA which traverses the 8 nm pore in about 200 s without any complicated folding phenomena. On average each traversal creates a current dip of 120 pA at the applied voltage of 120 mV. The translocation time vs the contour length of the dsDNA is plotted in Figure 5.20(c). Surprisingly, it was found that long dsDNA molecules display a nonlinear, power-law dependence with an exponent = 1.27.

5.7 Biosensors

Modern, productive, and advancing societies function due to improved lifestyle and health of the constituents obtained through the major improvements in environmental factors such as cleaner water, air, and food. Furthermore, health and life science-related improvements have also contributed to the growth of our society. One scientific tool that has enabled these improvements is the use of sensors for detecting chemical and biological molecules that are critical to the functioning of humans and other societal constituents. Among the various sensors and detectors that have been developed for identifying chemical and biological molecules, biosensors find a special place.

Biosensors are devices typically used to detect target biomolecules such as proteins, enzymes, or nucleic acids though they can have other applications such as the detection of chemical contaminants in water (also see Chapter 6). Biomolecules or chemical contaminants can be sensed or detected through a

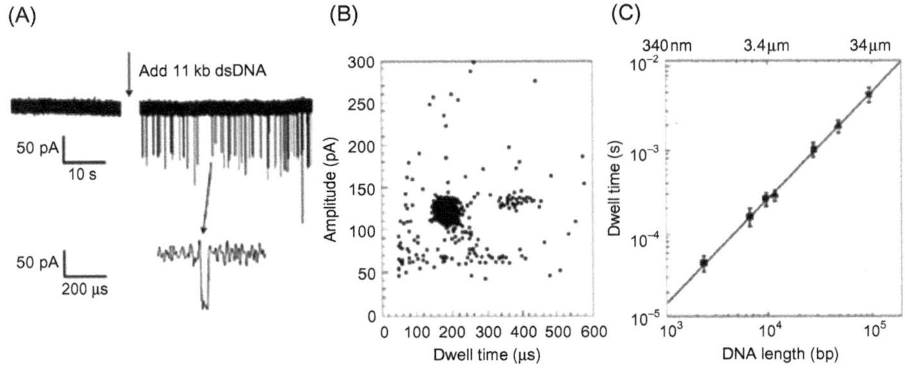

FIGURE 5.20

The figure shows data obtained from a blockage current measurement for DNA sequencing. The current spikes are proportional to the size and base-pair chemistry. Nanopores in other applications have been used to identify single-pair mismatches in base-pairs. (A) © 2007 Macmillan Publishers Ltd, reproduced with permission from [74]; (C) © 2005 American Chemical Society, reproduced with permission from [75].

variety of mechanisms but generally detection involves an interaction between the target molecule and a transducer to produce a measurable signal indicating the presence and concentration of that target molecule within a sample. In the case of a mixture of target molecules, the biosensor would enable the ability to discern between different species. Therefore, an ideal biosensor will provide specific information on the type and amount of the biomolecules present in a test sample.

Detection can be performed by a variety of methods. Common optical methods include fluorescence [76–79], optical cavity resonators [80–82], and surface plasmon resonance or SPR [83–89] in which a surface-based chemical reaction corresponds to a change in the refractive index and thus a shift in the optical signal. Electrochemical impedance spectroscopy or EIS [90–92] detects changes in surface or nanostructure impedance due to surface reactions or transport. In other electrically based techniques a surface-based reaction corresponds to a change in a measured electrical signal such as current [93–96], resistance [97–99], capacitance [100], or conductance [101–107] of the test sample. In addition, detection methods could also rely on mechanical changes induced due to adsorption of target molecules or analytes to cantilevers [108–110] or nanowires [106], changes to inherent biomolecular charge in the case of field effect transistors (FETs), sometimes also classified as microarray-type biosensors [111–116], or electrochemical changes such as those arising from redox reactions inducing variations in current flow [117,118]. Therefore, biosensors form a complex area of research given the diversity of target molecules, need for low false positives, and variety of detection platforms available [119–123]. Various detection methods are summarized in Table 5.4. Micro/nanofluidic-based biosensors incorporate one such detection device and may also contain different elements to control fludic transport as summarized in Table 5.2 above.

In addition, target biomolecules are typically found in some type of solution, and may require use of microarrays to immobilize target species on the surface of the biosensor [85, 112]. The solvents are usually aqueous solutions to preserve the natural state, or after processing may be organic materials. The sensor should be able to perform the function regardless of background solvents and interfering agents. Consequently, many different types of biosensors have been developed. Our focus will be on microfluidic and nanofluidic biosensors.

As discussed above, most biosensors rely on a target molecule adhering to a detector molecule and providing a measurable signal. Consequently, several different types of sensors have been developed for the detection of target molecules. In the next few paragraphs a broad-brushed discussion is presented for biosensors ranging from relatively common materials such as urea [124] to biomarkers for various autoimmune disorders and cancer. Our discussions do not target any specific category of biosensors since the focus here is on demonstrating the use of microfluidics and nanofluidics as a widely applicable platform for biosensing applications.

The most common substrates for microfluidic and nanofluidic devices are silicon, glass, and a variety of polymers. Silicon has the advantage of being widely used by the semiconductor industry, so its properties are well studied and fabrication methods are well developed. However, silicon is a semiconductor, so a SiO_2

Table 5.4 Summary of Detection Mechanisms for Various μ-Na*fl* Biosensors

Detection Method	Detection Mechanism	Measurable Response to Target Analyte	References
Mechanical	Cantilevers	Bending of cantilever or change in resonant frequency	[108–110]
	Nanowires	Change in surface charge through protonation/deprotonation	[106]
Electrical	Field effect transistors	Changes in inherent biomolecular charge	[111, 114–116, 125]
	Nanowires	Changes in electrical signal such as current, resistance, capacitance or conductance	[97–107]
Optical	Fluorescence	Intensity of fluorescent signal	[76–78]
	Optical cavity resonator	Shift in resonance wavelength proportional to change in mass	[80–82]
	Surface plasmon resonance	Shift in refractive index	[83,84, 86–89, 126]

Reprinted with permission from [47].

layer is thermally grown on the surface as an insulating layer for most silicon devices that require electrical isolation. This is particularly important for sensors that employ electrical-based detection methods since electrical effects in the bulk silicon substrate could potentially interfere with detection or provide stray noise in the sensor. Silicon with a thermally grown thin layer of SiO_2 has been used as a transducer where an Au/Cr contact was connected to the Si chip to monitor impedance changes caused by variation in charge distribution due to binding of DNA [103]. Glass has the advantage of being chemically stable and is transparent, which is particularly attractive when optical detection methods such as fluorescence or SPR are used. However, depending on the type of glass there can be a large amount of impurities or trace materials that may potentially interfere with either device fabrication or sensor operation. For example, soda lime glass is one of the cheapest and most commonly used forms of glass, but can contain a large amount of contaminants such as aluminum. Pyrex or borosilicate glass is often used in microfluidic and nanofluidics devices but is more expensive than soda lime. The best optical properties would be achieved in either fused-silica or quartz but these types of materials are significantly more expensive. One of the challenges with glass is the difficulty in bonding device layers to create sealed channels. Often high temperatures and/or large electric fields may be needed and therefore fabrication methods for bio-materials have to be adapted to meet the device bonding requirements. Polymers are attractive materials for substrates because of their low cost, disposability, and ability to adapt to several bio-material and diagnostic applications [127]. For example, a polymer-based disposable device capable of detecting 0.25–10 ppm of anesthetic propofal within 60 s has been already demonstrated (Figure 5.21) [124]. Poly(dimethyl siloxane) (PDMS) is the most commonly used polymer for fabrication of fluidic channels [76, 77, 84, 129] toward biosensor applications though PMMA is also used [130]. Conducting polymers such as poly(phenylene vinylene) (PPV) [96], and its soluble derivatives along with polyaniline (PAN) have been used for electronic components [101]. The main challenges in working with polymers relate to the relatively lower Young's modulus for these materials making polymer channels more prone to collapse and deformation than either silicon or glass. Furthermore, many polymers also show nonspecific adsorption of bio-materials and can therefore pose contamination challenges for biosensors.

A wide range of proteins have been studied including human chorionic gonadotropin [77, 87, 89], which is a biomarker for tumors, β-amyloid, an important biomarker for Alzheimer's disease caused by the cleaving of the amyloid precursor protein [131], the cancer biomarker alpha-fetoprotein [99], and interleukin-6, which is a biomarker for autoimmune disorders [86, 132]. Other target molecules include microRNAs [95, 99], a biomarker for cancer, DNA [88, 93, 98, 103, 107, 127], herpes simplex-2 virus infection [78], and common drinking water contaminates including *E. coli* [118], pesticides [95, 108], synthetic estrogen in river water [100], and heavy metals [76, 134–136].

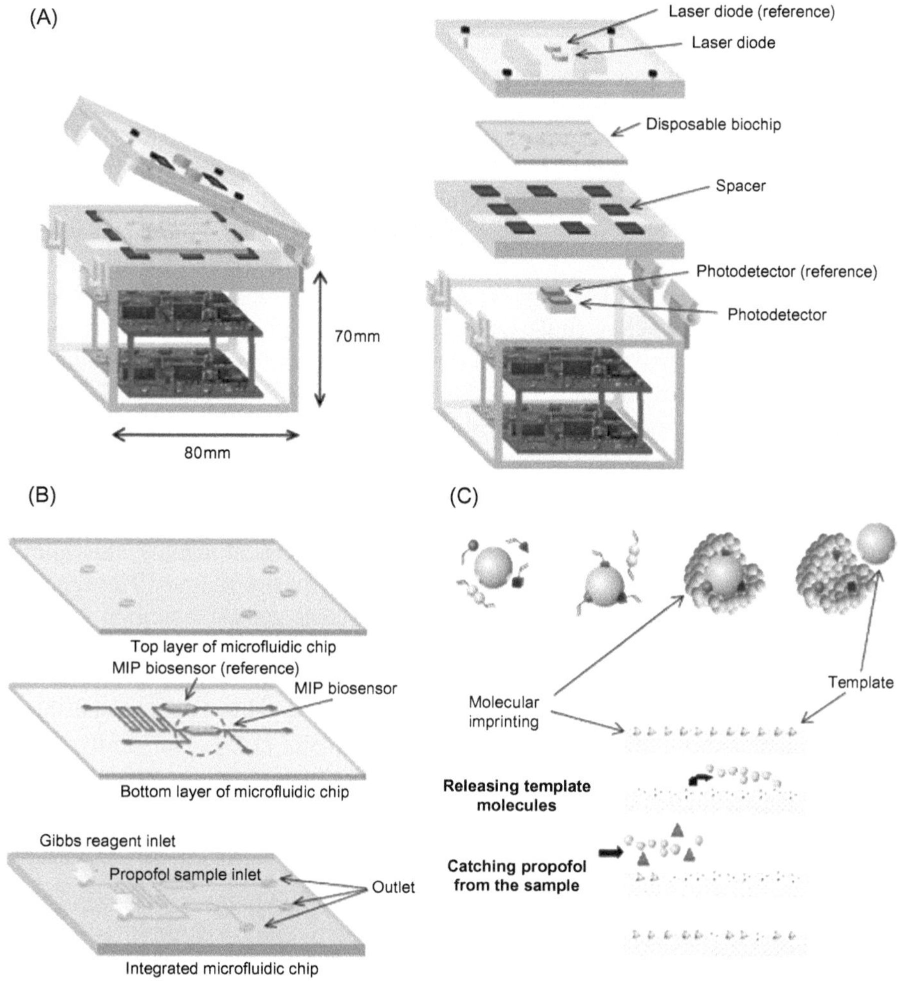

FIGURE 5.21

A disposable microfluidic biosensor that uses optical detection methods to detect the presence of anesthetic propofol. The device is compact and highly sensitive with a detection limit of 0.25 ppm. Reproduced with permission from [48].

Many detection methods rely on target molecule labeling and optical measurements. One such method is surface plasmon resonance imaging. In one study, target miRNA molecules were adsorbed onto a single-stranded locked nucleic acid microarray where poly(A) tails are added through surface reaction. DNA-coated Au nanoparticles bind to poly(A) tails which can then be analyzed using SPRI (surface plasmon resonance imaging). The percent reflectivity was found to vary linearly with concentration at concentrations of 10–500 fM. This study achieved a sensor that was 50 times more sensitive than a previously reported fluorescence-based microarray [137], with a detection limit of 10 fM (5 attomoles, dissolved in 500 μl) [83].

Methods that require labeling of target molecules and optical detection methods are limited by the ability of the fluorophore labels to bind with the correct molecules. In addition, quenching from the excitation light or from environmental effects limits the reliability of optical-based measurements. For these reasons, researchers have begun to develop label-free detection methods that do not rely on optical measurements. Recently silicon-based nanowires have been used for the detection of microRNA (miRNA), a nonprotein coding RNA that regulates gene expression and has been implicated as a biomarker for certain types of cancer and neurological disorders [97]. MiRNA cells bind to peptide nucleic acids (PNA) which are immobilized on the surface of the nanowires to serve as receptors for the miRNA. Changes in resistance measured before and after the introduction of the sample indicate that miRNA has bonded to the PNA on the surface of the nanowire. This detection method allows for label-free direct detection of miRNA. Since the method is based on changes in electrical signals, this allows for quantitative determination of the miRNA present in the sample. The reported detection limit is 1 fM. A potential limitation is that certain common proteins such as BSA also bind to PNA and may affect resistance signals though effects were not found to be significant in a recent study [97]. Streaming potential and Debye screening may also affect electrical signals as previously discussed; however, researchers are working to minimize these effects [125]. The similar study was done using silicon nanowires to detect DNA down to 10 fM [107]. Similar detection method was developed using polymer-based nanowires. In this study, concentration of MiRNA could be quantified in the range of 10 fM–20 pM with a detection limit of 5.0 fM [101].

Electrophoretic separation was used in a microfluidic analysis system to detect the presence of anthrax in 750 nl of mouse blood within 30 min (Figure 5.22) [129]. In another study, a nanofluidic device is used to localize gold nanoparticles (60 nm) and increase concentration of target molecules near the entrance of the channel. Since the channel is only 40 nm tall the particles become trapped at the entrance. Capillary forces drive the target molecules through the nanoparticle clusters where the target molecules bind creating an area that has a high concentration of target molecules. Surface-enhanced Raman spectroscopy, which depends on the excitation of molecules by infrared light, is then used to detect target molecule concentration down to 11.5 pM. The method is particularly attractive because it can distinguish between toxic $A\beta$ and harmless forms [131].

While many biosensors focus on the detection of biomarkers for certain diseases, biosensors also have application for detecting environmental contaminates. The Environmental Protection Agency had set the maximum concentration allowance for the pesticide dichlorvos at 5×10^{-10} M for drinking water. A carbon nanotube-based biosensor reduced the detection limit from 1 nM to 1 pM in

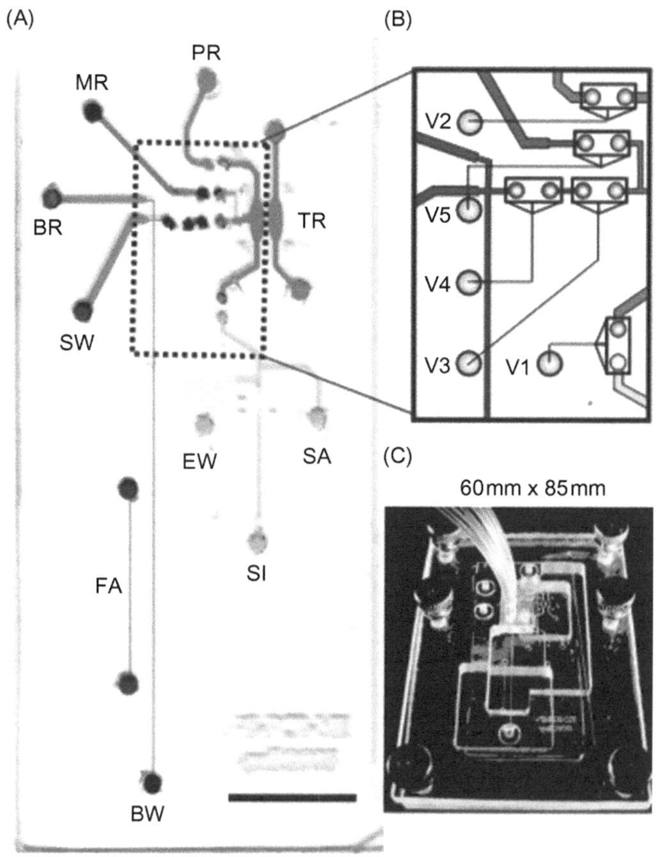

FIGURE 5.22

A multilayer microfluidic device that employs electrophoretic separation to detect anthrax in whole mouse blood. Reproduced with permission from [48].

10 min with a linear range of detection from 1 pM to 1 μM. It is believed that the decrease in the detection limit is due to the selective concentration of the pesticide inside the carbon matrix [95]. A further concern for clean drinking water is the presence of heavy metals such as arsenic, which can occur naturally or from materials like uranium, potentially arising from waste armaments and other industrial and military applications. The United States and Europe set the permissible level of arsenic at 10 μg/l while in other parts of the world it is 50 μg/l. Microfluidics-based biosensors are capable of detecting arsenic concentrations of 10 μg/l in 75–120 min in contrast to 120–180 μg/l using fluorescence-based techniques [76]. Biosensors have also been developed for the detection of other heavy metals [134–136], synthetic estrogen [94] and *E. coli* [118], which are all concerns for water contamination.

For all biosensors non-specific adsorption of biomolecules is a concern. Permselective membranes are one method for dealing with this issue. The principle of ion permselectivity is widely used in glucose biosensors. Glucose sensors are based on the rate of oxidation of glucose by dioxygen which is measured by the formation of hydrogen peroxide. A permselective polymer layer is used in amperometric glucose biosensors to avoid interferences of the detection of hydrogen peroxide by ascorbate, urate, and acetaminophen [138] found in blood samples. Permselective membranes are typically made of Nafion or materials such as (3-mercaptopropyl) trimethoxysilane (MPTMS), with studies indicating that MPTMS is a better permselective material since signals from interfering molecules were reduced when compared with Nafion-coated biosensors [139]. Nafion permselective membranes are also used in the glucose sensors to prevent reduction of a sensor signal in electric-based sensors caused by the dependence of the signal strength on buffer concentration. As the buffer concentration increases, the signal strength decreases. The use of the permselective membranes near the sensing surface reduces the influence of the buffer concentration [140]. In another study Nafion was used as the nanochannel in a microfluidic-based biosensor. The nanofluidic concentration polarization effect was used to preconcentrate the sample before detection. The detection limit for this device was 2.72 nM with a sample size of 25 μl and a detection time of 15 min, including the preconcentration step [77]. A recent study shows that nonspecific adsorption can be reduced by increasing self-assembled monolayer incubation time, decreasing surface roughness and in the case of gold electrodes, optimizing the directional alignment of the gold crystal [141]. The device used to perform this study is shown in Figure 5.23. Another method of dealing with the nonspecific binding of proteins exploits the Vroman effect. The Vroman effect is when a weak affinity protein bound to the surface of a substrate is displaced by a strong affinity protein. The

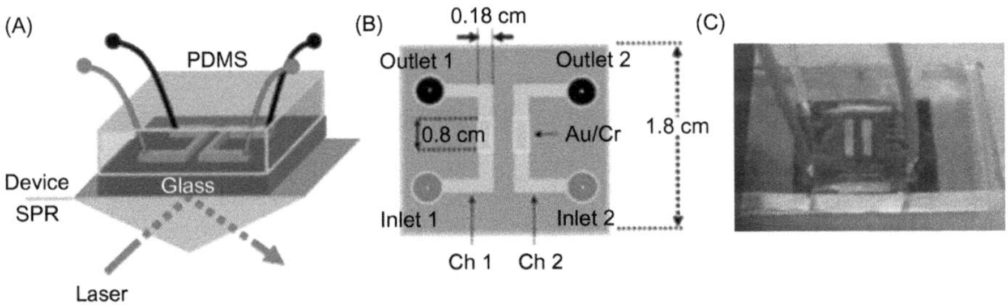

FIGURE 5.23

Schematic of a device used for studying methods to reduce nonspecific adsorption. SPR is used to detect binding of target molecules. Fluidic channels are fabricated in a PDMS slab and bonded with a glass cover to allow optical access for SPR detection. Fluid inlet/outlet tubes are inserted through the PDMS slab [140]. Reproduced with permission from [48].

reverse process does not occur. A protein's affinity to bind depends on the size of the protein, the concentration of proteins in the solution, the binding substrate and the pH of the solution. The reactions occur so that the total entropy of the system is minimized, which is why certain proteins will replace others already bound on a surface while the reverse process will not occur [142]. The Vroman effect allows a target protein with a strong affinity to be detected in the presence of several other lower affinity proteins. Displacement of proteins initially adsorbed to a surface result in a shift in reflectivity that can be measured by SPR [84].

Microfluidic biosensors offer the possibility of creating portable and potentially disposable detection devices. Recently, an electrochemical microfluidic biosensor was developed for the detection of DNA. The device uses an array of microelectrodes as a transducer and ferro/ferricyanide entrapping liposomes as signal reporters to quantify the concentration of DNA in the range of 1–50 fM in 6 min using only a simple potentiostat [133]. This type of sensing mechanism was used in the design of a miniaturized electrochemical detection system with an integrated potentiostat. The device is fully portable and relatively inexpensive (<$50) [93], when compared to some other microfluidics devices that can easily cost well over a few hundred dollars. This varies from fluorescence microscopy since it does not require expensive, bulky detection equipment and image processing software, though the detection limit for fluorescence-based detection has been reported to be as low as 0.2 fM [143]. In addition, devices have been developed that have six channels in parallel and use on capillary forces to fill the channels in 8–47 min. This could potentially lead to a single device that can be used to detect several different target molecules at once [78]. Researchers have also made advances towards fabricating reusable devices by applying a low (0.9 V) DC voltage to reset the sensing surface [84].

5.8 Nanomedicine and nanobiotechnology

One of the reasons the scientific community is excited about advances in nanofluidics and nanotechnology, centers on the possibility of manipulating processes, especially in biological systems, at the molecular level. This possibility has led to the rapid development of new materials, methods, and systems for applications in biomedicine [144–148]. A recent review explains the concept of nanomedicine and some of the early advances in applying the principles of nanotechnology to medicine [149]. Nanomaterials, such as silica or gold nanoparticles, quantum dots, functionalized micro- and nanocantilevers, and artificial nanopores, are being increasingly used for detection of DNA [150–154], RNA [155–157], and proteins [158–161]. In some cases, the asymmetry within the artificial nanostructures is being exploited to mimic the behavior of natural systems such as ion-channels [162–164]. Multiple fabrication techniques, for example e-beam lithography, focused ion-beam milling, and nanoimprint lithography, have been used to construct nanostructures for trapping and transporting DNA and focusing of proteins using dielectrophoresis [165]. Using well-defined nanochannels from 30–500 nm DNA and proteins have been separated by the steric effects that exploit the principle of entropic minimization in the transport of biomacromolecules within confined geometries [161]. Nanostructures and single nanopores have been used to reduce the entropic contribution to the total free energy of DNA, thereby modulating entry into a separation channel according to the oligonucleotide length [166, 167].

Silicon has been extensively used as a material for building systems for applications in biochemical systems at the nanoscale due to its versatility in microfabrication [168]. However, chemical modification can impart the novel properties of the solid–liquid interface that can be usefully exploited. Many materials and methods are currently being researched for drug-delivery applications [169] including dendrimers, such as poly(amido amine) or PAMAM [170]. Dendrimers have recently been attached to the confined surfaces within microfluidic devices and show a possible method for attachment of PAMAM dendrimers to nanofluidic devices [171]. The development of systems relying on bacteria as molecular motors for pumping small volumes of liquids has also been investigated [172].

5.9 Instrumentation and platforms at the microscale and nanoscale

5.9.1 Micro-gas chromatography: a case study for a lab-on-a-chip system

In the sections above, we have extensively discussed a variety of applications largely targeted at liquid-based μ-Na*fl* systems. In this section, we present a detailed overview of a micro-gas chromatograph to show not only the versatility of microsystems and nanosystems but also demonstrate how a standard analytical chemistry technique has been miniaturized to develop a portable system. The challenges, opportunities, and key results are all discussed.

Timely identification and measurement of gas constituents are essential in many areas of science and technology and to human activities. The detection of undesirable compounds in indoor and outdoor air, exhaust, fumes/plumes, closed spaces, battlefield, and many other environments is important to security and safety, climate change, food processing/monitoring, emission control in industrial plants, pharmaceutical screening, health care, and air quality monitoring. Toxic industrial compounds (TICs), volatile organic compounds (VOCs), semi-volatile organic compounds (SVOCs), explosives and narcotics, and chemical/biological agents are a few classifications of the vapors/gases of interest. In each application, an aim is to identify and quantify some or all of the chemical species in a mixture. A considerable number of gas-sensing technologies have been developed and pursued in last few decades, and a decent review of gas sensors is given elsewhere

[173]. Regardless of the sensing principle (electrochemical, optical, mass-based, etc.), selectivity remains a challenge, and gas chromatography (GC) is known to be the best technology to improve the selectivity of the gas sensors.

The miniaturization of analytical systems such as GC has drawn significant attention for more than 30 years as a means of reducing instrument size and analysis cost for the field deployable and portable applications. The first report of a microfabricated GC (μGC) system consisting of a silicon substrate with an etched column, gas sample loop, and resistance–temperature detector (RTD) was made in 1979 at Stanford University [174]. Since then, to address the need for toxic gaseous substance analysis in the field, chemical processing monitoring, and autonomous vapor analysis for environmental monitoring, several miniaturized GC systems have been developed in the national laboratories and universities [175–179]. Among these systems, the μChemLab® developed at Sandia National Laboratories and a hybrid microfabricated GC (μGC) at the University of Michigan are the state-of-the-art systems that utilize the MEMS and microfabrication technologies and incorporate individual components such as a microfabricated adsorbent preconcentrator, a dry-etched silicon separation column, and an integrated surface acoustic-wave sensor array [180,181]. Very recently, University of Illinois and Cbana Labs Inc. jointly developed a portable gas analyzer that consisted of a micro-flame ionization detector (μFID) and a μGC, and all the peripherals to operate μGC/FID were housed in a lunchbox of $20 \times 24 \times 10\ cm^3$ [182]. Likewise, large efforts in developing portable GCs with micromachined components have been made in the past two decades. Some notable companies include Agilent Technologies [183], SLS Microtechnology of German (now acquired by Elster GmbH) [184], C2V in the Netherlands [185], and Defiant Technologies of USA [186]. Some photographs of the exemplary miniature GC systems (only in the system level) are shown in Figure 5.24 [186].

The operation principle of a conventional GC system is depicted in Figure 5.25. In GC, a sample analyte is first collected and then injected into a separation channel or column, which is typically coated with a thin polymeric film, known as the stationary phase. Once a mixture of the analytes is separated into individual analyte bands, each band would be eluted to GC detectors for sensing and producing chromatograms. The three key components in a μGC system are sampling/injection units, microcolumns, and micro gas detectors, and a more detailed description of each component follows.

5.9.1.1 Micropreconcentrators

Many environmentally toxic chemicals and CWAs can be hazardous to humans at concentrations below 1 ppb, but many portable gas detectors have difficulties in detecting such a trace concentration of toxic vapors. In line with developing a microscale/portable detector with the highest possible sensitivity for a high-performance μGC system, a microscale (or MEMS) preconcentrator (μPC) can provide an alternative means to improve the detection limit of the detector. The role of the μPC is to capture the vapors of very low concentrated (<1 ppb) analytes

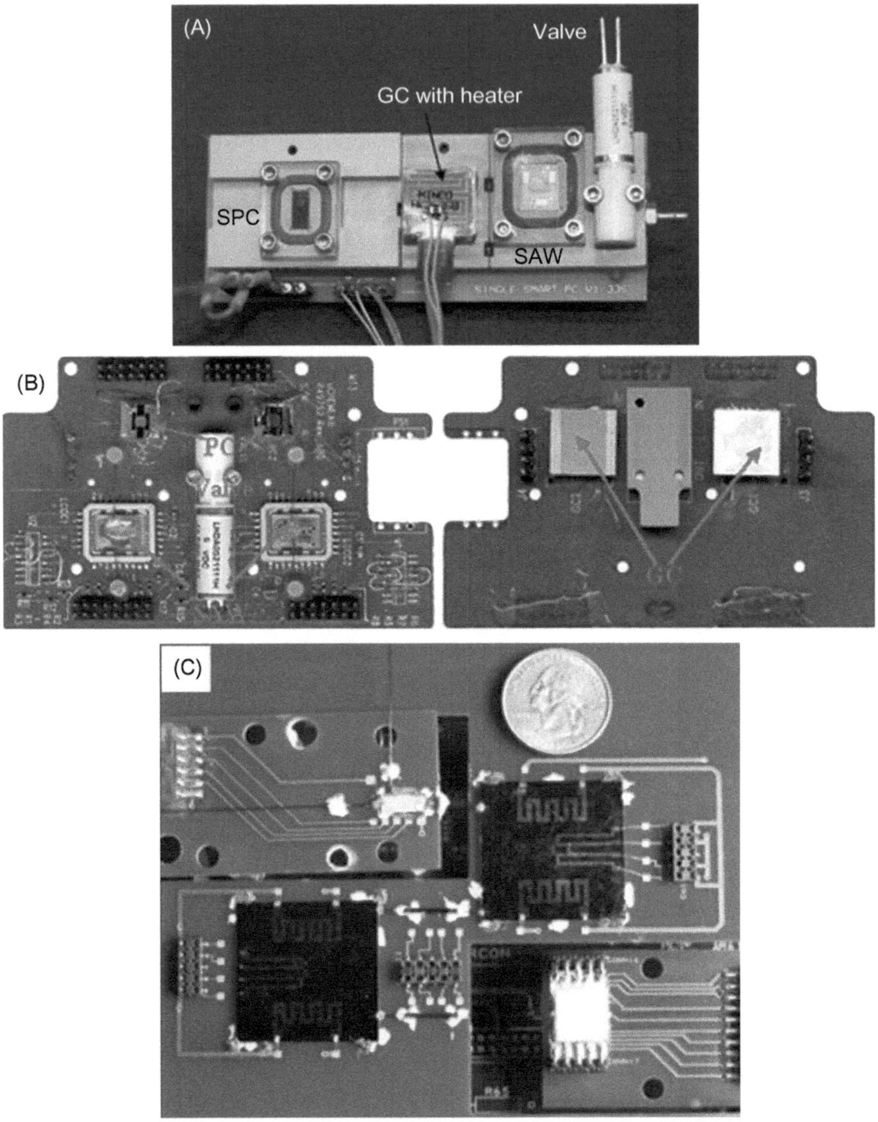

FIGURE 5.24

Photographs of (A) a test fixture for Sandia's μGC system with a micropreconcentrator, microcolumn and its heater, SAW array sensor, and Lee minivalve (© 2008 IEEE. Reproduced with permission from [187]), (B) hybrid integration of GC components on a circuit board for Sandia's μGC (© 2006 IEEE. Reproduced with permission from [181]), (C) integrated microanalytical system from University of Michigan (© 2011 American Chemical Society. Reproduced with permission from [189]).

from ambient and to thermally desorb them into a much smaller volume in a microfabricated separation column, thereby increasing the effective concentration to facilitate detection as well as providing a sharp injection plug to promote high-speed chromatographic separations (see Figure 5.26) [180].

Conventional PC (also called microtrap) is comprised of glass or stainless steel capillary tubes packed with one or more granular adsorbent materials. These capillary tubes suffer from a larger dead volume and slow heating due to their larger thermal mass. Microfabrication technology can overcome these limitations by making the size of PC small, which can significantly reduce the dead volume and thermal mass [187]. The design aspects in developing microscale PCs are related to the operational constraints such as limited (preferably less than a few mW) power consumption and fast (less than a few seconds) sampling time. The μPC must contain sufficient adsorbent mass (surface area) to ensure quantitative trapping of vapors from the sample stream, but small enough to be rapidly heated to ensure complete desorption and to minimize the desorbed-vapor bandwidth [188].

While columns and detectors for μGC systems have been widely investigated, the microscale preconcentrator has been relatively little studied. The μChemLab® at Sandia National Laboratories employed a freestanding SiNx microhotplate onto which the polymer adsorbent material was spray-coated (see Figure 5.27(a)) [187]. In their modified version of the PC, three-dimensional silicon structures were suspended on the central portion of the membrane to increase the SA/V up to a factor of 20 and to distribute heat more uniformly (see Figure 5.27(b)) [181]. On the other hand, Tian et al. from the University of Michigan used freestanding, high-aspect-ratio Si microstructures as thick microheaters and packed the PC with the carbon-particle-based adsorbents [189, 190]. Single-stage and multistage μPCs demonstrated a high adsorption capacity and an ability to capture compounds of a wide range of volatility (see Figure 5.27(c)) [189, 190]. Researchers at the University of Illinois integrated four microvalves with their μPC (sub-μL-microchannel filled with an array of microposts) so that sampling, preconcentrating,

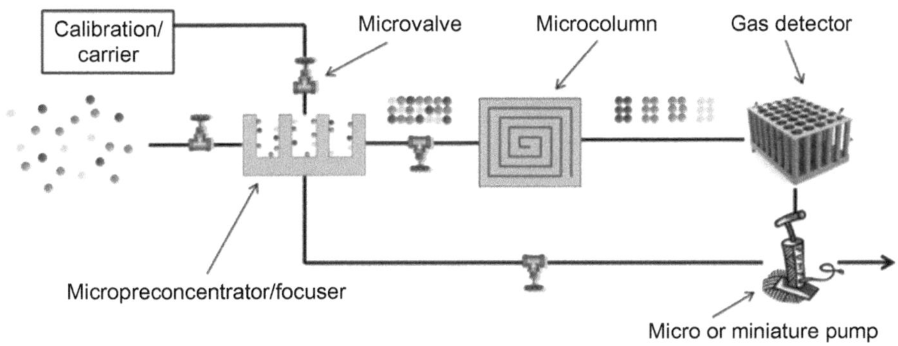

FIGURE 5.25

A simplified diagram of a typical gas chromatography system.

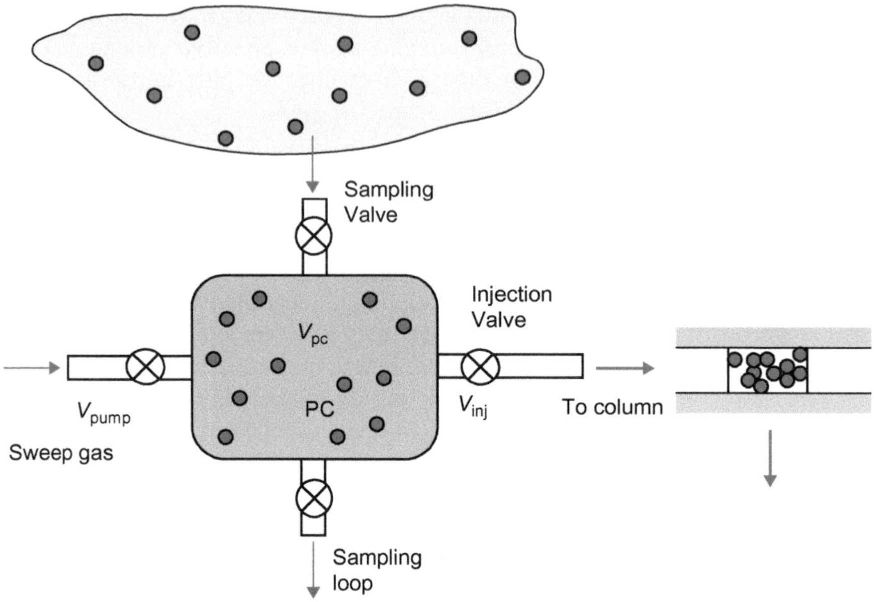

FIGURE 5.26

A diagram of the preconcentrator connected with valves: sampling a low concentrated vapor (analyte) into a small volume chamber packed with high surface-area adsorbents and later sweeping the thermally desorbed analytes into a column.

and injection could be performed in a single chip with monolithic integration (see Figure 5.27(d)) [191–193].

5.9.1.2 Microcolumns

When a carrier gas transports the sample mixture across the column, the individual components in the mixture spend different amounts of time (i.e., retention) absorbed in the stationary phase, causing separation to take place. The retention time of each component depends on several variables including its vapor pressure and polarity, column temperature, carrier gas speed, and the types of the stationary phase. Column selectivity can be further enhanced by using either mixed stationary phases or by coupling in series two wall-coated open tubular columns with each having a different stationary phase [194]. The state-of-the-art columns are made of fused-silica tubes whose inner walls are coated with the stationary phase.

In their first micromachined GC, Terry et al. used wet etching to create a spiral microchannel on a silicon wafer and anodically bonded it to a Pyrex glass substrate [174]. Since then, microcolumns have been mostly fabricated by wet anisotropic or deep reaction ion etching of silicon and bonding with glass. For example, Lambertus et al. from University of Michigan fabricated 150 μm wide,

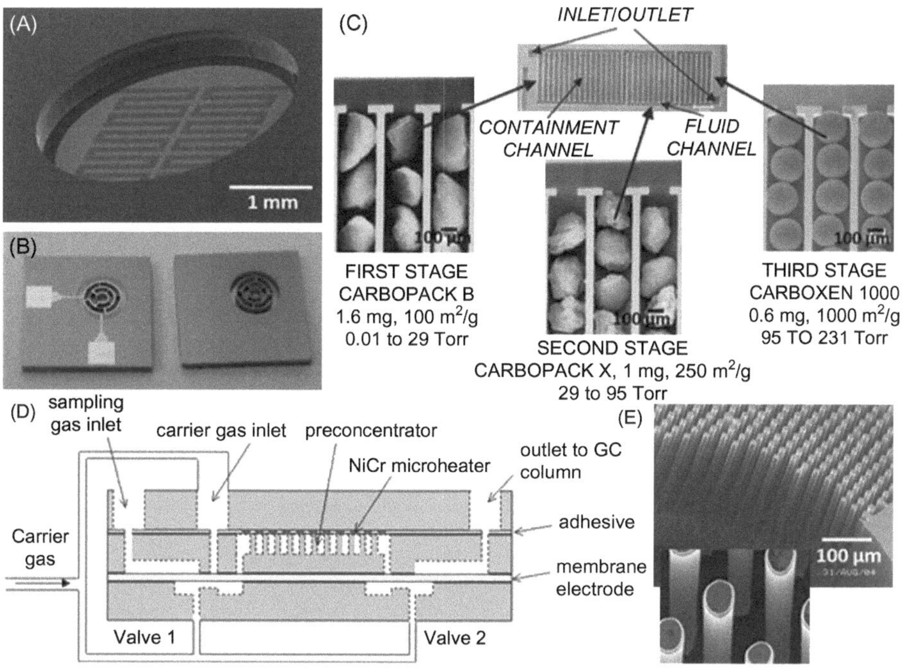

FIGURE 5.27

(A) DRIE-etched microhotplate preconcentrator developed by the Sandia National Laboratory: Pt serpentine heater fabricated on free-standing silicon nitride membrane, (B) Sandia's micropreconcentrator modified with 3D Si channels perpendicular to the substrate surface, (© 2006 IEEE. Reproduced with permission from [181]), (C) three-stage microfabricated preconcentrator focuser using a microheater packed with three carbon adsorbents by the University of Michigan (© 2003 IEEE. Reproduced with permission from [195]), (D) a schematic of the fully integrated μPC with 4 microvalves for sampling and injection developed by the University of Illinois, (E) the μPC filled with an array of microposts.

240 μm deep, and 3 m long microcolumns in silicon and sealed it with a Pyrex glass cover with anodic bonding (see Figure 5.28(a–b)) [178]. These columns were dynamically coated with nonpolar dimethylpolysiloxane or mildly polar phase trifluoropropylmethylpolysiloxane, producing the theoretical plates of 4600 to 8200 for nonpolar phase and 3500 to 5500 for mildly polar phase, respectively. Noh et al. from Georgia Institute Technology developed parylene-based GC microcolumns to achieve low thermal mass for rapid thermal cycling. DRIE etched channels of 100 μm in width, 350 μm in depth, and 1 m in length were used as a mold to create free-standing parylene channels, which was laminated with another parylene layer prior to release in KOH solution (see Figure 5.28(c–d)) [178]. Radadia et al. from the University of Illinois developed

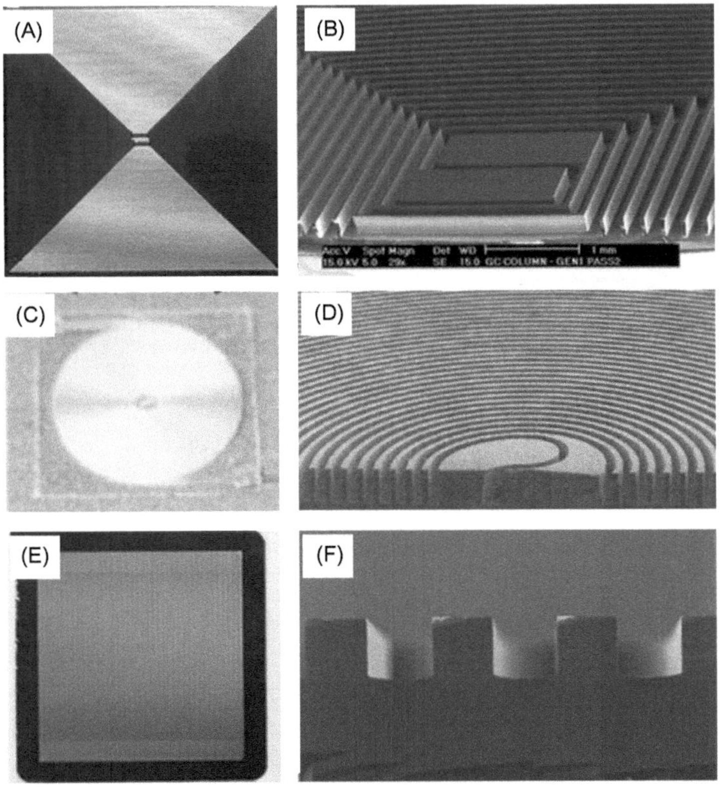

FIGURE 5.28

(A–B) A photograph of 3-m long microcolumn of 150 μm by 240 μm in cross-section and its SEM image (University of Michigan) (© 2004 American Chemical Society. Reproduced with permission from [178]), (C-D) a photograph of a 1-m long free-standing parylene column and the SEM image of a DRIE etched silicon microchannel used to mold the parylene film (Georgia Tech) (© 2002 IEEE. Reproduced with permission from [176]), (E–F) a photograph of 3-long serpentine microcolumn and its SEM image (University of Illinois) (© 2008 American Chemical Society. Reproduced with permission from [196]).

the 3-m-long serpentine microcolumn for organophosphate separation as shown in Figure 5.28(e–f) [179] and compared the separation performance of the microcolumns in terms of the channel geometry—serpentine, circular-spiral and square-spiral [197]. They found that the serpentine columns exhibited higher separation plate numbers, i.e., lower band broadening, due to thinner and more uniform stationary phase coating and favored hydrodynamics in the serpentine channels.

5.9.1.3 Micro gas sensors for micro GC

The gold standard for universal gas detection is the mass spectrometry (MS) system, where an analyte is ionized by an electron beam and the resulting charged particles travels through a magnetic field and gets sorted on the basis of mass-to-charge ratio. It is highly sensitive and very fast and can detect every gas compound, but typically entails large energy for ionization, very high vacuum, and enough space for acceleration, all of which make the miniaturization of MS extremely challenging. Therefore, miniaturization efforts have been focused on other popular GC detectors including flame ionization detectors (FIDs), thermal conductivity detectors (TCDs), and surface acoustic wave (SAW) sensors.

5.9.1.3.1 Micro flame ionization detectors

An FID is the most widely used GC detector and is an example of the ionization detector invented specifically for GC applications [198]. It is known that FIDs are the most effective detector for hydrocarbons and organic compounds. The column effluent entering an FID is burned in a small oxy-hydrogen flame, producing positively charged ions. These ions are traveling through a gap between two biased electrodes and are collected to form a current that becomes chromatographic signals. However, the conventional FID is too bulky and not compatible with the μGC in terms of fabrication and integration. In addition, high fuel gas consumption is required to maintain a stable flame, entailing a large gas supply and presenting a high explosion risk.

The benefit of miniaturization includes (i) less fuel and oxidant consumption due to the small chamber volume, (ii) the possible elimination of the high-pressure hydrogen storage when an electrolyzer is integrated, and (iii) small combustion area and much lower overall temperature allowing wider packaging materials, less energy usage, and integration with other μGC components in a close proximity. Zimmermann et al. reported the first kind of a μFID using micromachining technologies to create a micro burner unit, i.e., a silicon channel encapsulated inside a glass-silicon-glass sandwich structure [199,200]. Kuipers and Muller proposed a μFID using a premixed flame with hydrogen and oxygen [201]. Kim et al. investigated the stability of diffusion flames in the encapsulated microchannels with different geometries (as shown in Figure 5.29) and developed the μFID with a folded flame structure to increase ionization and collection efficiencies [202]. Kim's μFID featured the sensitivity of 40 mC/gC, which is twice higher than the best performing μFID reported so far [182].

5.9.1.3.2 Micro thermal conductivity detector (μTCD)

FIDs are unfortunately not sensitive to inorganic compounds such as H_2O, CO, CO_2, HCl, NO_2, and other permanent gases. Therefore, TCDs are essential for fixed-gas detection because no substitute has the same ease of use and stability. In addition, microfabricated thermal conductivity detectors (μTCDs) have been extensively developed as a favorite μGC detector because of its ease and

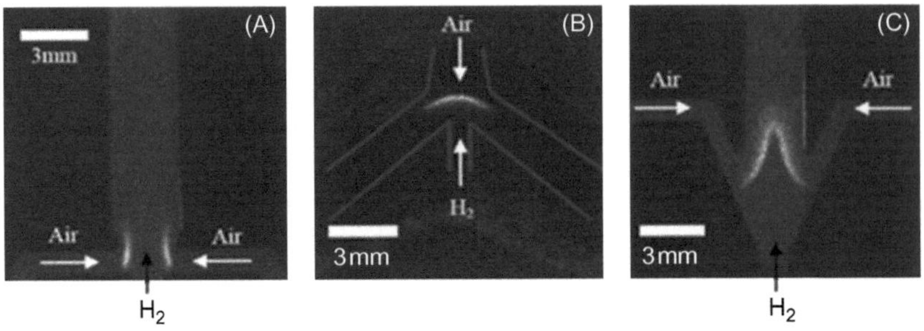

FIGURE 5.29

Images of flames taken with an iCCD camera. Hydrogen and air flows are at the same rate, 80 ml/min. (A) Two stable flames are observed due to the high diffusivity of hydrogen. (B) One stable flame is observed by configuring the channels in a way to make the counter-flow of air and hydrogen stream. (C) When hydrogen and air channels are meeting with a 150° angle, a stable single flame is observed [202].

Source [201]: Kim CH, El-Leathy AM , Xu F, Faeth GM. Soot surface growth and oxidation in laminar diffusion flames at pressures of 0.1e1.0 atm

simplicity in fabrication. TCDs consist of one or more heating elements in two gas streams: the reference stream containing pure carrier gas and the sample stream containing the column effluent transported by the same carrier gas. If there is a detectable change in thermal conductivity of the gas in the sample stream, the temperature of the heating element changes accordingly and can be detected by the standard bridge circuit.

One of the earliest research efforts on silicon-based TCDs was made at Stanford University [174]. They developed μTCDs that were several orders of magnitude smaller than the conventional TCDs, and the commercial products from their work are currently available. Most of the subsequent research on μTCDs uses the similar device platform—a thin film of heating element deposited on a thin membrane suspended in the middle of a microchannel etched in a silicon wafer [203–205]. Figure 5.30 shows the exemplary structure of μTCDs. Nickel, copper, or tungsten and their alloys are typically used for a heating element due to their high temperature-resistance coefficients. The LPCVD SiN_x film is the most widely used support structure because of its structure stability as well as its compatibility with both wet anisotropic etching and dry etching during the release step.

5.9.2 Micro-scale impedance measurements for flow diagnostics at the nanoscale: a case study

In this section, we present another case study where a microscale impedance measurement system has been developed for probing nanoscale flows. As has been

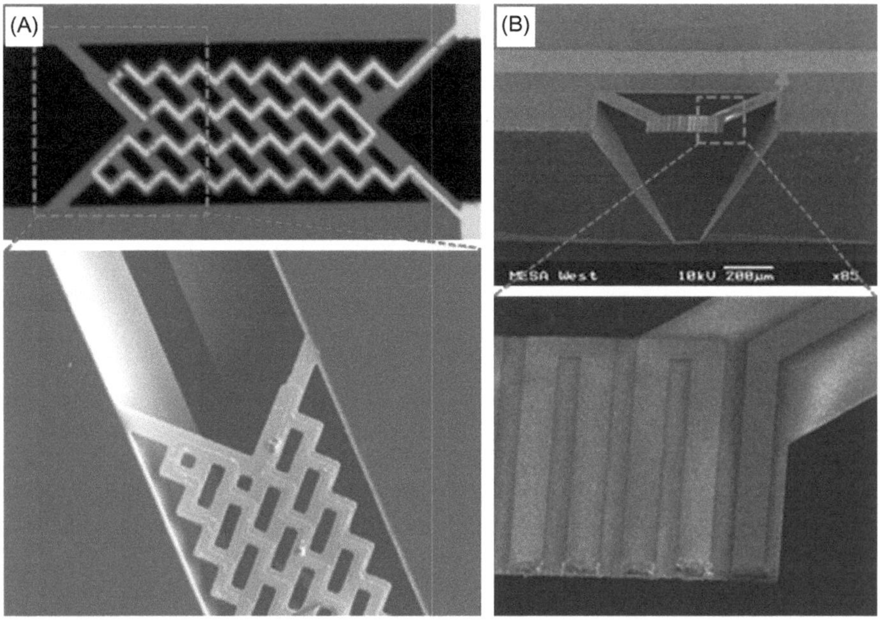

FIGURE 5.30

(A) A photograph and SEM image of thermally isolated, low mass SiN_x bridges (as a heater pad) suspended in a 150 μm wide Si channel. On top of the support structure is a 5 μm wide meandering Ni self-heating resistor (© 2009 IEEE. Reproduced with permission from [205]), (B) SEM images of the cross-sectional view of μTCD, revealing the meandered Pt heater on the SiN_x membrane suspended in the KOH-etched silicon microchannel (© 2007 Elsevier. Reproduced with permission from [204]).

evident throughout this chapter, for liquid-based devices aqueous salt solutions and system that can measure ionic transport can be very useful. Therefore, the ability to monitor ionic transport across nanopores is important for a variety of scientific and technological reasons, including the development of synthetic constructs mimicking ion-transport in biological channels, single molecule detection, and the development of molecular gates controlling separation of attomolar concentrations. Macroscale electrochemical impedance spectroscopy (EIS) typically uses samples and electrodes with several square centimeters of area and cannot have electrodes placed near the nanopores. Thus, detecting and characterizing ionic transport across a few nanopores at a known transmembrane potential is difficult, often yielding extremely poor signal to noise (S/N) ratio for systems such as molecular gates. In μ-EIS, the electrodes can be located relatively close to the nanoporous membrane. Also, with greatly reduced sensor size, small changes in ionic current can be detected with a reasonable S/N ratio. However, device integration challenges such as incorporating ideally polarized working and counter electrodes with an ideally nonpolarized on-chip reference electrode must be addressed. Also, the relative location and dimensions of the electrodes must allow for accurate transmembrane potential measurements. Thus, a μ-EIS sensor needed to be developed that can provide rapid and direct characterization of translocating ionic species through nanoporous membranes without the need for amplification, chemical modification, surface adsorption, or binding agents. The μ-EIS impedance data gives magnitude and phase information along with I-V characteristics, thus also providing insight into the interaction between the translocating ions and the electric double layer (EDL) within nanopores due to the interplay between the surface zeta potential and the ionic charge of the electrolyte. Developing a system in which nanopore impedance can be measured accurately will help develop in situ monitoring of nanopore electrokinetic flows.

A microfabricated multilayer μ-EIS system consisting of integrated nanoporous membranes, and working, counter, and reference electrodes is shown in Figure 5.31. The schematic represents an assembled device. A 1000 Å thick sputtered Au layer, serving as the cell electrodes, is patterned on a 100 mm Si wafer, and a polyimide layer is spin-coated to serve the dual role of providing electrical isolation for electroplating the Ag layer and adhesion between the die stack. Double-sided photolithography and DRIE yield four 18×16 mm Si dies that are used to assemble the multilayer stack with the nanoporous membrane sandwiched in between. Device assembly requires several alignment and bonding steps. The various layers are bonded using a commercial adhesive and the process for applying and curing the adhesives has been reported previously. The electrodes are located $\sim$500 μm from the membrane for accurate measurements of voltage drop across the membrane. A planar Ag/AgCl reference electrode is incorporated to provide on-chip reference. The formation of the reference electrode with the appropriate chemical structure is verified by XPS (Figure 5.32).

Impedance data is collected for varying salt concentrations of monobasic potassium dihydrogen phosphate (KH_2PO_4) in an aqueous solution. For validation of the μ-EIS measurement, impedance measurements are conducted on a macroscale 3-electrode electrochemical cell with 800 nm pores in polycarbonate track etched (PCTE) membranes. Figure 5.33 shows a comparative Nyquist plot for the 10 nm and the 800 nm pore diameter PCTE membranes at 100 mM salt concentration. The 10 nm pores display a mass transport limit indicated by the 45° line at low frequencies. At 1 mM salt with overlapping EDLs, the concentration impedance for the 10 nm pores increases by almost 2.4 times at low frequency (<1 Hz) and the imaginary part of impedance shows an increase of about 31% in comparison to the 100 mM case. Equivalent circuit modeling was carried out to estimate a possible electrical circuit that would better elucidate the physics of nanopore ionic transport. The presence of Warburg impedance indicates that hindered ionic transport occurs in the 10 nm pore, and its absence till about 0.1 Hz for the 800 nm at 100 mM indicates normal electrokinetic flows without interacting EDLs. This paper demonstrates the fabrication and testing for a μ-EIS sensor, and the use of such a sensor to evaluate electrokinetic ionic transport across nanopores.

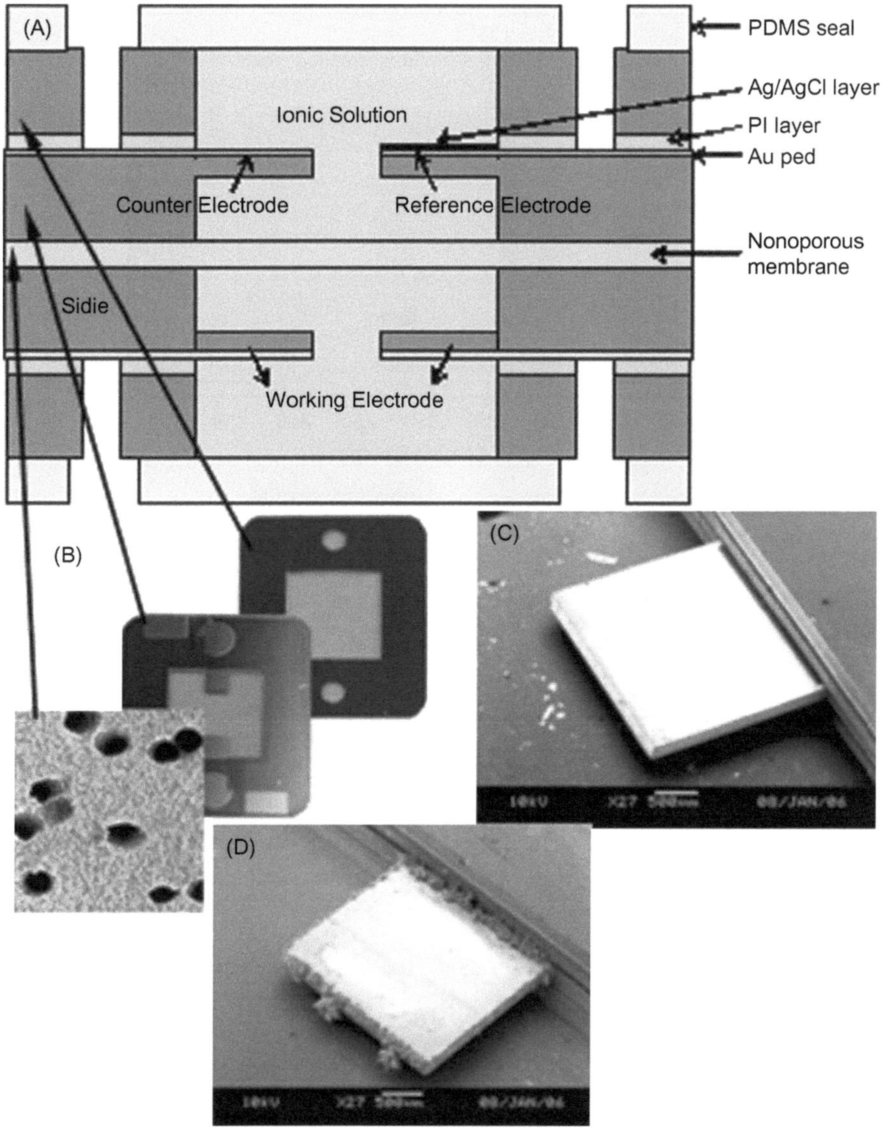

FIGURE 5.31

A microfabricated multilayer μ-EIS system consisting of integrated nanoporous membranes, and working, counter, and reference electrodes (Reproduced with permission from [206]).

It should be noted that the two case studies discussed in this section demonstrate a few aspects of integrated microsystems and nanosystems. First, these systems comprise several components. Second, fabrication and assembly of these

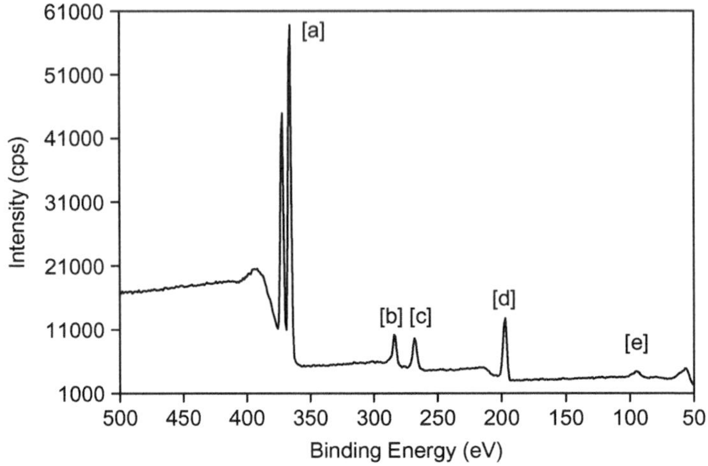

FIGURE 5.32

The formation of the reference electrode with the appropriate chemical structure is verified by XPS. Here you see the XPS spectrum of a Ag/AgCl reference electrode. Some of the key peaks are shown in the spectrum. [a] Ag(3d), [b] C(1s), [c] Cl(2s), [d] Cl(2p), and [e] Au(4f) (Reproduced with permission from [206]).

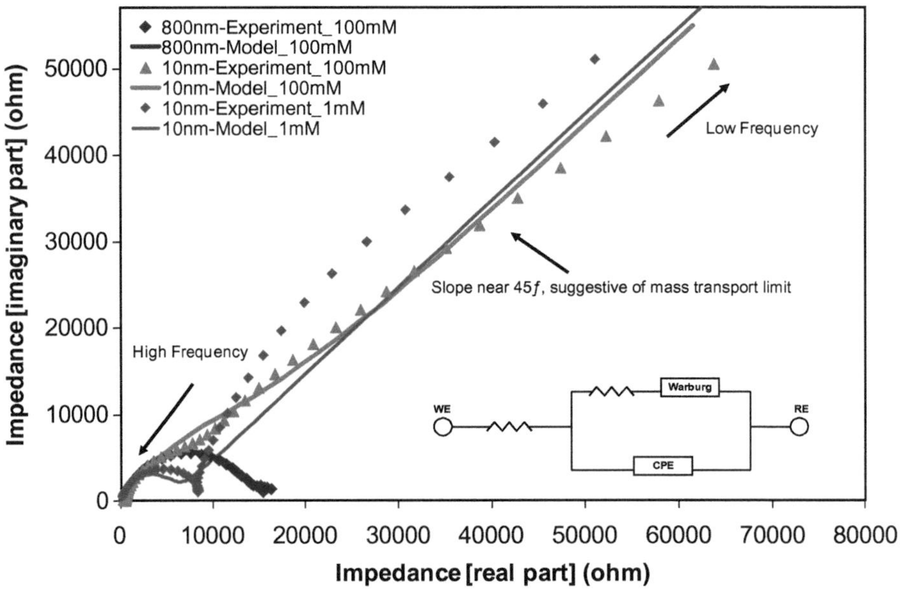

FIGURE 5.33

A comparative Nyquist plot for the 10 nm and the 800 nm pore diameter PCTE membranes at 100 mM salt concentration (Reproduced with permission from [206]).

components is a rigorous engineering task and needs sufficient development and research time. Third, several characterization steps may be needed to verify component performance and eventual device operation. Overall, the chapter provided a broad-brushed overview of many different enabling technologies and systems for microfluidic and nanofluidic devices and systems. In the next chapter, we continue with the theme of showcasing the breadth of applications possible by discussing areas related to energy and environmental research that relies on microfluidic and nanofluidic systems.

References

[1] Nam S-W, Rooks MJ, Kim K-B, Rossnagel SM. Ionic field effect transistors with sub-10 nm multiple nanopores. Nano Lett 2009;9:2044–8.

[2] Yan R, Liang W, Fan R, Yang P. Nanofluidic diodes based on nanotube heterojunctions. Nano Lett 2009;9:3820–5.

[3] Prakash S, Karacor MB, editors. Surface mediated flows in glass nanofluidic channels. The 13th Solid-State Sensors, Actuators and Microsystems Workshop. Hilton Head Island, SC: Transducer Research Foundation; 2010.

[4] Cho SK, Moon H, Kim C-J. Creating, transporting, cutting, and merging liquid droplets by electrowetting-based actuation for digital microfluidic-based circuits. J Microelectromech Syst 2003;12:70–80.

[5] Wheeler AR, Moon H, Bird CA, Loo RRO, Kim C-J, Loo JA, et al. Digital microfluidics with in-line sample purification for proteomics analysis with MALDI-MS. Anal Chem 2005;77:534–40.

[6] Prakash S, Karacor MB, Banerjee S. Surface modification in microsystems and nanosystems. Surf Sci Rep 2009;64:233–54.

[7] Yanik AA, Huang M, Artar A, Chang T-Y, Altug H. Integrated nanoplasmonic-nanofluidic biosensors with targeted delivery of analytes. Appl Phys Lett 2010;96:021101.

[8] Karnik R, Fan R, Yue M, Li D, Yang P, Majumdar A. Electrostatic control of ions and molecules in nanofluidic transistors. Nano Lett 2005;5(5):943–8.

[9] Schasfoort RBM, Schlautmann S, Hendrikse J, van den Berg A. Field-effect flow control for microfabricated fluidic networks, Science 1999;286:942–54.

[10] Kuo T-C, Cannon Jr DM, Chen Y, Tulock JJ, Shannon MA, Sweedler JV, et al. Gateable nanofluidic interconnects for multilayered microfluidic separation systems. Anal Chem 2003;75:1861–7.

[11] Kuo T-C, Cannon Jr DM, Shannon MA, Bohn PW, Sweedler JV. Hybrid three-dimensional nanofluidic/microfluidic devices using molecular gates. Sens Actuators A 2003;102:223–33.

[12] Schasfoort RBM, Schlautmann S, Hendrikse J, van den Berg A. Field-effect flow control for microfabricated fluidic networks. Science 1999;286:942–5.

[13] Karnik R, Castelino K, Majumdar A. Field-effect control of protein transport in a nanofluidic transistor circuit. Appl Phys Lett 2006;88:123114.

[14] Daiguji H, Oka Y, Shirono K. Nanofluidic diode and bipolar transistor. Nano Lett 2005;5:2274–80.

[15] Vlassiouk I, Siwy ZS. Nanofluidic diode. Nano Lett 2007;7(3):552–6.

[16] He Y, Gillespie D, Boda D, Vlassiouk I, Eisenberg RS, Siwy ZS. Tuning transport properties of nanofluidic devices with local charge inversion. J Am Chem Soc 2009;131:5194 52-2.

[17] Pardon G, van der Wijngaart W, editors. Electrostatic gating of ion and molecule transport through a nanochannel-array membrane. Solid-State Sensors, Actuators and Microsystems Conference (Transducers). Beijing, China: IEEE; 2011.

[18] Oh KW, Rong R, Ahn CH. In-line micro ball valve through polymer tubing. Proc microTAS 2001 2001;407–8.

[19] Fu C, Rummler Z, Schomburg W. Magnetically driven micro ball valves fabricated by multilayer adhesive film bonding. J Micromech Microeng 2003;13:S96–102.

[20] Oh KW, Rong R, Ahn CH. Miniaturization of pinch-type valves and pumps for practical micro total analysis system integration. J Micromech Microeng 2005;15:2449–55.

[21] Oh KW, Ahn CH. A review of microvalves. J Micromechan Microeng 2006;16: R13–39.

[22] Goll C, Bacher W, Bustgens B, Maas D, Menz W, Schomburg W. Microvalves with bistable buckled polymer diaphragms. J Micromech Microeng 1996;6:77–9.

[23] Goll C, Bacher W, Bustgens B, Maas D, Ruprecht R, Schomburg WK. An electrostatically actuated polymer microvalve equipped with a movable membrane electrode. J Micromech Microeng 1997;7:224–6.

[24] Bae B, Han J, Masel RI, Shannon MA. A bidirectional electrostatic microvalve with microsecond switching performance. J Microelectromech Syst 2007c;16(6):1461–71.

[25] Han J, Flachsbart B, Masel RI, Shannon MA. Micro-fabricated membrane gas valves with a non-stiction coating deposited by C4F8/Ar plasma. J Micromech Microeng 2008;18(9):095015.

[26] Han J, Yeom J, Mensing G, Joe D, Masel RI, Shannon MA. Surface energy approach and AFM verification of the (CF)n treated surface effect and its correlation with adhesion reduction in microvalves. J Micromech Microeng 2009;19(8):085017.

[27] Kirby BJ, Shepodd TJ, Hasselbrink EF. Voltage-addressable on/off microvalves for high-pressure microchip separations. J Chromatogr 2002;979:147–54.

[28] Kohl M, Skrobanek KD, Miyazaki S. Development of stress-optimised shape memory microvalves. Sens Actuators A 1999;72:243–50.

[29] Kohl M, Dittmann D, Quandt E, Winzek B. Thin film shape memory microvalves with adjustable operation temperature. Sens Actuators A 2000;83:214–9.

[30] Kohl M, Dittmann D, Quandt E, Winzek B, Miyazaki S, Allen DM. Shape memory microvalves based on thin films or rolled sheets. Mater Sci Eng A 1999;273-5:784–8.

[31] Reynaerts D, Peirs J, Van Brussel H. An implantable drug-delivery system based on shape memory alloy micro-actuaries. Sens Actuators A 1997;61:455–62.

[32] Pemble CM, Towe BC. A miniature shape memory alloy pinch valve. Sens Actuators A 1999;77:145–8.

[33] Neagu CR, Gardeniers JGE, Elwenspoek M, Kelly JJ. An electrochemical active valve. Electrochim Acta 1997;42: 3367–13.

[34] Neagu CR, Gardeniers JGE, Elwenspoek M, Kelly JJ. An electrochemical microactuator: principle and first results. J Microelectromech Syst 1996;5:2–9.

[35] Yang Z, Maeda R. A world-to-chip socket for microfluidic prototype development. Electrophoresis 2002;23:3474–8.

[36] Yang Z, Maeda R. Socket with built-in valves for the interconnection of microfluidic chips to macro constituents. J Chromatogr A 2003;1013:29–33.

[37] Zhu L, Meng DD, Kroodsma N, Yeom J, Shannon MA, editors. An integrated microfluidic self-regulating and self-circulating hydrogen generator for fuel cells. Tech Dig The 15th Int Conf on Solid-State Sensors, Actuators and Microsystems (Transducers 09). Denver, CO: IEEE; Jun. 21–Jun. 25 2009.
[38] Yang EH, Han SW, Yang SS. Fabrication and testing of a pair of passive bivalvular microvalves composed of p + silicon diaphragms. Sensors Actuators A 1996;57:75–8.
[39] Wu Y, Misra S, Karacor MB, Prakash S, Shannon MA. Dynamic response of AFM cantilevers to dissimilar functionalized silica surfaces in aqueous electrolyte solutions. Langmuir 2010;26(22):16963–72.
[40] Pollack MG, Shenderov AD, Fair RB. Electrowetting-based actuation of droplets for integrated microfluidics. Lab Chip 2002;2:96–101.
[41] Cannon Jr. DM, Kuo T-C, Bohn PW, Sweedler JV. Nanocapillary array interconnects for gated analyte injections and electrophoretic separations in multilayer microfluidic architectures. Anal Chem 2003;75:2224–30.
[42] Chang I-H, Tulock JJ, Liu J, Cannon Jr DM, Bohn PW, Sweedler JV, et al. Miniaturized lead sensor based on lead-specific DNAzyme in a nanocapillary interconnected microfluidic device. Environ Sci Technol 2005;39(10):3756–61.
[43] Tulock JJ, Shannon MA, Bohn PW, Sweedler JV. Microfluidic separation and gateable fraction collection for mass-limited samples. Anal Chem 2004;76:6419–25.
[44] Prakash S, Piruska A, Gatimu EN, Bohn PW, Sweedler JV, Shannon MA. Nanofluidics: systems and applications. IEEE Sens J 2008;8(5):441–50.
[45] Prakash S, Pinti M, Bellman K. Variable Cross-section nanopores fabricated in silicon nitride membranes using a transmission electron microscope. J Micromech Microeng 2012;22:067002.
[46] Shannon MA, Flachsbart BR, Iannacone JM, Wong K, Cannon Jr DM, Fa K, et al., editors. Nanofluidic interconnects within a multilayer microfluidic chip for attomolar biochemical analysis and molecular manipulation. Proceedings of the 3rd Annual International EMBS Special Topic Conference on Microtechnologies in Medicine and Biology. Kahuhu, HI: IEEE; 2005.
[47] Prakash S, Pinti M, Bhushan B. Theory, fabrication and applications of microfluidic and nanofluidic biosensors. Philos T Roy Soc A 2012;370:2269–303.
[48] Wernette DP, Swearingen CB, Cropek DM, Lu Y, Sweedler JV, Bohn PW. Incorporation of a DNAzyme into Au-coated nanocapillary array membranes with an internal standard for Pb(II) sensing. Analyst 2006;131:41–7.
[49] Gong M, Flachsbart BR, Shannon MA, Bohn PW, Sweedler JV. Fluidic communication between multiple vertically segregated microfluidic channels connected by nanocapillary array membranes. Electrophoresis 2008;29:1237–44.
[50] Gong M, Kim BY, Flachsbart BR, Shannon MA, Bohn PW, Sweedler JV. An on-chip fluorogenic enzyme assay using a multilayer microchip interconnected with a nanocapillary array membrane. IEEE Sensors J 2008;8(5):601–7.
[51] Wang Y, Bhushan B, Maali A. Atomic force microscopy measurement of boundary slip on hydrophilic, hydrophobic, and superhydrophobic surfaces. J Vac Sci Technol 2009;27:754–60.
[52] Wang Z, King TL, Branagan SP, Bohn PW. Enzymatic activity of surface-immobilized horseradish peroxidase confined to micrometer- to nanometer-scale structures in nanocapillary array membranes. Analyst 2009;134:851–9.

[53] Boukany PE, Morss A, Liao W-C, Henslee B, Jung H, Zhang X, et al. Nanochannel electroporation delivers precise amounts of biomolecules into living cells. Nat Nanotechnol 2011;6:747–54.
[54] Kim BY, Yang J, Gong M, Flachsbart BR, Shannon MA, Bohn PW, et al. Multidimensional separation of chiral amino acid mixtures in a multilayered three-dimensional hybrid microfluidic/nanofluidic device. Anal Chem 2009;81(7):2715–22.
[55] Lokuge I, Wang X, Bohn PW. Temperature-controlled flow switching in nanocapillary array membranes mediated by poly(N-isopropylacrylamide) polymer brushes grafted by atom transfer radical polymerization. Langmuir 2007;23(1):305–11.
[56] Schibel AEP, Heider EC, Harris JM, White HS. Fluorescence microscopy of the pressure-dependent structure of lipid bilayers suspended across conical nanopores. J Am Chem Soc 2011;133:7810–5.
[57] Spencer MG, Flachsbart BR, Yasunaga T, Kuo T-C, Sweedler JV, Bohn PW, Shannon MA. Design and fabrication of multiple nanofluidic interconnects for three-dimensional microanalytical systems. In: Micro total analysis systems. Monterey, CA; 2001.
[58] Khandurina J, Jacobson SC, Waters LC, Foote RS, Ramsey JM. Microfabricated porous membrane structure for sample concentration and electrophoretic analysis. Anal Chem 1999;71:1815–9.
[59] Bayley H, Martin CR. Resistive-pulse sensing: from microbes to molecules. Chem Rev 2000;100:2575–94.
[60] Seo J, Lean MH, Kole A. Membrane-free microfiltration by asymmetric inertial migration. Appl Phys Lett 2007;91:033901–3.
[61] Yamada M, Nakashima M, Seki M. Pinched flow fractionation: continuous size separation of particles utilizing a laminar flow profile in a pinched microchannel. Anal Chem 2004;76:5465.
[62] Fa K, Tulock JJ, Sweedler JV, Bohn PW. Profiling pH gradients across nanocapillary array membranes connecting microfluidic channels. J Am Chem Soc 2005;127:13928–33.
[63] Kuo T-C, Kim H-K, Cannon Jr DM, Shannon MA, Sweedler JV, Bohn PW. Nanocapillary arrays effect mixing and reaction in multilayer fluidic structures. Angew Chem 2004;116:1898–901.
[64] Zhang Y, Timperman AT. Integration of nanocapillary arrays into microfluidic devices for use as analyte concentrators. Analyst 2003;128:537–42.
[65] Iannacone JM, Jakubowski JA, Bohn PW, Sweedler JV. A multilayer poly(dimethylsiloxane) electrospray ionization emitter for sample injection and online mass spectrometric detection. Electrophoresis 2005;26:4684–90.
[66] Kim BY, Swearingen CB, Ho JA, Romanova EV, Bohn PW, Sweedler JV. Direct immobilization of Fab' in nanocapillaries for manipulating mass-limited samples. J Am Chem Soc 2007;129:7620–6.
[67] Wernette DP, Mead C, Bohn PW, Lu Y. Surface immobilization of catalytic beacons based on ratiometric fluorescent DNAzyme sensors: a systematic study. Langmuir 2007;23(18):9513–21 Epub 2007 Aug 4.
[68] Wang YC, Stevens AL, Han J. Million-fold preconcentration of proteins and peptides by nanofluidic filter. Anal Chem 2005;77:4293–9.
[69] Piruska A, Baragan SP, Minnis AB, Wang Z, Cropeck DM, Sweedler JV, et al. Electrokinetic control of fluid transport in gold-coated nanocapillary array membranes in hybrid nanofluidic-microfluidic devices. Lab Chip 2010;10:1237–44.

[70] Shannon MA, Bohn PW, Elimelech M, Georgiadis JG, Marinas BJ. Science and technology for water purification in the coming decades. Nature 2008;452(20):301–10.
[71] Kim SJ, Ko SH, Kang KH, Han J. Direct seawater desalination by ion concentration polarization. Nat Nanotechnol 2010;5:297–301.
[72] Dhathathreyan A. Real-time monitoring of invertase activity immobilized in nanoporous aluminum oxide. J Phys Chem B 2011;115:6678–82.
[73] Bowski L, Saini R, Ryu DY, Vieth WR. Kinetic modeling of hydrolysis of sucrose by invertase. Biotechnol Bioeng 1971;13:641–56.
[74] Dekker C. Solid-state nanopores. Nature Nanotech 2007;2:209–15.
[75] Storm AJ, Storm C, Chen J, Zandbergen J, Joanny J-F, Dekker C. Fast DNA translocation through a solid-state nanopore. Nano Lett 2005;5:1193–7.
[76] Buffi N, Merulla D, Beutier J, Barbaud F, Beggah S, Lintel HV, et al. Development of a microfluidics biosensor for agarose-bead immobilized Escherichia coli bioreporter cells for arsenite detection in aqueous samples. Lab Chip 2011;11:2369.
[77] Lee JH, Han J. Concentration-enhanced rapid detection of human chorionic gonadotropin as a tumor marker using nanofluidic preconcentrator. Microfluid Nanofluid 2010;9:973–9.
[78] Zubair A, Burbelo PD, Vincent LG, Iadarola MJ, Smith PD, Morgan NY. Microfluidic LIPS serum antibody detection: demonstration of a rapid test for HSV-2 infection. Biomed Microdevices 2011.
[79] Gervais L, Hitzbleck M, Delamarche E. Capillary-driven multiparametric microfluidic chips for one-step immunoassays. Biosens Bioelectron 2011;27:64–70.
[80] Hu J, Sun X, Agarwal A, Kimerling LC. Design guidelines for optical resonator biochemical sensors. J Opt Soc Am B 2009;26:1032–41.
[81] Fukuyama M, Yamatogi S, Ding H, Nishida M, Kawamoto C, Amemiya Y, et al. Selective detection of antigen-antibody reaction using Si ring optical resonators. Jpn J Appl Phys 2010;49: 04DL9.
[82] White IM, Zhu H, Suter JD, Fan X, Zourob M. Label-free detection with the liquid core optical ring resonator sensing platform. Methods Mol Biol 2009;503:139–65.
[83] Fang S, Lee HJ, Wark AW, Corn RM. Attomole microarray detection of MicroRNAs by nanoparticle-amplified SPR imaging measurements of surface polyadenylation reactions. J Am Chem Soc 2006;128(43):14044–6.
[84] Choi S, Chae J. Surface plasmon resonance biosensor based on Vroman effect: towards cancer biomarker detection. IEEE 15th International Mixed-Signals, Sensors, and Systems Test Workshop; Scottsdale, AZ; 2009. p. 1–3.
[85] Metzger J, Pv Landenberg, Kehrel M, Buhl A, Lackner KJ, Luppa PB. Biosensor analysis of 2-glycoprotein I—reactive autoantibodies: evidence for isotype-specific binding and differentiation of pathogenic from infection-induced antibodies. Clin Chem 2007;53(6):1137–43.
[86] Chou T-H, Chuang C-Y, Wu C-M. Quantification of interleukin-6 in cell culture medium using surface plasmon resonance biosensors. Cytokine 2010;51:107–11.
[87] Piliarik M, Bockova M, Homola J. Surface plasmon resonance biosensor for parallelized detection of protein biomarkers in diluted blood plasma. Biosens Bioelectron 2010;26:1656–61.
[88] Malic L, Veres T, Tabrizian M. Nanostructured digital microfluidics for enhanced surface plasmon resonance imaging. Biosens Bioelectron 2011;26:2053–9.

[89] Springer T, Piliarik M, Homola J. Real-time monitoring of biomolecular interactions in blood plasma using a surface plasmon resonance biosensor. Anal Bioanal Chem 2010;398:1955–61.
[90] Pei R, Cheng Z, Wang E, Yang X. Amplification of antigen-antibody interactions based on biotin labeled protein-streptavidin network complex using impedance spectroscopy. Biosens Bioelectron 2001;16(6):355–61.
[91] Vitarelli MJ, Prakash S, Talaga DS. Determining nanocapillary geometry from electrochemical impedance spectroscopy using a variable topology network circuit model. Anal Chem 2011;83(2):533–41.
[92] Rodriguez MC, Kawde AN, Wang J. Aptamer biosensor for label-free impedance spectroscopy detection of proteins based on recognition-induced switching of surface charge. Chem Comm 2005;4267–9.
[93] Kwakye S, Goral VN, Baeumner AJ. Electrochemical microfluidic biosensor for nucleic acid detection with integrated minipotentiostat. Biosens Bioelectron 2006;21:2217–23.
[94] Martinez NA, Schneider RJ, Messina GA, Raba J. Modified paramagnetic beads in a microfluidic system for the determination of ethinylestradiol (EE2) in river water samples. Biosens Bioelectron 2010;25:1376–81.
[95] Sotiropoulou S, Caniotakis NA. Lowering the detection limit of the acetylcholinesterase biosensor using a nanoporous carbon matrix. Anal Chim Acta 2005;530:199–204.
[96] Chen L, McBranch DW, Wang H-L, Helgeson R, Wudl F, Whitten DG. Highly sensitive biological and chemical sensors based on reversible fluorescence quenching in a conjugated polymer. PNAS 1999;96(22):12287–92.
[97] Zhang G-J, Chua JH, Chee R-E, Agarwal A, Wong SM. Label-free direct detection of MiRNAs with silicon nanowire biosensors. Biosens Bioelectron 2009; 24. Epub 2504-2508.
[98] Javanmard M, Davis RW. A microfluidic platform for electrical detection of DNA hybridization. Sens Actuators B Chem 2011;154:22–7.
[99] Maeng JH, Lee BC, Ko YJ, Cho W, Ahn Y, Cho NG, et al. A novel microfluidic biosensor based on an electrical detection system for alpha-fetoprotein. Biosens Bioelectron 2008;23:1319–25.
[100] Moreno-Hagelsieb L, Lobert PE, Pampin R, Bourgeois D, Remacle J, Flandre D. Sensitive DNA electrical detection based interdigitated Al/Al_2O_3 microelectrodes. Sens Actuator B 2004;98:269–74.
![101] Fan Y, Chen X, Trigg AD, Tung C-h, Kong J, Gao Z. Detection of MicroRNAs using target-guided formation of conducting polymer nanowires in nanogaps. J Am Chem Soc 2007;129:5437–43.
[101] Iqbal SM, Balasundaram G, Ghosh S, Bergstrom DE, Bashir R. Direct current electrical characterization of ds-DNA in nanogap junctions. Appl Phys Lett 2005;86:153901.
![103] Macanovic A, Marquette C, Polychronakos C, Lawrence MF. Impedance-based detection of DNA sequences using a silicon transducer with PNA as the probe layer. Nucleic Acids Res 2004;32(e20):.
[104] Durand NFY, Renaud P. Label-free determination of protein–surface interaction kinetics by ionic conductance inside a nanochannel. Lab Chip 2009;9:319–24.
[105] Chen R, Bangsaruntip S, Drouvalakis KA, Kam NWS, Shim M, Li Y, et al. Noncovalent functionalization of carbon nanotubes for highly specific electronic biosensors. PNAS 2003;100(9):4984–9.

[106] Cui Y, Wei Q, Park H, Lieber CM. Nanowire nanosensors for highly sensitive and selective detection of biological and chemical species. Science 2001;293:1289.
[107] Hahm J, Lieber CM. Direct ultrasensitive electrical detection of DNA and DNA sequence variations using nanowire nanosensors. Nano Lett 2004;4(1):51–4.
[108] Raiteri R, Grattarola M, Butt M, Skladal P. Micromechanical cantilever-based biosensor. Sens Actuators B Chem 2001;79:115–26.
[109] Waggoner PS, Tan CP, Craighead HG. Microfluidic integration of nanomechanical resonators for protein analysis in serum. Sens Actuators B Chem 2010;150: 550–5.
[110] Hwang KS, Lee S-M, Kim SK, Lee JH, Kim TS. Micro- and nanocantilever devices and systems for biomolecule detection. Annu Rev Anal Chem 2009;2:77–98.
[111] Lee S, Bhushan B. Biological molecules in therapeutic nanodevices. In: Bhushan B, editor. Spinger handbook of nanotechnology. 3rd ed. Berlin: Springer-Verlag; 2010. p. 453–84.
[112] Olthius W, Bergveld P, Kruise J. The exploitation of ISFETs to determine acid-base behaviour of proteins. Electrochim Acta 1997;43:3483–8.
[113] Schoning MJ, Poghossian A. Recent advances in biologically sensitive field-effect transistors (BioFETS). Analyst 2002;127:1137–51.
[114] Bhushan B. MEMS/NEMS and BioMEMS/BioNEMS: materials, devices, and biomimetics. In: Bhushan B, editor. Springer handbook of nanotechnology. 3rd ed. Berlin: Springer-Verlag; 2010. p. 1663–740.
[115] Bhushan B, Tokachichu D, Keener MT, Lee SC. Nanoscale adhesion, friction, and wear studies of biomolecules on silicon-based surfaces. Acta Biomater 2006;2:39–49.
[116] Bhushan B, Tokachichu DR, Keener MT, Lee SC. Morphology and adhesion of biomolecules on silicon based surfaces. Acta Biomater 2005;1:327–41.
[117] Wang J. Towards genoelectronics: electrochemical biosensing of DNA hybridization. Chem Eur J 1999;5:1681–5.
[118] Mannoor MS, Zhang S, Link AJ, McAlpine MC. Electrical detection of pathogenic bacteria via immobilized antimicrobial peptides. PNAS 2010;107(45):19207–12.
[119] Tsai H-KA, Moschou EA, Kulinsky L, Daunert S, Madou MJ. Integrating biosensors and drug delivery: a step closer toward scalable responsive drug delivery systems. Adv Mater 2009;21:656–60.
[120] Watts AS, Urbas AA, Moschou EA, Gavalas VG, Zoval JV, Madou MJ, et al. Centrifugal microfluidics with integrated sensing microdome optodes for multiion detection. Anal Chem 2007;79:8046–54.
[121] Nelson WC, Peng I, Lee G-A, Loo JA, Garrell RL, Kim C-J. Incubated protein reduction and digestion on an electrowetting-on-dielectric digital microfluidic chip for MALDI-MS. Anal Chem 2010;82(23):9932–7.
[122] Shah GJ, Veale JL, Korin Y, Reed EF, Gritsch HA, Kim C-J. Specific binding and magnetic concentration of CD8 + T-lymphocytes on electrowetting-on-dielectric platform. Biomicrofluidics 2010;4(4):12.
[123] Lang HP, Hegner M, Gerber C. Cantilever array sensors. Mater Today 2005;4:30–6.
[124] Singh M, Verma N, Garg AK, Redhu N. Urea biosensors. Sens Actuator B 2008;134:345–51.
[125] Bunimovich YL, Shin YS, Yeo WS, Amori M, Kwong G, Heath JR. Quantitative real-time measurements of DNA hybridization with alkylated nonoxidized silicon nanowires in electrolyte solution. J Am Chem Soc 2006;128:16323–31.

[126] Chen X, Matsumoto N, Hu Y, Wilson GS. Electrochemically mediated electrodeposition/ electropolymerization to yield a glucose microbiosensor with improved characteristics. Anal Chem 2002;74:368–72.
[127] Siegrist J, Gorkin R, Bastien M, Stewart G, Peytavi R, Kido H, et al. Validation of a centrifugal microfluidic sample lysis and homogenization platform for nucleic acid extraction with clinical samples. Lab Chip 2010;10:363–71.
[128] Hong CC, Chang PH, Lin CC, Hong CL. A disposable microfluidic biochip with on-chip molecularly imprinted biosensors for optical detection of anesthetic propofol. Biosens Bioelectron 2010;25:2058–64.
[129] Easley CJ, Karlinsey JM, Bienvenue JM, Legendre LA, Roper MG, Feldman SH, et al. A fully integrated microfluidic genetic analysis system with sample-in–answer-out capability. PNAS 2006;103(51):19272–7.
[130] Chikkaveeraiah BV, Liu H, Mani V, Papadimitrakopoulos F. A microfluidic electrochemical device for high sensitivity biosensing: detection of nanomolar hydrogen peroxide. Electrochem Commun 2009;11:819–22.
[131] Chou IH, Benford M, Beier HT, Cote GL. Nanofluidic biosensing for beta-amyloid detection using surface enhanced Raman spectroscopy. Nano Lett 2008;8 (6):1729–35.
[132] Messina GA, Panini NV, Martinez NA, Raba J. Microfluidic immunosensor design for the quantification of interleukin-6 in human serum samples. Anal Biochem 2008;380:262–7.
[133] Goral VN, Zaytseva NV, Baeumner AJ. Electrochemical microfluidic biosensor for the detection of nucleic acid sequences. Lab Chip 2006;6:414–21.
[134] Zhang X-B, Kong R-M, Lu Y. Metal ion sensors based on DNAzymes and related DNA molecules. Ann Rev Anal Chem 2011;4:104–28.
[135] Wang Z, Lee JH, Lu Y. Fluorescent sensor for Hg2 + in aqueous solution based on structure-switching DNA. Chem Comm 2008;2008:6005–7.
[136] Lan T, Furuya K, Lu Y. A highly selective lead sensor based on a classic lead DNAzyme. Chem Comm 2010;46:3896–8.
[137] Shingara J, Keiger K, Shelton J, Laosinchai-Wolf W, Powers P, Conrad R, et al. An optimized isolation and labeling platform for accurate microRNA expression profiling. RNA 2005;11:1461–70.
[138] Zhang Y, Hu Y, Wilson GS, Moatti-Sirat D, Poitout V, Reach G. Elimination of the acetaminophen interference in an implantable glucose sensor. Anal Chem 1994;66:1183–8.
[139] Matsumoto T, Furusawa M, Fujiwara H, Matsumoto Y, Ito N. A micro-planar amperometric glucose sensor unsusceptible to interference species sens actuators. B Chem 1998;49:68–72.
[140] Yang YL, Tseng TF, Lou SL, editors. Using MPTMS as Permselective Membrane of Biosensors. 29th Annual IEEE EMBS. Lyon, France; August 23–26, 2007.
[141] Soldatkin AP, El'skaya AV, Shul'ga AA, Jdanova AS, Dzyadevich SV, Jaffrezic-Renault N, et al. Glucose sensitive conductometric biosensor with additional nafion membrane:reduction of influence of buffer capacity on the sensor response and extension of dynamic range. Anal Chim Acta 1994;288:197–203.
[142] Vroman L, Adams AL. Findings with the recording ellipsometer suggesting rapid exchange of specific plasma proteins at liquid solid interfaces. Surf Sci 1969;16:438–46.

[143] Zaytseva NV, Montagna RA, Baeumner AJ. Microfluidic biosensor for the serotype-specific detection of dengue virus. Anal Chem 2005;77:7520–7.

[144] Jain KK. The role of nanobiotechnology in drug discovery. Drug Discovery Today 2005;10:1435–42.

[145] Jain KK. Nanotechnology in clinical laboratory diagnostics. Clinica Chim Acta 2005;358:37–54.

[146] Leary SP, Liu CY, Apuzzo MLJ. Toward the emergence of nanoneurosurgery: Part II-Nanomedicine: Diagnostics and imaging at the nanoscale level. Neurosurgery 2006;58:805–23.

[147] Leary SP, Liu CY, Yu C, Apuzzo MLJ. Toward the emergence of nanoneurosurgery: Part I-Progress in nanoscience, nanotechnology, and the comprehension of events in the mesoscale realm. Neurosurgery 2005;57:606–34.

[148] Langer R, Peppas NA. Advances in biomaterials, drug delivery, and bionanotechnology. AIChE J 2004;49(12):2990–3006.

[149] Freitas RA. What is nanomedicine? Nanomed NBM 2005;1:2–9.

[150] Tegenfeldt JO, Prinz C, Cao H, Huang RL, Austin RH, Chou SY, et al. Micro- and nanofluidics for DNA analysis. Anal Bioanal Chem 2004;318(7):1678–92.

[151] Lin H, Musik MD, Nicewarner SR, Salinas FG, Benkovic SJ, Natan MJ, et al. Colloidal au-enhanced surface plasmon resonance for ultrasensitive detection of DNA hybridization. J Amer Chem Soc 2000;122:9071–7.

[152] Mirkin CA, Letsinger RL, Mucic RC, Storhoff JJ. A DNAbased method for rationally assembling nanoparticles into macroscopic materials. Nature 1996;382:.

[153] Nam JM, Stoeva SI, Mirkin CA. Bio-bar-code-based DNA detection with PCR-like sensitivity. J Amer Chem Soc 2004;126:5932–3.

[154] Park SJ, Taton TA, Mirkin CA. Array-based electrical detection of DNA with nanoparticle probes. Science 2002;295:1503–6.

[155] Arntz Y, Seelig JD, Lang HP, Zhang J, Hunziker P, Ramseyer JP, et al. Label-free protein assay based on a nanomechanical cantilever array. Nanotechnology 2003;14:86–90.

[156] Fritz J, Baller MK, Lang HP, Rothuizen H, Vettiger P, Meyer E, et al. Translating biomolecular recognition into nanomechanics. Science 2000;288:316–8.

[157] Hagan MF, Majumdar A, Chakraborty AK. Nanomechanical forces generated by surface grafted DNA. J Phys Chem B 2002;106:10163–73.

[158] Georganopoulou DG, Chang L, Nam JM, Thaxton CS, Mufson EJ, Klein WL, et al. Nanoparticle-based detection in cerebral spinal fluid of a soluble pathogenic biomarker for alzheimer's disease. Pro Nat Acad Sci USA 2005;102:2273–6.

[159] Hirsch LR, Jackson JB, Lee A, Halas NJ, West JL. A whole blood immunoassay using gold nanoshells. Anal Chem 2003;75:2377–81.

[160] Nam JM, Thaxton CS, Mirkin CA. Nanoparticle-based bio-bar codes for ultrasensitive detection of proteins. Science 2003;301:1884–6.

[161] Rosi NR, Mirkin CA. Nanostructures in biodiagnostics. Chem Rev 2005;105:1547–62.

[162] Man J, Fu J. Biomolecular separation by steric hindrance using nanofluidic filters. Proc Ann Int Conf IEEE Eng Med Biol 2004;:2611–4.

[163] Siwy Z, Fulinski A. Fabrication of a synthetic nanopore ionpump. Phys Rev Lett 2002;89:19803-1–19803-4.

[164] Siwy Z, Heins E, Harrell CC, Kohli P, Martin CR. Conical nanotube ion-current rectifiers: the role of surface charge. J Am Chem Soc 2004;126:10850–1.

[165] Chou C-F, Gu J, Wei Q, Liu Y, Gupta R, Nishio T, et al. Nanopatterned structures for biomolecular analysis toward genomic and protemic applications. In: Proc. SPIE; 2005. p. 183–92.

[166] Cao H, Tegenfeldt JO, Austin RH, Chou SY. Gradient nanostructures for interfacing microfluidics and nanofluidics. App Phys Lett 2002;81:3058–61.

[167] Ho C, Qiao R, Heng JB, Chatterjee A, Timp RJ, Aluru NR, et al. Electrolytic transport through a synthetic nanometerdiameter pore. Proc Nat Acad Sci 2005;102:10445–50.

[168] Gardeniers H, Berg, AVD. Integrated micro-and nanofluidics: Silicon revisited. In: Proc. Amer. Soc. Mech. Eng., Fluid Engineering Division; 2002. p. 1199–204.

[169] Hughes GA. Nanostructure-mediated drug delivery. Nanomed Nanotech Biol Med 2005;1:22–30.

[170] Aulenta F, Hayes W, Rannard S. Dendrimers: a new class of nanoscopic containers and drug delivery. Eur Polym J 2003;39:1741–71.

[171] Prakash S, Long TM, Selby JC, Moore JS, Shannon MA. 'Click' modification of silica surfaces and glass microfluidic channels,". Anal Chem 2007;79:1661–7.

[172] Tung S, Kim J-W, Malshe AP, Lee CC, Pooran R. A cellular motor driven microfluidic system. In: Proc. 12th Int. Conf. Solid-State Sensors, Actuators, and Microsyst. 2003. p. 678–81.

[173] Janata J, Josowicz M, DeVaney DM. Chemical sensors. Anal Chem 1994;66:207R–228RR.

[174] Terry SC, Jerman JH, Angell JB. A gas chromatographic air analyzer fabricated on a silicon wafer. IEEE Trans Electron Devices 1979;26:1880–6.

[175] Overton EB, Carney KR. New horizons in gas chromatography: field applications of microminiaturized gas chromatographic techniques. Trends Anal Chem 1994;13:258–62.

[176] Frye-Mason G, Kottenstette RJ, Lewis PR, Heller EJ, Manginell RP, Adkins DR, et al., editors. Hand-held miniature chemical analysis system for detection of trace concentrations of gas phase analytes. Proceedings of Micro Total Analysis Systems Workshop. Enschede, Netherlands; 2000.

[177] Noh H-S, Hesketh PJ, Frye-Mason GC. Parylene gas chromatographic column for rapid thermal cycling. J Microelectromechan Syst 2002;11(6):718–25.

[178] Lambertus G, Elstro A, Sensening K, Potkay J, Agah M, Scheuering S, et al. Design, fabrication, and evaluation of microfabricated columns for gas chromatography. Anal Chem 2004;76:2629–37.

[179] Radadia AD, Masel RI, Shannon MA, Jerrell JP, Cadwallader KR, Micromachined GC. Columns for fast separation of organophosphonate and organosulfur compounds. Anal Chem 2008;80(11):4087–94.

[180] Lu C-J, Steinecker WH, Tian W-C, Oborny MC, Nichols JM, Agah M, et al. First-generation hybrid MEMS gas chromatograph. Lab Chip 2005;5:1123–31.

[181] Lewis PR, Manginell P, Adkins DR, Kottenstette RJ, Wheeler DR, Sokolowski SS, et al. Recent advancements in the gas-phase MicroChemLab. IEEE Sensors 2006;6 (3):784–95.

[182] Bae B, Kim J, Yeom J, Chen Q, Ray C, Shannon M, editors. Development of a portable gas analyzer using a micro-gas chromatograph/flame ionization detector (micro-GC/FID) for NASA's environmental missions. 42nd International

Conference on Environmental Systems (ICES); 2012 15–19 July 2012. San Diego, CA: American Institute of Aeronautics and Astronautics.

[183] Agilent Technologies, 490 Micro GC; 2012.

[183] Elster microGC. Available from: <http://www.elster-instromet.com/en/micro_GC_technology.html>; 2012.

[185] C2V Micro GC. Available from: <http://www.c2v.nl[182]/index.shtml>; 2012.

[186] Defiant Technologies. 2012.

[187] Manginell RP, Frye-Mason GC, Kottenstette RJ, Lewis PR, Wong CC, editors. Microfabricated planar preconcentrator. Solid-State Sensor and Actuator Workshop. Hilton Head Island, SC, USA; 2000.

[188] Tian W-C, Pang SW, Lu C-J, Zellers ET. Microfabricated preconcentrator-focuser for a microscale gas chromatograph. J Microelectromechan Syst 2003;12:264–72.

[189] Tian W-C, Pang SW. Freestanding microheaters in Si with high aspect ratio microstructures. J Vacuum Sci Technol B 2002;20:1008–12.

[190] Tian W-C, Pang SW. Thick and thermally isolated Si microheaters for microfabricated preconcentrators. J Vacuum Sci Technol B 2003;12:274–9.

[191] Bae B, Yeom J, Masel RI, Shannon MA, editors. A five-microvalve fully integrated preconcentrator. IEEE Sensors Conference; Atlanta, GA; 2007b.

[191] Bae B, Yeom J, Radadia AD, Masel RI, Shannon MA, editors. A fully-integrated MEMS preconcentrator for rapid gas sampling. The 14th International Conference on Solid State Sensors, Actuators and Microsystems: Transducers '07; 2007a; Lyon, France: IEEE.

[193] Yeom J, Field CR, Bae B, Masel RI, Shannon MA. The design, fabrication and characterization of a silicon microheater for an integrated MEMS gas preconcentrator. J Micromechan Microeng 2008;18:125001.

[194] Veriotti T, Sacks R. A tandem column ensemble with an atmospheric pressure juction-point vent for high-speed GC with selective control of peak-pair separation; Anal Chem 2001; 73(4):813–819.

[195] Tian W-C, Chan HKL, Lu C-J, Pang SW, Zellers ET. Multi-stage microfabricated preconcentrator-focuser for microscale gas chromatography system. J Microelectromechan Syst 2005;14:498–507.

[196] Radadia AD, Morgan RD, Masel RI, Shannon MA. Partially buried microcolumns for micro-gas analyzers. Anal Chem 2009;81(9):3471–7.

[197] Radadia AD, Salehi-Khojin A, Masel RI, Shannon MA. The effect of microcolumn geometry on the performance of micro-gas chromatography columns for chip scale gas analyzers. Sens Actuators B Chem 2010;150:456–64.

[198] McNair HM, Miller JM. Basic gas chromatography. New York, NY: John Wiley & Sons; 1997.

[199] Zimmermann S, Krippner P, Vogel A, Muller J. Miniaturized flame ionization detector for gas chromatography. Sens Actuators B Chem 2002;83:285–9.

[200] Zimmermann S, Wischhusen S, Muller J. Micro flame ionization detector and micro flame spectrometer. Sens Actuators B Chem 2000;63:159–66.

[201] Kuipers WJ, Müller J. A planar micro-flame ionization detector with an integrated guard electrode. J. Micromech. Microeng. 2008;18:064015.

[202] Kim CH, El-Leathy AM, Xu F, Faeth GM. Soot surface growth and oxidation in laminar diffusion flames at pressures of 0.1e1.0 atm. Combust Flame 2004;136:191–207.

[203] Wu YE, Chen K, Chen CW, Hsu KH. Fabrication and characterization of thermal conductivity detectors (TCDs) of different flow channel and heater designs. Sens Actuators A Phys 2002;100:37–45.

[204] Cruz D, Chang JP, Showalter SK, Gelbard F, Manginell RP, Blain MG. Microfabricated thermal conductivity detector for the micro-ChemLab™. Sens Actuators B Chem 2007;121:414–22.

[205] Kaanta BC, Jonca AJ, Chen H, Zhang X. Temperature distribution on thermal conductivity detectors for flow rate insensitivity. Sens Actuators Phys 2011;167:146–51.

[206] Prakash S, Yeom J, Jin N, Adesida I, Shannon MA. Characterization of ionic transport at the nanoscale. Proceedings of the institution of mechanical engineers, Part N. J Nanoeng Nanosyst 2007;220(2):45–52.

[207] Swaminathan VV, Gibson LR, Pinti M, Prakash S, Bohn PW, Shannon MA. Ionic transport in nanocapillary array membranes. J Nanopart Res 2012;14:951–65.

Select Bibliography

Abbasi F, Mirzadeh H, Katbab AA. Modification of polysiloxane polymers for biomedical applications: a review. Polym Int 2001;50(12):1279–87.

Abgrall P, Lowb L-N, Nguyen N-T. Fabrication of planar nanofluidic channels in a thermoplastic by hot-embossing and thermal bonding. Lab Chip 2007;7:520–2.

Abgrall P, Nguyen NT. Nanofluidic devices and their applications. Anal Chem 2008;80 (7):2326–41.

Abgrall P, Nguyen NT. Nanofluidics. Artech House: Norwood, MA; 2009.

Adiga SP, Jin CM, Curtiss LA, Monteiro-Riviere NA, Narayan RJ. Nanoporous membranes for medical and biological applications. WIREs Nanomed Nanobiotechnol 2009;1:568–81.

Agerbaek ML, Keiding K. Streaming potential during cake filtration of slightly compressible particles. J Colloid Interface Sci 1995;169:255–342.

Ali M, Tahir MN, Siwy ZS, Neumann R, Tremel W, Ensinger W. Hydrogen peroxide sensing with horseradish peroxidase-modified polymer single conical nanochannels. Anal Chem 2011;83:1673–80.

Alvarez M, Calle A, Tamayo J, Lechuga L, Abad A, Montoya A. Development of nanomechanical biosensors for detection of the pesticide DDT. Biosens Bioelectron 2003;18:649–53.

Ashkin A, Dziedzic JM. Internal cell manipulation using infrared laser traps. Proc Nat Acad Sci USA 1989;86:7914–8.

Auroux PA, Iossifidis D, Reyes DR, Manz A. Micro total analysis systems. 2. Analytical standard operations and applications. Anal Chem 2002;74(12):2637–52.

Ayhan H, Cicek H, Tuncel SA. Investigation of surface properties of biodegradable albumin microspheres via phagocytosis phenomena. J Bioact Compat Pol 2003;18 (4):273–82.

Bang ZL, MacDonald NC, An RIE. Process for submicron, silicon electromechanical structures. international conference on solid-state sensors and actuators; San Francisco: CA 1991. p. 520–3.

Bard AJ, Faulkner LR. Electrochemical methods: fundamentals and applications. 2nd ed. John Wiley & Sons; 2004.

Bhushan B. Biomimetics: lessons from nature. Philos T Roy Soc A 2009;367:1445–86.

Bhushan B. Nanotribology, nanomechanics, and materials characterization. In: Bhushan B, editor. Springer handbook of nanotechnology. 3rd ed. Berlin: Springer-Verlag; 2010. p. 789–856.

Bhushan B. Springer handbook of nanotechnology. 3rd ed. Heidelberg: Springer-Verlag; 2010.

Bhushan B, Kasai T, Kulik G, Barbieri L, Hoffman P. AFM study of perfluorosilane and alkylsilane self-assembled monolayers for anti-stiction in MEMS/NEMS. Ultramicroscopy 2005;105:176–88.

Bhushan B, Pan Y. Role of electric field on surface wetting of polystyrene surface. Langmuir 2011;27:9425–9.

Bliss CL, McMullin JN, Backhouse CJ. Rapid fabrication of a microfluidic device with integrated optical waveguides for DNA fragment analysis. Lab Chip 2007;7(10): 1280–7.

Bogunia-Kubik K, Sugisaka M. From molecular biology to nanotechnology and nanomedicine. BioSystems 2002;65(2-3):123–38.

Bohn PW. Nanoscale control and manipulation of molecular transport in chemical analysis. Ann Rev Anal Chem 2009;2:279–96.

Burgmayer P, Murray RW. An ion gate membrane: electrochemical control of ion permeability through a membrane with an embedded electrode. J Am Chem Soc 1982;104:6139–40.

Burgreen D, Nakache FR. Electrokinetic flow in ultrafine capillary slits. J Phys Chem 1964;68(5):1084–91.

Buyukserin F, Kohli P, Wirtz MO, Martin CR. Electroactive nanotube membranes and redox-gating. Small 2007;3:266–70.

Cao H, Tegenfeldt JO, Chou SY. Gradient nanostructures for interfacing microfluidics and nanofluidics. Appl Phys Lett 2002;81(16):3058–60.

Cao H, Yu Z, Wang J, Tegenfeldt JO, Austin RH, Chen E, et al. Fabrication of 10 nm enclosed nanofluidic channels. Appl Phys Lett 2002;81(1):174–6.

Castelino K, Satyanarayana S, Sitti M. Manufacturing of two and three-dimensional micro/ nanostructures by integrating optical tweezers with chemical assembly. Robotica 2005;23(4):435–9.

Cervera J, Patricio R, Manzanares JA, Mafe S. Incorporating ionic size in the transport equations for charged nanopores. Microfluid Nanofluid 2010;9:41–53.

Chatterjee AN, Cannon Jr DM, Gatimu EN, Sweedler JV, Aluru NR, Bohn PW. Modeling and simulation of ionic currents in three-dimensional microfluidic devices with nanofluidic interconnects. J Nanopart Res 2005;7:507–16.

Chen IJ, White IM. High-sensitivity electrochemical enzyme linked assay on a microfluidic interdigitated microelectrode. Biosens Bioelectron 2011;26:4375–81.

Choi S, Chae J. A regenerative biosensing surface in microfluidics using electrochemical desorption of short-chain self-assembled monolayer. Microfluid Nanofluid 2009;7: 819–27.

Chowdiah P, Wasan DT, Gidaspow D. On the interpretation of streaming potential data in nonaqueous media. Colloids Surf 1983;7:291–9.

Chun K-Y, Mafe S, Ramirez P, Stroeve P. Protein transport through gold-coated, charged nanopores: effects of applied voltage. Chem Phys Lett 2006;418:561–4.

Chun K-Y, Stroeve P. Protein transport in nanoporous membranes modified with self-assembled monolayers of functionalized thiols. Langmuir 2002;18:4653–8.

Ciardelli G, Rechichi A, Sartori S, D'Acunto M, Caporale A, Peggion E, et al. Bioactive polyurethanes in clinical applications. Polym Adv Technol 2006;17(9-10):786–9.

Cirino NM, Musser KA, Egan C. Multiplex diagnostic platforms for detection of biothreat agents. Expert Rev Mol Diagn 2004;4:841–57.

Conlisk AT. The debye-huckel approximation: its use in describing electroosmotic flow in micro- and nanochannels. Eletrophoresis 2005;26(10):1896–912.

Conlisk AT. Essentials of micro- and nanofluidics: with applications to the biological and chemical sciences. Cambridge, UK: Cambridge University Press; 2012.

Conlisk AT, Kumar A, Rampersaud A. Ionic and biomolecular transport in nanochannels. Nanoscale Microscale Thermophys Eng 2007;11(1–2):177–99.

Conlisk AT, McFerran J, Zheng Z, Hansford D. Mass transfer and flow in electrically charged micro- and nano-channels. Anal Chem 2002;74(9):2139–50.

Contento NM, Branagan SP, Bohn PW. Electrolysis in nanochannels for in situ reagent generation in confined geometries. Lab Chip 2011;11:3634–41.

Danielsson B. Calorimetric biosensors. J Biotechnol 1990;15:187–200.

Datta S, Conlisk AT, Kanani DM, Zydney AL, Fissell WH, Roy S. Characterizing the surface charge of synthetic nanomembranes by the streaming potential method. J Colloid Interface Sci 2010;348(1):85–95.

De Campos AM, Sanchez A, Gref R, Calvo P, Alonso MJ. The effect of a PEG versus a chitosan coating on the interaction of drug colloidal carriers with the ocular mucosa. Eur J Pharm Sci 2003;20(1):73–81.

Delft KMvon, Eijkel JCT, Mijatovic D, Druzhinina TS, Rathgen H, Tas NR, et al. Micromachined Fabry-Perot interferometer with embedded nanochannels for nanoscale fluid dynamics. Nano Lett 2007;7(2):345–50.

Deng L, Wang Y, Shang L, Wen D, Wang F, Dong S. A sensitive NADH amd glucose biosensor tuned by visible light based on thionine bridged carbon nanotubes and gold nanoparticles multilayer. Biosens Bioelectron 2008;24:957–63.

Desai S, Bodas D, Patil KR, Patole M, Singh RP. Tailor-made functional surfaces: potential elastomeric biomaterials I. J Biomat Sci Polym E 2003;14(12):1323–38.

Donnan FG. Theory of membrane equilibria and membrane potentials in the presence of non-dialysing electrolytes: a contribution to physical-chemical physiology. J Membr Sci 1995;100:45–55.

Eijkel JCT, Bomer J, Tas NR, van den Berg A. 1-D nanochannels fabricated in polyimide. Lab Chip 2004;4:161–3.

Erikson D, Li D. Three-dimensional structure of electroosmotic flow over hetereogeneous surfaces. J Phys Chem 2003;107:12212–20.

Erikson D, Li D. Analysis of alternating electroosmotic flows in a rectangular microchannel. Langmuir 2003;19(2003):5421–30.

Erikson D, Mandal S, Yang AHJ, Cordovez B. Nanobiosensors: optofluidic, electrical and mechanical approaches to biomolecular detection at the nanoscale. Microfluid Nanofluid 2008;4(1-2):33–52.

Falconnet D, Csucs G, Grandin HM, Textor M. Surface engineering approaches to micropattern surfaces for cell-based assays. Biomaterials 2006;(16):3044–63.

Fällman E, Schedin S, Jass J, Andersson M, Uhlin BE, Axner O. Optical tweezers based force measurement system for quantitating binding interactions: system design and application for the study of bacterial adhesion. Biosens Bioelectron 2004;19(11):1429–37.

Faucheux N, Schweiss R, Lutzow K, Werner C, Groth T. Self-assembled monolayers with different terminating groups as model substrates for cell adhesion studies. Biomaterials 2004;25:2721–30.

Fievet P, Szymczyk A, Aoubiza B, Pagetti J. Evaluation of three methods for the characterisation of the membrane-solution interface: streaming potential, membrane potential and electrolyte conductivity inside pores. J Membr Sci 2000;168:87–100.

Flachsbart BR, Wong K, Iannacone JM, Abante EN, Vlach RI, Rauchfuss PA, et al. Design and fabrication of a multilayered polymer microfluidic chip with nanofluidic interconnects via adhesive contact printing. Lab Chip 2006;6:667–74.

Franks W, Tosatti S, Heer F, Seif P, Textor M, Hierlemann A. Patterned cell adhesion by self-assembled structures for use with a CMOS cell-based biosensor. Biosens Bioelectron 2007;22(7):1426–33. PubMed PMID: ISI:000244385400033.

Fredrickson CK, Fan ZH. Macro-to-micro interfaces for microfluidic devices. Lab Chip 2004;4:526–33.

Freeman GR, March NH. Triboelectricity and some associated phenomena. Mat Sci Eng 1999;15:1454–8.

Pardon G, van der Wijngaart W. Electrostatic gating of ion and molecule transport through a nanochannel-array membrane. Proc IEEE Solid-State Sens Actuators Microsyst Conf (Transducers). Beijing; 2011. p. 1610–3.

Golden JP, Floyd-Smith TM, Mott DR, Ligler FS. Target delivery in a microfluidic immunosensor. Biosens Bioelectron 2007;22:2763–7.

Gray JJ. The interaction of proteins with solid surfaces. Curr Opin Struc Biol 2004;14:110–5.

Grayson ACR, Shawgo RS, Johnson AM, Flynn NT, Li YW, Cima MJ, et al. A BioMEMS review: MEMS technology for physiologically integrated devices. P IEEE 2004; 92(1):6–21.

Guo LJ, Cheng X, Chou CF. Fabrication of size-controllable nanofluidic channels by nanoimprinting and its application for DNA stretching. Nano Lett 2004;4(1):69–73.

Han J, Fu J, Schoch RB. Molecular sieving using nanofilters: past, present and future. Lab Chip 2008;8(1):23–33.

Han J, Turner SW, Craighead HG. Entropic trapping and escape of long DNA molecules at submicron size constriction. Phys Rev Lett 1999;83(8):1688–91.

Hatakeyama T, Murakami K, Miyamoto Y, Yamasaki N. An assay for lectin activity using microtiter plate with chemically immobilized carbohydrates. Anal Biochem 1996;237 (2):188–92.

Henniker J. Triboelectricity in polymers. Nature 1962;196:474.

Herr AE, Molho JI, Drouvalakis KA, Mikkelsen JC, Utz PJ, Santiago JG, et al. On-chip coupling of isoelectric focusing and free solution electrophoresis for multidimensional separations. Anal Chem 2003;75:1180–7.

Herrmann M, Roy E, Veres T, Tabrizian M. Microfluidic ELISA on non-passivated PDMS chip using magnetic bead transfer inside dual networks of channels. Lab Chip 2007;7 (11):1546–52.

Hoang HT, Segers-Nolten IM, Berenschot JW, Boer MJd, Tas NR, Haneveld J, et al. Fabrication and interfacing of nanochannel devices for single-molecule studies. J Micromech Microeng 2009;19:065017.

Holt JK, Park HG, Wang Y, Stadermann M, Artyukhin AB, Grigoropoulos CP, et al. Fast mass transport through sub-2-nanometer carbon nanotubes. Science 2006;312: 1034–7.

Holtz B, Wang YN, Zhu XY, Guo A. Denaturing and refolding of protein molecules on surfaces. Proteomics 2007;7(11):1771–4.

Huang SS, Yin YF. Transport and separation of small organic molecules through nanotubules. Anal Sci 2006;22:1005–9.

Huisman IH, Pradanos P, Calvo JI, Hernandez A. Electroviscous effects, streaming potential, and zeta potential in polycarbonate track-etched membranes. J Membr Sci 2000;178:79–92.

Hulteen JC, Jirage KB, Martin CR. Introducing chemical transport selectivity into gold nanotubule membranes. J Am Chem Soc 1998;120:6603–4.

Israelachvili JN. Intermolecular and surface forces. 2nd ed. San Diego: Academic Press; 1992.

Jackman RJ, Wilbus JL, Whitesides GM. Fabrication of submicrometer features on curved substrates by microcontact printing. Science 1995;269(5224):664–6.

Jagerszki G, Gyurcsanyi RE, Hofler L, Pretsch E. Hybridization-modulated ion fluxes through peptide-nucleic-acid-functionalized gold nanotubes. A new approach to quantitative label-free DNA analysis. Nano Lett 2007;7:1609–12.

Jirage KB, Hulteen JC, Martin CR. Effect of thiol chemisorption on the transport properties of gold nanotubule membranes. Analy Chem 1999;71:4913–8.

Kaanta BC, Chen H, Lambertus G, Steinecker WH, Zhdaneev O, Zhang X, editors. High sensitivity micro-hermal conductivity detector for gas chromatography. Proc of the 22nd IEEE Int Conf on Micro Electro Mechanical Systems (MEMS'09); 25–29 January 2009; Sorrento, Italy: IEEE.

Kaholek M, Lee W-K, LaMattina B, Caster KC, Zauscher S. Fabrication of stimulus-responsive nanopatterned polymer brushes by scanning-probe lithography. Nano Lett 2004;4(2):373–6.

Kasemo B. Biological surface science. Surf Sci 2002;500:577–656.

Kastner JF, editor. Online gas quality measurement technique based on optical and thermal gas properties. Natural Gas Quality, Energy Measurement, and Interchangeability. Chicago, IL, USA; 2005.

Kemery PJ, Steehler JK, Bohn PW. Electric field mediated transport in nanometer diameter channels. Langmuir 1998;14:2884–9.

Kim DR, Lee CH, Zheng X. Probing flow velocity with silicon nanowire sensors. Nano Lett 2009;9(5):1984–8.

Kim SJ, Wang Y-C, Lee JH, Jang H, Han J. Concentration polarization and nonlinear electrokinetic flow near a nanofluidic channel. Phys Rev Lett 2007;99:044501.

Kirby BJ. Micro- and Nanoscale Fluid Mechanics: Cambridge; 2010.

Kohli P, Harrell CC, Cao Z, Gasparac R, Tan W, Martin CR. DNA-functionalized nanotube membranes with single-base mismatch selectivity. Science 2004;305:984–6.

Krenkova J, Foret F. Immobilized microfluidic enzymatic reactors. Electrophoresis 2004;25(21-22):3550–63..

Ku J-R, Lai S-M, Ileri N, Ramirez P, Mafe S, Stroeve P. pH and ionic strength effects on amino acid transport through Au-nanotubule membranes charged with self-assembled monolayers. J Phys Chem C 2007;111(7):2965–73.

Kuo T-C, Sloan LA, Sweedler JV, Bohn PW. Manipulating molecular transport through nanoporous membranes by control of electrokinetic flow: effect of surface charge density and debye length. Langmuir 2001;17:6298–303.

Lao KP, Yao NK, Kuo TS, editors. Sub-60nm nanofluidic channels fabricated by glass-glass bonding. 28th IEEE EMBS Annual International Conference; 2006 Aug 30-Sept 3; New York, NY, USA.

Laser DJ, Santiago JG. A review of micropumps. J Micromech Microeng 2004;14: R35–64.

Lee I, Lee JW, Greenbaum E. Biomolecular electronics: vectorial arrays of photosynthetic reaction centers. Phys Rev Lett 1997;79(17):3294–7.

Lee J, Yeon YK, Kim J, Kim Y, Jo K. Roof-collapsed PDMS mask for nanochannel fabrication. Bull Korean Chem Soc 2011;32(1):33.

Lee SB, Martin CR. Electromodulated molecular transport in gold-nanotube membranes. J Am Chem Soc 2002;124:11850–1.

Li D. Electrokinetics in microfluidics. Elsevier; 2004.

Lu Y, Liu JW. Functional DNA nanotechnology: emerging applications of DNAzymes and aptamers. Curr Opin Biotechnol 2006;17:580–8.

Luong JHT, Make KB, Glennon JD. Biosensor technology: technology push versus market pull. Biotechnol Adv 2008;26:492–500.

Lyklema J. Fundamentals of interface and colloid science. Academic Press; 1995.

Madou M. Fundamentals of microfabrication. Boca Raton, FL: CRC Press; 1997.

Madou MJ. Fundamentals of microfabrication and nanotechnology. CRC Press; 2012.

Mala GM, Yang C, Li D. Electrical double layer potential distribution in a rectangular microchannel. Colloid Surf A 1998;135:109–16.

Manginell RP, Adkins DR, Moorman MW, Hadizadeh R, Copic D, Porter DA, et al. Mass-sensitive microfabricated chemical preconcentrator. J Microelectromech Syst 2008;17 (6):1396–407.

Mannion JT, Reccius CH, Cross JD, Craighead HG. Conformational analysis of single DNA molecules undergoing entropically induced motion in nanochannels. Biophys J 2006;90:4538–45.

Manz A, Graber N, Widmer HM. Miniaturized total analysis systems: a novel concept for chemical sensing. Sens Actuators B 1990;1:244–8.

Mao P, Han J. Fabrication and characterization of 20 nm planar nanofluidic channels by glass–glass and glass–silicon bonding. Lab Chip 2005;5:837–44.

Marrian CRK, Tennant DM, Nanofabrication. J. Vac Sci Technol 2003;21(5):207–15.

Martin CR, Nishizawa M, Jirage K, Kang M. Investigations of the transport properties of gold nanotubule membranes. J Phys Chem B 2001;105:1925–34.

Martin CR, Nishizawa M, Jirage K, Kang MS, Lee SB. Controlling ion-transport selectivity in gold nanotubule membranes. Adv Mater 2001;13:1351–62.

Masliyah JH, Bhattacharjee S. Electrokinetic and colloid transport phenomena. John Wiley & Sons; 2006.

Matsumoto K, Kawakami T, Nakao M, Hatamura Y, Kitamori T, Sawada T, editors. IEEE MEMS. 1998.

Mazumder M, Bhushan B. Propensity and geometrical distribution of surface nanobubbles: effect of electrolyte, roughness, pH, and substrate bias. Soft Matter 2011.

Mesquida P, Blanco EM, McKendry RA. Patterning amyloid peptide fibrils by AFM charge writing. Langmuir 2006;22:9089–91.

Moritz J, Dieny B, Nozieres JP, Landis S, Lebib A, Chen Y. Domain structure in magnetic dots prepared by nanoimprint and e-beam lithography. J Appl Phys 2002;91 (10):7314–6.

Mulero R, Prabhu AS, Freedman KJ, Kim MJ. Nanopore-based devices for bioanalytical applications. J Assoc Lab Autom 2010;15(3):243–52.

Napoli M, Eijkel JCT, Pennathur S. Nanofluidic technology for biomolecule applications: a critical review. Lab Chip 2010;10:957–85.

Nemes P, Knolhoff AM, Rubakhin SS, Sweedler JV. Metabolic differentiation of neuronal phenotypes by single-cell capillary electrophoresis-electrospray ionization-mass spectrometry. Anal Chem 2011;83:6810–7.

Nguyen NT, Wereley ST. Fundamentals and applications of microfluidics. Artech House; 2002.

Nguyen N-T, Wu Z. Micromixers—a review. J Micromech Microeng 2005;15:R1–6.

Nishizawa M, Menon VP, Martin CR. Metal nanotubule membranes with electrochemically switchable ion-transport selectivity. Science 1995;268:700–2.

Noroozi Z, Kido H, Peytavi R, Nakajima-Sasaki R, Jasinkas A, Micic M, et al. A multiplexed immunoassay system based upon reciprocating centrifugal microfluidics. Rev Sci Instrum 2011;82.

Nystrom M, Lindstrom M, Matthiasson E. Streaming potential as a tool in the characterization of ultrafiltration membranes. Colloids Surf 1989;36:297–312.

O'Brien II MJ, Bisong P, Ista LK, Rabinovich EM, Garcia AL, Sibbett SS, et al. Fabrication of an integrated nanofluidic chip ising interferometric lithography. J Vac Sci Technol 2003;21(6):2941–5.

Oukhaled A, Cressiot B, Bacri L, Pastoriza-Gallego M, Betton J-M, Bourhis E, et al. Dynamics of completely unfolded and native proteins through solid-state nanopores as a function of electric driving force. ACS Nano 2011;5(5):3628–38.

Palchetti I, Mascini M. Nucleic acid biosensors for environmental pollution monitoring. Analyst 2008;133:846–54.

Parka S-M, Huhb YS, Craigheada HG, Erickson D. A method for nanofluidic device prototyping using elastomeric collapse. PNAS 2009;106(37):15549–54.

Perera DM, Neluni T, Ito T. Cyclic voltammetry on recessed nanodisk-array electrodes prepared from track-etched polycarbonate membranes with 10-nm diameter pores. Analyst 2010;135:172–6.

Persson F, Thamdrup LH, Mikkelsen MBL, Jaarlgard SE, Skafte-Pedersen P, Bruus H, et al. Double thermal oxidation scheme for the fabrication of SiO_2 nanochannels. Nanotechnology 2007;18(245301).

Pijanowska DG, Remiszewska E, Lysko JM, Jazwinski J, Torbicz W. Immobilization of bioreceptors for microreactors. Sens Actuator B-Chem 2003;91(1-3):152–7.

Pinti M, Prakash S, editors. A two-step wet etch process for the facile fabrication of hybrid micro-nanofluidic devices. ASME International Mechanical Engineering Congress and Exposition. Denver, CO; 2011 November 11–17.

Piruska A, Branagan SP, Cropek DM, Sweedler JV, Bohn PW. Electrokinetically driven fluidic transport in integrated three-dimensional microfluidic devices incorporating gold-coated nanocapillary array membranes. Lab Chip 2008;8:1625–31.

Piruska A, Gong M, Sweedler JV, Bohn PW. Nanofluidics in chemical analysis. Chem Soc Rev 2010;39:1060–72.

Plecis A, Schoch RB, Renaud P. Ionic transport phenomena in nanofluidics: experimental and theoretical study of the exclusion-enrichment effect on a chip. Nano Lett 2005;5(6):1147–55.

Powell MR, Cleary L, Davenport M, Shea KJ, Siwy ZS. Electric-field-induced wetting and dewetting in single hydrophobic nanopores. Nature Nanotechnol 2011;6:798–802.

Prakash S, Pinti M, Bellman K. Variable cross-section nanopores fabricated in silicon nitride membranes using a transmission electron microscope. J Micromech Microeng 2012;22.

Prakash S, Yeom J, Shannon MA, editors. A microfabricated impedance sensor for ionic transport in nanopores. 11th Solid-state sensors, actuators, and microsystems workshop. Hilton Head Island, SC; 2006.

Probstein RF. Physicochemical hydrodynamics: an introduction. 2nd ed. John Wiley & Sons; 2003.

Qiang L, Vaddiraju A, Rusling JF, Papadimitrakopoulos F. Highly sensitive and reusable Pt-black microfluidic electrodes for long-term electrochemical sensing. Biosens Bioelectron 2010;26(2):682–8.

Qiao R, Aluru NR. Ion concentration and velocity in nanochannel electroosmotic flows. J Chem Phys 2003;118:4692–701.

Qiao R, Aluru NR. Charge inversion and flow reversal in a nanochannel electroosmotic flow. Phys Rev Lett 2004;92(19):198301.

Qiao R, Georgiadis JG, Aluru NR. Differential ion transport induced electroosmosis and internal recirculation in heterogeneous osmosis membranes. Nano Lett 2006;6: 995–9.

Qin D, Xia Y, Whitesides GM. Soft lithography for micro- and nanoscale patterning. Nat Protoc 2010;5(3):491–502.

Qu W, Li DA. Model for overlapped EDL fields. J Colloid Interface Sci 2000;224:397–407.

Raiteri R, Grattarola M, Butt M, Skladal P. Micromechanical cantilever-based biosensor. Sens Actuators B Chem 2001;79:115–26.

Renkin EM. Filtration, diffusion, and molecular sieving through porous cellulose membranes. J General Physiol 1954;38(2):225–43.

Rice CL, Whitehead R. Electrokinetic flow in a narrow cylindrical capillary. J Phys Chem 1965;69(11):4017–24.

Rogers K. Principles of affinity-based biosensors. Mol Biotechnol 2000;14:109–29.

Sadr R, Yoda M, Gnanaprakasam P, Conlisk AT. Velocity measurements inside the diffuse electric double layer in electro-osmotic flow. Appl Phys Lett 2006; 89(4):044103.

Sakaguchi M, Kashiwabara HA. Generation mechanism of triboelectricity due to the reaction of mechaniradicals with mechanoions which are produced by mechanical fracture of solid polymer. Colloid Poly Sci 1992;270:621–6.

Sauberlich S, Klee D, Richter EJ, Hocker H, Spiekermann H. Cell culture tests for assessing the tolerance of soft tissue to variously modified titanium surfaces. Clin Oral Implan Res 1999;10(5):379–93.

Schmickler W. Interfacial electrochemistry. Oxford; 1996.

Schoch RB, Han J, Renaud P. Transport phenomena in nanofluidics. Rev Modern Phys 2008;80:839–83.

Schroder K, Meyer-Plath A, Keller D, Besch W, Babucke G, Ohl A. Plasma-induced surface functionalization of polymeric biomaterials in ammonia plasma. Contrib Plas Phys 2001;41(6):562–72.

Schroeder DV. An introduction to thermal physics. San Francisco, CA: Addison Wesley Longman; 2000.

Shannon MA, Bohn PW, Elimelech M, Georgiadis JG, Marinas BJ. Science and technology for water in coming decades. Nature 2009;452:301–10.

Sharmaa S, Johnsonb RW, Desai TA. XPS and AFM analysis of antifouling PEG interfaces for microfabricated silicon biosensors. Biosens Bioelectron 2004;20(2):227–39.

Sheehan PE, Whitman LJ. Detection limits for nanoscale biosensors. Nano Lett 2005;5 (4):803–7.

Sparreboom W, van den Berg A, Eijkel JCT. Principles and applications of nanofluidic transport. Nat Nanotechnol 2009;4:713–20.

Sparreboom W, van den Berg A, Eijkel JCT. Transport in nanofluidic systems: a review of theory and applications. J Phys 2010;12.

Squires TM, Messinger RJ, Manalis SR. Making it stick: convection, reaction and diffusion in surface-based biosensors. Nat Biotechnol 2008;26(4):417–26.

Srinivasan V, Pamula V, Pollack M, Fair R. A digital microfluidic biosensor for multianalyte detection. IEEE 2003;227–330.

Stavis C, Clare TL, Butler JE, Radadia AD, Carre R, Zeng H, et al. Functionalization of thin-film diamond for highly stable and selective biological interfaces. P Natl Acad Sci USA 2011;108:983–8.

Stern E, Wagner R, Sigworth FJ, Breaker R, Fahmy TM, Reed MA. Importance of the Debye screening length on nanowire field effect transistor sensors. Nano Lett 2007;7(11):3405–9.

Svoboda K, Block SM. Biological applications of optical forces. Ann Rev Biophys Biom 1994;23:247–85.

Swaminathan VV, Gibson LR, Pinti M, Prakash S, Bohn PW, Shannon MA. Ionic transport in nanocapillary membrane systems. J Nanoparticle Res 2012;14.

Szymczyk A, Aoubiza B, Fievet P, Pagetti J. Electrokinetic phenomena in homogeneous cylindrical pores. J Colloid Interface Sci 1999;216:285–96.

Talaga DS, Li J. Single-molecule protein unfolding in solid state nanopores. J Am Chem Soc 2009;131(26):9287–97.

Tamaki E, Hibara A, Kima HB, Tokeshi M, Kitamori T. Pressure-driven flow control system for nanofluidic chemical process. J Chromatogr A 2006;1137:256–62.

Tang KC, Liao E, Ong WL, Wong JDS, Agarwal A, Nagarajan R, et al. Evaluation of bonding between oxygen plasma treated polydimethyl siloxane and passivated silicon. JPhCS 2006;34:155–61.

Tas NR, Berenschot JW, Mela P, Jansen HV, Elwenspoek M, van den Berg A. 2D-Confined nanochannels fabricated by conventional micromachining. Nano Lett 2002;2 (9):1031–2.

Verbridge SS, Edel JB, Stavis SM, Moran-Mirabal JM, Allen SD, Coates G, et al. Suspended glass nanochannels coupled with microstructures for single molecule detection. J Appl Phys 97(12):124317–4.

Vijayendran RA, Motsegood KM, Beebe DJ, Leckband DE. Evaluation of a three-dimensional micromixer in a surface-based biosensor. Langmuir 2003;19: 1824–8.

Vinogradova OI, Belyaev AV. Wetting. Roughness and hydrodynamic slip, PSP Review Volume. Singapore: Pan Stanford Publishing; 2010.

Wang LP, Shao PG, van Kan JA, Ansari K, Bettiol AA, Pan XT, Wohland T, et al. Fabrication of nanofluidic devices utilizing proton beam writing and thermal bonding techniques. Nucl Instrum Meth B 2007;260:450–4.

Wang Y, Bhushan B. Boundary slip and nanobubble study in micro/nanofluidics using atomic force microscopy. Soft Matter 2010;6:29–66.

Wang Y, Bhushan B. Liquid microdroplet sliding on hydrophobic surfaces in the presence of electric field. Langmuir 2010;26:4013–7.

Wang Y, Wang Y, Chen K, Li B. Non equilibrium molecular dynamics simulation of electrokinetic effects on heterogenous ionic transport in nano-channel. Chem Eng Sci 2011;66:2807–16.

Woias P. Micropumps—past, progress and future prospects. Sens Actuators B 2005;105:28–38.

Wu LQ, Ghodssi R, Elabd YA, Payne GF. Biomimetic pattern transfer. Adv Funct Mater 2005;15(2):189–95.

Xuan X. Ion separation in nanofluidics. Electrophoresis 2008;29:3737–43.

Yang WS, Auciello O, Butler JE, Cai W, Carlisle JA, Gerbi J, et al. DNA-modified nanocrystalline diamond thin-films as stable, biologically active substrates. Nat Mater 2002;1(4):253–7.

Yeom J, Shannon MA. Detachment lithography of photosensitive polymers: a route to fabricating three-dimensional structures. Adv Funct Mater 2009;19:1–7.

Zhang G-J. MicroRNA and cancer. Methods Protocols 2011.

Zhu H, Holl M, Ray T, Bhushan S, Meldrum DR. Characterization of deep wet etching of fused silica glass for single cell and optical sensor deposition. J Micromech Microeng 2009;19(065013):8pp.

CHAPTER

Energy and Environmental Applications

6

CHAPTER OUTLINE

6.1 Introduction

Microfluidic and nanofluidic (μ-Na*fl*) systems have created several distinct areas of applications. One such area is that of energy and environment-related technical challenges. In this chapter, we discuss several examples for the technical challenges solved by microfluidic and nanofluidics either as smaller components in engineered systems or as complete systems.

6.2 Combustion-based devices

Gas-phase fluidic systems that rely on combustion have been explored for power generation systems. Several studies have been conducted on such portable

S. Prakash & J. Yeom: Nanofluidics and Microfluidics. DOI: http://dx.doi.org/10.1016/B978-1-4377-4469-9.00006-8

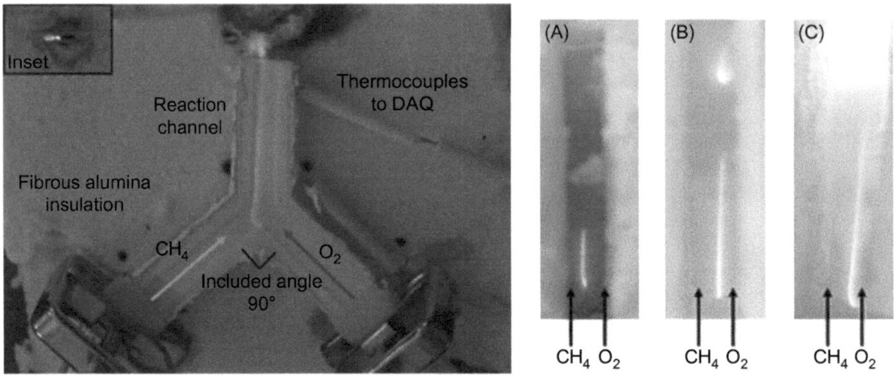

FIGURE 6.1

The left panel shows a digital photograph of a sub-millimeter combustor in operation for flame studies. The right panel shows three examples of flame structure observed in the reaction channel with conventional photography. © 2009 American Chemical Society. Reproduced with permission from [1].

systems including work on fundamentals of flame structure and heat transfer in confined microscale spaces where research studies have shown the difficulties in sustaining homogeneous combustion using premixed flames in confined spaces with a large surface-area-to-volume ratio ($>10^3\ m^{-1}$). Flame studies by Prakash et al. [1] in alumina-based sub-millimeter combustors have led to establishing the fundamentals of microcombustion on sustaining homogeneous flames in confined spaces by managing a variety of parameters as discussed next. As an example, Figure 6.1 shows images of flame structure captured by conventional photography and Figure 6.2 shows high-speed flame imaging depicting more detail in the flame structure in contrast to conventional imaging techniques.

To overcome some of the difficulties due to structural failure, non-premixed or catalytic combustion has been used with laminar flows and minimal influence of turbulent mixing as $Re < 100$. In addition, the critical role played by surfaces in reducing or preventing quenching of essential radicals has also been discussed. Other studies have evaluated the role of heat transfer, stability criteria, mixing patterns, shear conditions, and a broad range of fuel and oxidizer parameters and boundary conditions affecting both steady-state flame structure and transient flames in confined spaces. These fundamental studies lay the groundwork for development of devices that are coupled to thermoelectrics for direct conversion of combustion heat to electricity, and in other applications such as thermophotovoltaic devices, and microengines (Figure 6.3).

In one of the early demonstrations of using a combustion-based microfluidic system, it was shown that a microcombustor was directly coupled to a thermoelectric element for heat-to-electricity conversion. The idea has undergone several iterations in the past 10–15 years and recently a silicon combustor with

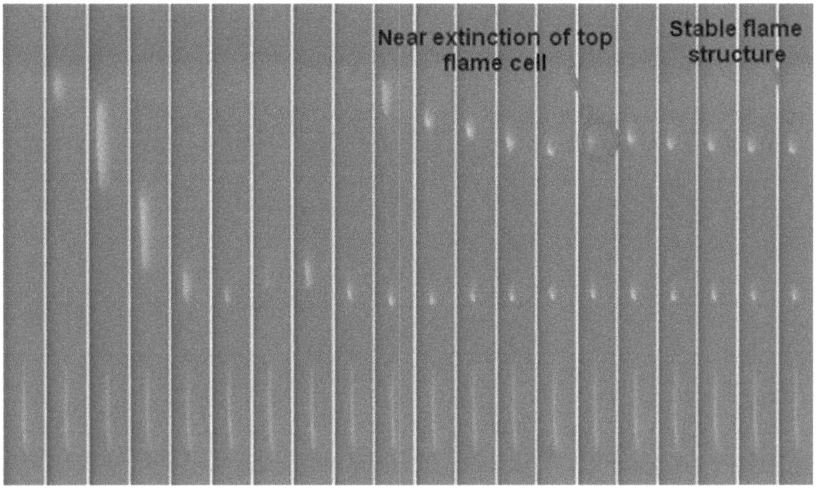

FIGURE 6.2

A high-speed image of the flame stabilization process in the alumina microchannel within the sub-millimeter combustor. Note the discontinuous flame structure observed in Figure 6.1 is formed through a repeated extinction–ignition process in the microchannel. During this extinction–ignition phase audible frequencies are heard and recorded. Complete details and analysis of the transient flames were done by Prakash et al. Reproduced with permission from [2].

an integrated thermopile was demonstrated as shown in Figure 6.4. In this system, catalytic combustion of butane was used with a flow rate in the 100–200 sccm range with overall combustor dimensions of $35 \times 28.5 \times 4\ mm^3$ at 5.5 g. Maximum power generated by the system was slightly over 7 W with conversion of heat to electricity at ~2.5% or less. In other similar approaches, combustion of liquid fuels such as butanol has also been explored for coupling to thermoelectrics. Methanol combustion led to efficiency of 1.1% with power generation on the order of 1 W. In all these systems, two major challenges remain. First, managing the heat flow is needed to minimize external convective and radiative losses as discussed extensively by Prakash et al. [1] in their work on importance of "boundary conditions" for microcombustion. Furthermore, as it is now known that radiative losses dominate heat loss, development of heat shield to mitigate radiation losses is critical. Note from Figure 6.4 that the thermal management system is several times larger in size than the power or heat generation module. Furthermore, the fuel and oxidizer sources are not coupled on the module either. Consequently, the second challenge lies in developing a completely integrated system with onboard fuel and exhaust management along with power generation.

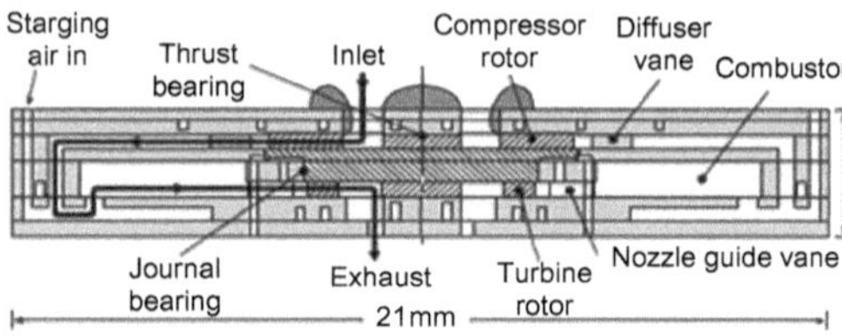

FIGURE 6.3

The picture on the left shows a microcombustor-powered thermophotovoltaic cell with cooling fins. The system is made of a SiC (silicon carbide) emitter, and a GaSb (gallium antimony) photovoltaic cell array. In a microcombustor of 0.113 cm^3 in volume, when the flow rate of hydrogen is 4.20 g/h, the micro-TPV system is able to deliver an electrical power output of 1.02 W, corresponding to an open-circuit electrical voltage of 2.28 V and a short-circuit current of 0.59 A. The picture on the right shows the schematic for a microturbine engine developed at MIT comprising a 6-wafer Si stack. Left: © 2004, AIP Publishing LLC. Reproduced with permission from [3]. Right: © 2000 IEEE. Reproduced with permission from [4]).

6.3 Microfuel cells

Batteries have been known as an ideal energy source for many solid-state devices, especially portable electronics, largely due to their concise form factor. Increased power demands of modern portable electronics; however, drive an extensive search for alternative technologies for portable power sources. Microcombustion-based systems were considered one such alternative as hydrocarbon fuels can provide the high-energy density of 40–50 MJ/kg (in comparison to 0.4–0.7 MJ/kg of state-of-the-art lithium-ion batteries), direct conversion of thermal energy to electricity has been widely investigated using combustion-based technology. However, at length scale smaller than 1 mm, combustion systems suffer from many problems and examples of research progress both at the fundamental level and device level discussed in Section 6.2.

Fuel cells (FCs), one of the oldest electrical energy conversion technologies convert chemical energy into electricity and have been scaled down into smaller systems. FCs intrinsically possess five to ten times the energy density of lithium-ion batteries, and its conversion efficiency currently is much higher than combustion of hydrocarbon fuels at microscale, suggesting need for further fundamental research at the microscale. From a simple scaling analysis by Nguyen and Chan in 2006 [5] it was pointed out that both advantages and drawbacks can result from miniaturization of FCs. First, dominant surface effects with miniaturization can

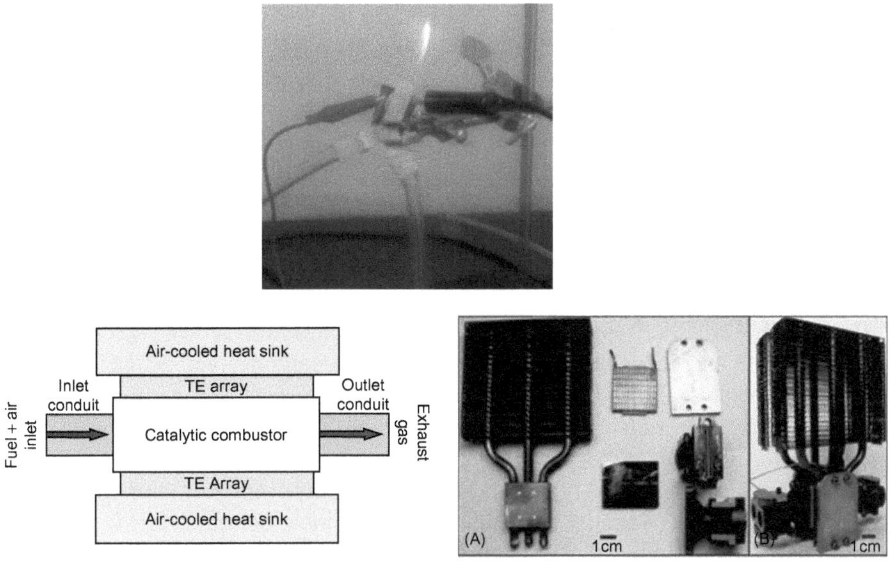

FIGURE 6.4

The top panel shows operation of an early stage laboratory scale microcombustor-thermoelectric converter during system development in the Microsystems and Nanosystems Laboratory at Ohio State University. A conventional chromel–alumel TE element was used to show viability of the concept. The lower panel shows a schematic for the catalytic combustor-thermoelectric converter and digital photograph of the system are shown. Note the size of the cooling array in contrast to the entire power generation system indicating the significant importance of developing a better thermal management system for these devices. © 2011 American Chemical Society. Reproduced with permission from [6].

potentially improve reaction conditions and increase the cell performance. Higher pressure drop with miniaturization can increase the pressure across the diffusion layer, helping diffusion of fuel and oxidant to the membrane electrode assembly (MEA) but require more energy for pumping. With miniaturization, the Peclet number (a ratio of convection to molecular diffusion) becomes smaller; thus fuel diffusion can be facilitated to the membrane in case of gaseous fuels (e.g., H_2). For micro direct methanol fuel cell (DMFC) in which methanol liquid is used as a fuel, the Peclet number is still observed to be too high ($\sim$100), so that the diffusion of fuel is limited decreasing the fuel utilization efficiency.

As with all other microfluidics and nanofluidics devices, micro-fuel cells are typically fabricated with the MEMS technologies, have microstructure with a critical dimension less than 100 μm, and generate less than 5 W of electricity. A variety of the fuel cell technologies have been subject to the worldwide miniaturization efforts. The interested reader is directed to specific review articles for the different types of the miniaturized FCs including micro-polymer electrolyte membrane (PEM) FCs [5,7], micro DMFCs [8], microbial FCs [9,10], and micro solid oxide fuel cells [11,12]. Among these, the PEM FCs and DMFCs have received most attention partly due to their low temperature operation. In the following sections, we will examine a few examples of how the micro-/nanofluidic approaches can help to solve the technical challenges and enhance the performance of the PEMFCs and DMFCs.

6.3.1 Laminar flow-based fuel cells

Many key challenges and issues with the micro-PEMFCs and DMFCs have to do with the use of PEMs—the solid polymer electrolyte membrane. The PEM separates the fuels in anode from the oxidants in cathode, and when in action, protons dissociated from the anode reaction are driven to the cathode side and react with oxygen molecules and the electrons arriving through the external circuit to form water molecules (oxygen reduction reaction). Several problems arise in operating the PEM-type FCs. First, the proton conductivity depends on the moisture content of the membrane. Nafion, the most common PEM, for example, needs to be hydrated to facilitate the proton transport across the membrane, posing the issue of the operating temperature and water management. In typical conditions, protons are accompanying water molecules when crossing, and thus the anode tends to be dried out while the cathode suffers from flooding. In addition, swelling and shrinkage of the membrane during water uptake and dehydration can deform the membrane, and the fatigue from the repeated mechanical deformation can lead to the component and packaging failure. Another issue is fuel crossover. In DMFC, methanol permeates across the membrane, which results in the degradation in cell performance and fuel utilization. Therefore, the conventional DMFCs are operated at relatively low methanol concentrations, entailing an auxiliary device to dilute the highly concentrated or neat methanol stored in a fuel reservoir [13].

In an effort to mitigate the water management issue and fuel crossover, it has been proposed for the solid PEMs to be replaced with the stable liquid–liquid interface between the two streams in the laminar flow-based device [14–16]. Figure 6.5A shows that two liquid streams, one containing a fuel and the other an oxidant, are merged into a single microfluidic channel and laminar flow between the electrodes placed on the opposing sidewalls of the channel. This membraneless laminar flow fuel cell (LFFC) takes advantage of the inherent low Reynolds number of the fluid flow at microscale—due to limited mixing and lack of turbulence, fuel crossover can only take place via diffusion and may be minimized by adjusting stream flow rates and channel dimensions [14]. Since the electrolyte is in a liquid form, anode dry out and cathode flooding are automatically eliminated. Another distinct merit of LFFCs is its flexibility in the type of fuels and working media that can be operated in the same configuration (with the proper catalysts tailored for each fuel type). Different fuels such as methanol, ethanol, formic acid, hydrazine, and sodium borohydride were delivered to the LFFC in either

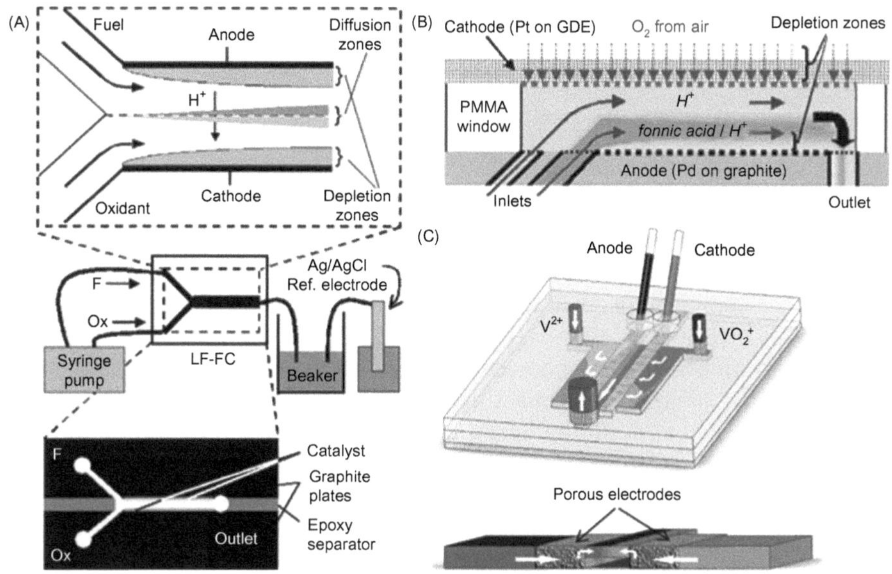

FIGURE 6.5

Schematics of (A) a membraneless laminar flow-based fuel cell (LFFC) and its experiment setup: a fuel and an oxidant stream flow laminarly in parallel between electrodes placed on opposing inside walls of a microfluidic channel [16], (B) an air-breathing LFFC with a porous gas diffusion electrode (exposed to ambient air) as the cathode [17], (C) the new LFFC architecture with flow-through porous electrodes, employing aqueous vanadium redox couple (V_2^+ and VO_2^+) [18]. (A) Reproduced by permission of The Electrochemical Society, from [16]; (B) © 2005 American Chemical Society. Reproduced with permission from [17]; (C) © 2008 American Chemical Society. Reproduced with permission from [18].

acidic or alkaline electrolyte media and produced more than 100 mW/cm^2 of the power density in the air-breathing mode [19]. With a slight modification to the common LFFC design (exposing a gas diffusion electrode to ambient air and using it as the cathode), the air-breathing microfluidic LFFC was also developed to increase the dissolved oxygen concentration in the cathode stream [17] (see Figure 6.5B).

One of the main issues in the operation of LFFC is a low level of fuel utilization per single pass. A fuel in the anode stream tends to pass the reaction zone unreacted (or unconsumed), and fuel utilization is limited by the geometrical constraints driven by the diffusion length of the fuel oxidizer in contrast to the physical device dimensions. Recently, a new LFFC architecture that incorporates the flow-through porous electrodes was developed to exploit the 3D large surface area inside the porous electrodes, enhancing the transport rates of fuels and oxidants and thus leading to much higher fuel utilization (see Figure 6.5C) [18].

The power density of 20 mW/cm^2 at 0.8 V with an active fuel utilization of 94% (for aqueous vanadium redox species) was achieved when the cell was operated at the flow rate of 1 μl/min. More detailed discussion on the design and performance of microfluidic LFFCs can be found elsewhere [5].

6.3.2 Inorganic–organic hybrid proton exchange membrane based on porous silicon

As discussed in the previous example, the Nafion membrane poses serious issues as a proton conductor in the micro-PEMFCs. The MEA—the heart of the PEMFC—is typically miniaturized using the standard microfabrication techniques. The Nafion membrane, however, is difficult to be patterned via photolithography, and bonding it to the silicon substrate is not straightforward. The large volumetric change during operation due to a change in the hydration level also makes the bonding interface unstable and susceptible to premature device failure. One idea is to use a thin, suspended porous silicon layer as a proton exchange membrane [20–22] in the work done at the University of Illinois at Urbana-Champaign. Nanoporous silicon can be readily prepared using anodic etching in hydrofluoric acid, and its porosity, tortuosity, and surface area are controlled by the anodization parameters (e.g., anodization current). The anodization process is compatible with the IC and MEMS fabrication, and because the PEM is based on porous Si, the construction of the MEA and further integration with other microfabricated fuel cell components can be greatly facilitated. And unlike the polymer-based PEMs, nanoporous Si may be stable at elevated temperature (>200°C).

Figure 6.6A shows the schematic of the micro-PEMFC with the nanoporous silicon membrane [23]. A thin silicon membrane is first suspended using anisotropic etching of silicon and then anodized to create the pores within the silicon membrane. The average pore size ranges from 5 to 7 nm, as shown in Figure 6.6B and C. The porous silicon membrane with the uniform thickness and open-ended pores is fabricated using the self-terminating, single-step anodization process developed as part of long-term efforts at the University of Illinois (see Figure 6.6D). In an acidic aqueous media, the silicon pore surfaces are terminated with the hydroxyl group, allowing for proton conduction [20]. While promoting high proton conductivity, the size of the pores, is too large to serve as a diffusion barrier to prevent fuel crossover and for the membrane to remain hydrated in a low humidity level. The self-assembled monolayer, 3-mercapto-propyltrimethoxysilane (MPTMS), is assembled on the modified pore surfaces and later oxidized to create $-SO_3H$ terminal group, which is the case in Nafion (Figure 6.6E). The size of the pores after the self-assembly is reduced to 3–5 nm (Figure 6.6F) and further decreased by depositing an ultrathin layer of silicon oxide at the mouth of each pore (Figure 6.6G). This inorganic membrane functionalized with an organic layer

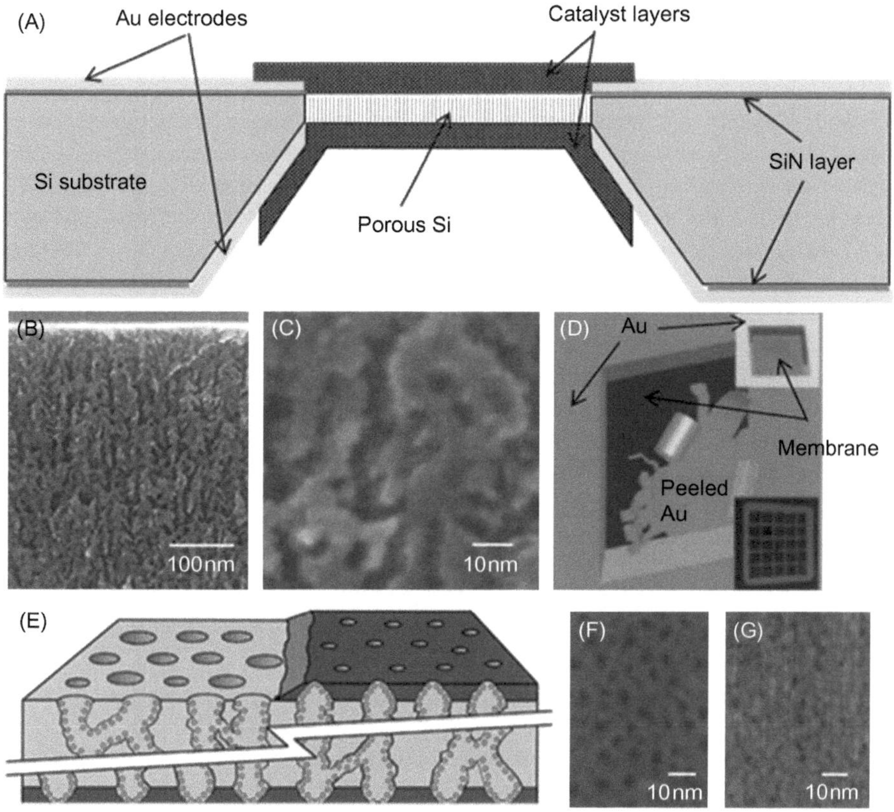

FIGURE 6.6

(A) A schematic of the micro-PEMFC with nanoporous silicon membrane, (B) and (C) cross-sectional SEM images of the nanoporous membrane fabricated using the anodization process, (D) a schematic of the Au layer peeling process for self-termination of the anodization and creating open-ended pores, (E) a schematic showing the sulfonated function groups (circles) in the pore sidewalls and the thin oxide layer to reduce the pore aperture, and the SEM images of the top view of the porous silicon membranes (F) before and (G) after the oxide deposition. © 2010 Macmillan Publishers Ltd. Reproduced with permission from [23].

to form an inorganic–organic porous membrane with asymmetric pore geometry helps to maintain its hydration level even at low humidity environment without sacrificing the overall proton conductivity [24]. The micro-PEMFC with the nanoporous silicon membrane produced an order of magnitude higher power density than that with the Nafion membrane in a dry operating condition.

6.3.3 Onboard hydrogen generators for microfuel cells

Although a vast amount of literature has been reported on efforts to miniaturize the PEMFC components, shrinking the size of the ancillary systems (pumps, valves, sensors, separators, control circuitry, etc.) has remained a challenge for most μ-Na*fl* devices. Microfuel cells suffer from similar limitations. It is especially relevant when developing microfuel cells that have the total size of a few mm^3 (μl) because the fuel delivery and control units are often larger than the MEA and consume significant power, thereby reducing the overall energy and power density of the fuel cell system [25]. From the system-level perspective, liquid fuels (methanol, formic acid, etc.) pose the problems of fuel delivery, product separation, and water management; and the devices of handling these issues are difficult to be down-scaled and integrated into millimeter-scale FCs. Hydrogen-based micro-PEMFCs may work well if onboard hydrogen generators can deliver hydrogen with minimal intervention and power consumption. Unlike pressurized or liquefied hydrogen tanks, metal and chemical hydrides are easier to store and handle and have relatively high theoretical hydrogen volumetric density of 60–150 kg H_2/m^3 [26]. Hydrolysis of these hydrides would generate hydrogen. Therefore, the development of the simplified auxiliary units, preferably in a completely passive means, for hydrogen generation and delivery would be highly desirable. Several microfluidics/nanofluidics-enabled approaches to controlling hydrogen generation at microscale are introduced in the following.

The first design as shown in Figure 6.7A utilizes a microfluidic self-regulating mechanism to control the hydrogen generation rate upon the load [27,28]. The micro hydrogen generator (MHG) unit consists of a metal hydride compartment, a water compartment, and an MEMS control layer, which separates two chambers by the meandering microfluidic channel. The hydride chamber partially packed with CaH_2 (allowing for byproduct volume expansion) also serves as the microreactor to generate hydrogen when water is delivered to the hydride chamber by capillary force. If the hydrogen generation rate exceeds the hydrogen consumption rate in the micro-PEMFC, the hydride chamber becomes pressurized, filling the microchannel with hydrogen and stopping a further water flow into the hydride chamber. Therefore, when the fuel cell is in idle, no excessive hydrolysis occurs. A different self-regulating mechanism for MHG has been proposed using a membrane microvalve as shown in Figure 6.7B. The hydride chamber and water reservoir are separated by the membrane that has many via holes. The control mechanism is similar to the previous design, but instead of the water meniscus in the microchannel acting as a virtual valve, when the chamber is pressurized due to a higher generation, the membrane is deflected toward the water port, blocking the water flow and ceasing the further hydrogen generation. The membrane valves and the fabricated MHG are shown in Figure 6.7C and D.

Though the self-regulating MHG systems have been successfully developed and integrated into the micro-PEMFC, they are ultimately non-rechargeable and

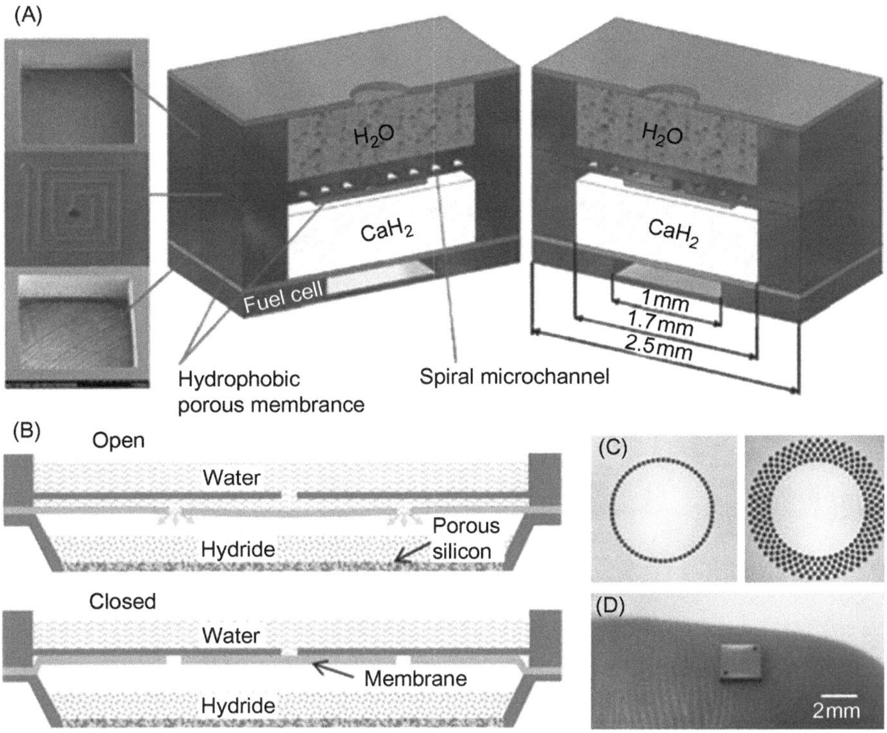

FIGURE 6.7

(A) A schematic of the integrated micro power source based on the self-regulating MHG and micro silicon fuel cell with SEM images of the water reservoir, the microchannel control layer, and the hydride chamber, (B) A schematic and operating principle of the membrane valve-based self-regulating MHG, (C) the polyimide membrane with an array of via holes for water diffusion, (D) a photograph of a $3 \times 3 \times 1$ mm^3 fully integrated microfuel cell . (A): © 2008, Reproduced with permission from Elsevier, from [28]; © 2008, Reproduced with permission from Elsevier from [27]; C-D© 2010, Reproduced with permission from Elsevier from [24].

only work as a solution for disposable systems. Fuel circulation and exchange with a rechargeable reservoir is essential for the high degree of fuel utilization and extended operating time. To develop a fully integrated MHG unit with self-circulation and self-regulation mechanism, hydrogen generation from a catalytic hydrolysis of alkaline hydride solutions is utilized [30]. A passive pumping, micro-valve, and micro gas–liquid separator are all integrated into a single microchannel device— (i) the hydrogen gas bubbles generated from catalytic hydrolysis are employed to circulate the hydride solution in a microchannel loop and (ii) a nanoporous hydrophobic membrane used to separate the hydride solution and hydrogen gas serves as a venting port regulated by the hydrogen pressure [31].

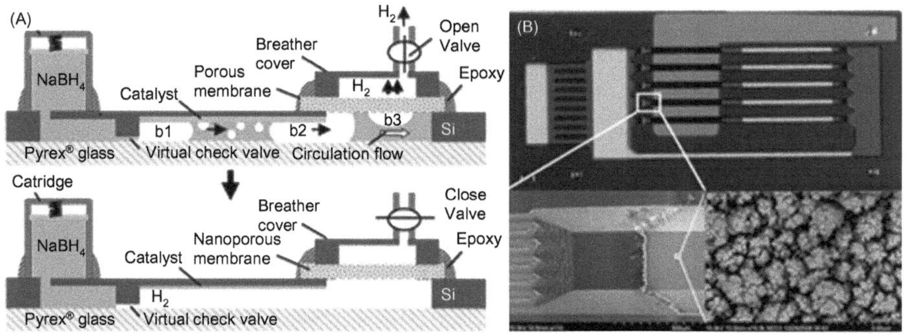

FIGURE 6.8

(A) Schematic of the microfluidic hydrogen generator. Top: the self-circulation mechanism. Bottom: the self-regulation mechanism. (B) Optical and SEM images of the fabricated silicon chip with V-shaped virtual check valves, reaction channels, venting windows, and Pt black catalyst . © 2009 IEEE. Reproduced with permission from [32].

Figure 6.8 shows the working principle of the self-circulation and self-regulation in the microfluidic hydrogen generator. The rightward net flow to circulate a hydride solution is introduced by: rightward growth of hydrogen bubble (b1) which is restricted by virtual check valve (smaller channels) on its left; rightward transportation of hydrogen bubble (b2) which is captured by the hydrophobic membrane; symmetric collapsing of hydrogen bubble (b3) by removing it through the nanoporous hydrophobic membrane. When hydrogen is not being consumed by the fuel cell or the outlet valve is closed, a hydride solution will be forced out of the catalyst channel to stop the reaction. Faster hydrogen consumption during high load (high current) will decrease the hydrogen pressure and increase the venting rate, resulting in a higher pumping rate. Therefore, the on-demand supply of hydrogen to the fuel cell can be achieved without any power consumption.

Another type of fuel cell being miniaturized but suffering from low power density is the microbial fuel cell, which converts biological activity at the electrode for charge transfer toward electricity generation. An example of a three-chamber microbial fuel cell is discussed below in the section on water desalination.

6.4 Electrokinetic energy conversion

Converting mechanical (e.g., pressure drop) to electrical energy (e.g., streaming current) through electrokinetic energy conversion or generation (EEC or EKEG) has drawn the renewed interest owing to the recent advancements in nanotechnology. Colloidal scientists discovered nearly 200 years ago that when an electrolyte in a fine capillary is subject to pressure difference, a streaming current and

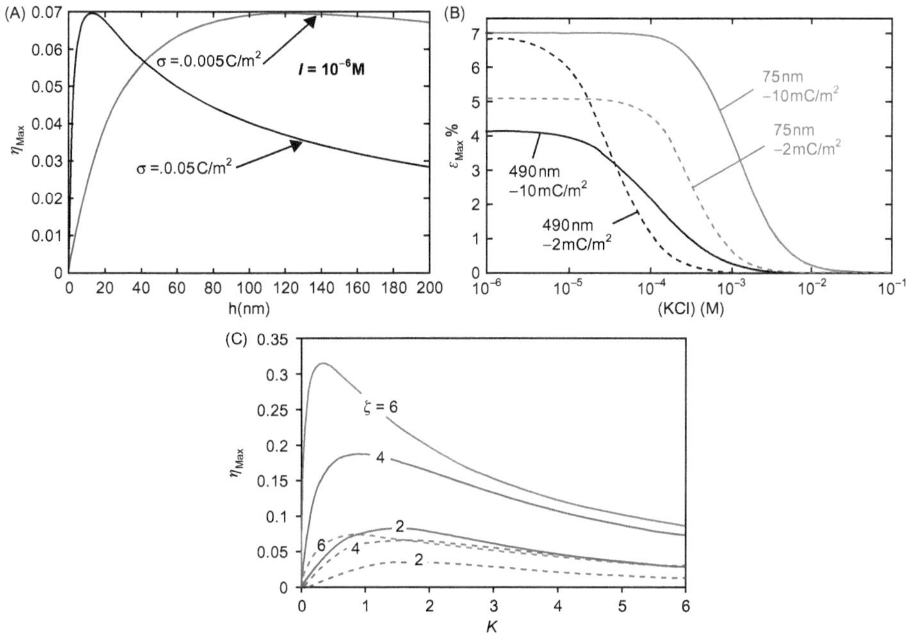

FIGURE 6.9

The figure presents theoretical calculations for efficiency of energy conversion in electrokinetic devices. (A) shows a calculation by the Conlisk group at Ohio State University predicting efficiency as a function of nanochannel depth for two different surface charge densities at a fixed ionic strength for the electrolyte [33]. (B) is from the Dekker group presenting a combination of calculated and measured data for a variety of nanochannels as a function of electrolyte concentrations. © 2006 American Chemical Society. Reproduced with permission from [34]. (C) presents similar calculations as used for the first two but a slip length for nanochannels was included by the Xuan group [35].

an electrical potential difference are created between the two ends of the capillary. The interest in such devices arises out of the ability to use waste mechanical pressure (e.g., exit stream from waste water facilities or pressure generated in wheels of vehicles during motion) to generate useful energy. Consequently, sometimes devices such as the EKEG or EEC are also referred to as energy harvesters.

A first thermodynamic analysis and proof-of-concept was demonstrated in 1964 by Osterle et al. The thermodynamic efficiency of the electrokinetic energy conversion systems has been studied since and theoretical estimates range from 1 to 70% based on geometric configuration, critical micro- or nanochannel dimensions, electrolyte type and concentration, surface charge density of the micro- or nanochannels, and the pressure applied (see Figure 6.9 for some examples of theoretical calculations). Li et al. [36] showed that the characteristic channel dimensions approach the nanometer regime, the electrical double layers of the channel walls can overlap, a the EEC is most efficient. In their mini review, Pennathur et al. [37] discussed the role for slip flow in nanochannels to further increase the efficiency of transforming mechanical pressure to a usable electrical potential. Despite these theoretical predictions, the power extracted from this process is still miniscule (~pW), and the efficiency from the best existing device employing nanochannels tops around a few percent (usually 1–3%). More recently, work by Wu, Prakash, and co-workers has shown that it is essential to manage the surface charge density and the relative hydrophobicity (or surface energy) of a surface to provide control over slip conditions exhibited by chemically modified surfaces (see Figure 6.10). The engineered boundary conditions can then be employed to build higher efficiency EEC or EKEG devices.

6.5 Sensors for water contamination

A major global technological challenge for the next century lies in developing new technologies for water purification. Membrane processes are an integral part of developing freshwater supplies through filtration and desalination and are discussed in the next section. However, another challenge lies in knowing the composition of water to determine levels of contaminants. Several different μ-Na*fl* sensors have been developed for contaminant detection in water. For example, catalytic DNA molecules or DNAzymes have been used for recognition of Pb^{2+}-mediated cleavage of DNA, which helps to build colorimetric sensors for measuring lead contamination in water. Recognition reactions are most frequently implemented as homogeneous reactions, however, DNAzyme-functionalized PMMA have been also incorporated into microfluidic devices for real-time detection of lead and arsenic (see Figure 6.11). PDMS-based microfluidic devices have been developed with integrated optical detection methods for measurement of algal fluorescence to quantify contamination in water due to herbicides or metal ions arising from agricultural run-offs. The bioassay allowed detection of the toxic effects of the herbicide Diuron on *Chlamydomonas reinhardtii* green algae with a concentration as low as 11 nM (Figure 6.12). In yet another example, an integrated microfluidic device with wireless communication to transmit sensor data has also been developed for parts per million-level sensitivity of the sensor toward phosphate detection in water.

6.6 Energy-efficient water purification

Recent reports and various publications have identified the importance of clean water not only for public health, but also for energy and security needs. Consider the following statistics that present a sobering picture for the current state of

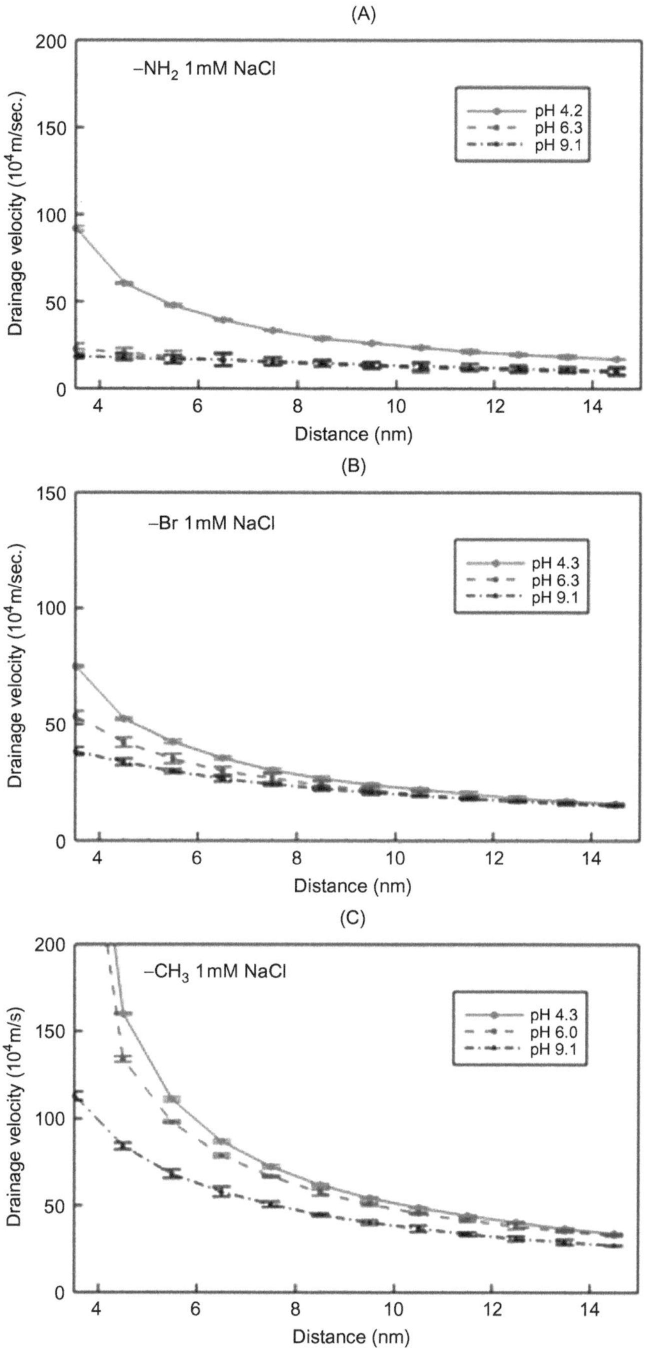

FIGURE 6.10

The figure depicts experimental measurement of drainage velocity for chemically modified surfaces for a fluid draining between an AFM tip and the modified surfaces. The measurements were done as function of solution pH and showed that hydrophilic surfaces (A) and (B) also show finite drainage velocity but the hydrophobic surface (C) shows the highest. Furthermore, it should be noted that the drainage velocity can be correlated to a surface slip velocity. © 2010 American Chemical Society. Reproduced with permission from [38].

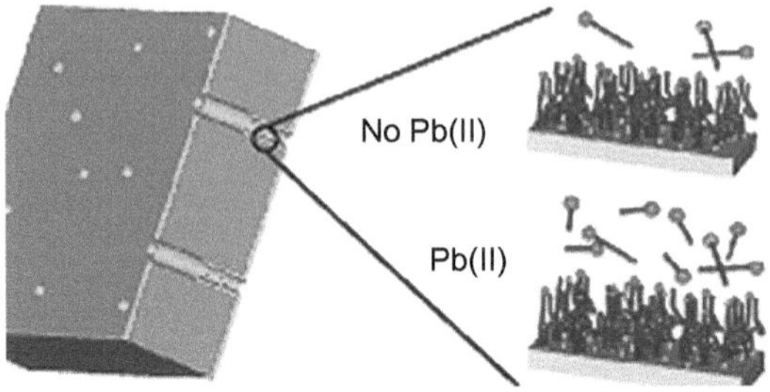

FIGURE 6.11

A schematic from work by the Bohn group at Notre Dame University showing DNAzyme-functionalized nanopores for detection of lead for applications in sensing heavy metal contaminants in water. © 2008 Macmillan Publishers Ltd. Reproduced with permission from [39].

water supplies and the importance for advancing the science and technology to increase freshwater supplies.

Among the many challenges that exist toward generating high-quality potable water, desalination of salt water has been identified as a key step toward making progress for developing sustainable sources for freshwater. However, to create effective desalination and subsequent distribution strategies, it is important to review how the water is distributed around the planet. The total available water on Earth is approximately 1.4×10^{21} l (or about 332,500,000 mi^3). Of this seemingly enormous water supply, over 99% of the water is currently inaccessible to human use, and more than 97% of the Earth's water exists in oceans, bays, seas, and saline aquifers as large reservoirs of salt water. In fact, several estimates place approximate supply of usable freshwater at ~0.7% or about 9.8×10^{18} l. While this may appear to be a large number, accounting for population increase, demands on freshwater for agriculture, industry, potentially changing climate, declining freshwater quality from worldwide contamination via industrial, municipal, and agricultural discharge, and increasing energy needs leads to a rather bleak picture for future availability of clean water for direct human use. Therefore, the ability to affordably and sustainably desalinate water can resolve many of the impending and projected water crises by providing an increased supply of freshwater from more abundant salt water sources.

A huge variety of water desalination methods exist. As per a review article in 2005, about 88% of the available desalination production capacity employs either reverse osmosis (RO) or multistage flash (MSF) distillation processes for freshwater production from either brackish or seawater. Other main technologies include multi-effect distillation (MED), vapor compression (VC) for seawater (~32,000 ppm of salt or higher; commonly at ~35,000 ppm), and electrodialysis (ED), which is more commonly used for brackish water. The primary energy requirement for MSF, MED, and VC is in the form of thermal energy, while RO requires primarily mechanical

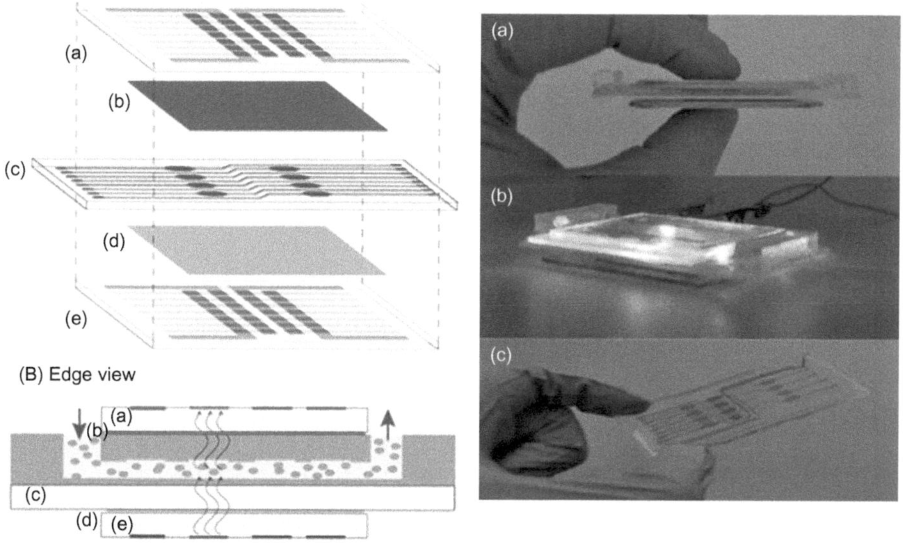

FIGURE 6.12

Schematic showing design and structure of fluorescence sensor and photographs for the device after fabrication and during testing as it is used for photo detection of algal fluorescence. Reproduced with permission from The Royal Society of Chemistry from [40].

energy, and ED requires primarily electrical energy. Other methods such as solar distillation, freezing, gas hydrate processes, membrane distillation, capacitive deionization, and ion exchange are used for desalination, but current technology levels for these processes find limited use and are not commercially viable on a worldwide scale for widespread implementation. Table 6.1 summarizes the energy consumption for some of the water desalination methods that are currently used.

We note here that the idea of generating freshwater from salt water is not new and dates back to the inventions of the Greek sailors around the fifth century with wool-based condensers with the osmosis-based filtration, and desalination methods that used polymeric membranes began gaining popularity in the mid-1900s. Next a discussion of some of the emerging bio- and nanotechnology approaches is presented.

6.6.1 Microfluidic and nanofluidic concentration polarization for desalination

Concentration polarization in most membrane-based water desalination systems is considered to be a loss term and immense resources and effort have been employed in better understanding and mitigating polarization effects. Recent research has revealed that the imbalances created by concentration polarization (enrichment and depletion regions) at the micro-nanochannel interface (see Chapter 3 for additional

Table 6.1 Summary of Energy Consumption and Pretreatment of Source Water for Some Common Water Desalination Methods

Process	MSF	MED/TVC	RO	ED*
Heat consumption (kJ/l)	290	145–390	–	–
Electricity consumption (kJ/l)	10.8–18	5.4–9	9–25.2	4.32–9
Total energy consumption (kJ/l)	300.8–308.8	150.9–399	9–25.2	4.32–9
Typical production capacity (m^3/day)	~76,000	~36,000	~20,000	~19,000
Conversion to freshwater	10–25%	23–33%	20–50%	80–90%
Pretreatment required	Little	Little	Demanding	Moderate

**Estimates based on brackish water, all other methods consider sea water.*

discussion of interfacial phenomena) and can be sustained for extended periods of time. Figure 6.13 shows an example where the interfacial development of concentration polarization was used to desalinate seawater. In this device, pretreatment was needed for the removal of Ca^{2+} ions and physical filtration for elimination of precipitation and large debris was needed before actual desalination. The membraneless system that was tested was primarily driven by electrokinetic flows and achieved a 99% salt rejection ratio. In addition to achieving seawater desalination, this system was shown to remove most solid particles, microorganisms, and biomolecules based on their charge. The energy consumption of this system was reported at approximately 5 W h/l for flow rates of 0.25 μl/min.

6.6.2 Advanced membranes

In approximately the past 5 years many new classes of membranes with functional nanoscale components or those inspired by biological systems have been developed. One kind of advanced membrane, comprises of aligned carbon nanotubes (CNTs), exhibits super-high flux for water transport. The measured fluxes were found to be higher by up to three orders of magnitude than those predicted by conventional Hagen–Poiseuille flow theory. These aligned CNT membranes were shown to have average tube diameters of 1.6 nm and ultrasmooth, hydrophobic walls which do not follow the basic "no-slip" condition, thereby letting fluids slide along the wall in an essentially frictionless configuration permitting high flux through the CNTs. While water flows nearly effortlessly through the aligned CNTs, the energy cost is expended in the water molecules entering the CNTs, and in the modification of the mouths of the CNTs to reject small salt ions, which increases the energy needed. The final energy usage for these membranes is still to be determined for concentrated salt solutions with a high rejection potential.

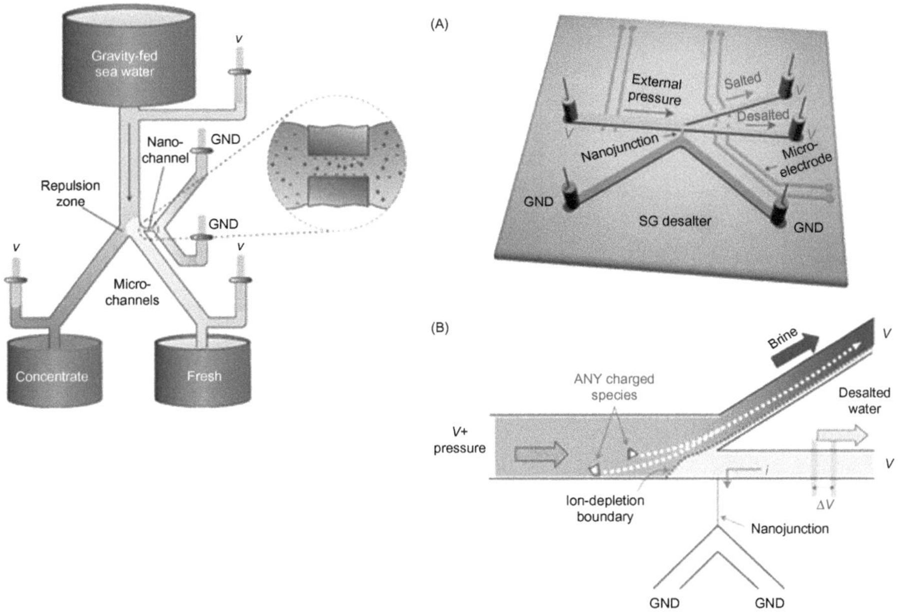

FIGURE 6.13

(A) Small scale setup for ion concentration polarization desalination of seawater. (B) Schematic of microchannel–nanochannel prototype tested for water desalination using concentration polarization. © 2010 Macmillan Publishers Ltd. Reproduced with permission from [41].

6.6.3 Microbial desalination cells

Microbial FCs convert biowaste to electricity through microbial activity by generating electrons available for harvesting in an external load circuit. In one approach, a three-cell system was converted into a water desalination system (Figure 6.14). It was shown that the three-cell design with an integrated anion exchange membrane (AEM) and cation exchange membrane (CEM) could generate a small amount of potential (typically less than 1 V) and desalinate water. The bacteria grow on the anode side and discharge protons into the water; however, the protons cannot pass through the AEM, so negatively charged ions from the salinated water flow through the AEM to balance the positive charges produced. A similar process takes place on the cathode side, except protons are consumed, requiring positively charged ions from the salinated water to cross the CEM to correct the charge imbalance. The amount of salt removed from each concentration was at least 88% ± 2% for the 5 g/l case and up to 94% ± 3% for the 20 g/l case. Comparison of the electrons harvested and NaCl removed revealed a charge transfer efficiency of almost 100% due to negligible effects of electrolysis (due to the low current generated) and insignificant back diffusion of ions from the electrode chambers to the desalination chamber.

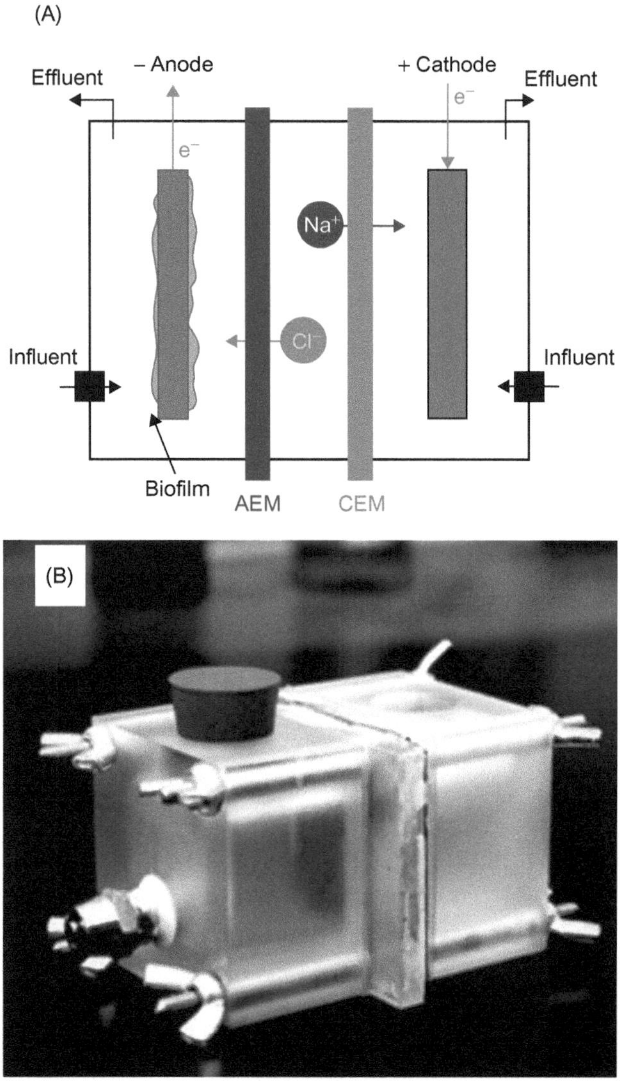

FIGURE 6.14

(A) A basic schematic representing a three-chamber cell design for coupling a microbial fuel cell with a water desalination system. (B) A photograph of the Microbial desalination cell prototype used in a three-chamber configuration. © 2009 American Chemical Society. Reproduced with permission from [42].

6.7 Asymmetry-driven ion pumps

As with nanotechnology, biology has also provided clues to developing advanced water desalination systems by following the ideas of ion transport in biological ion channels and ion pumps. While existential proof for the working of these systems have been around in living organisms for a long time, implementing technologies for practical systems have continued to be a challenge due to several gaps in engineering these systems due to a lack of mechanistic understanding of how to manipulate ion transport in artificial systems. In particular, mimicking ion-channel gating behavior for transport and control of ions has generated significant interest leading to the development of fluidic transistors and conical nanopores for ion pumps for the potassium ion. One advantage of bioinspired systems is that the fundamental processes occur at the nanoscale and engineering these systems could allow for rapid development of bionanotechnology as a core area for water desalination research. Further, with most biological systems operating near thermodynamic energy minima, such systems hold the promise of delivering the most energy-efficient water desalination systems ever conceived. However, still significant questions remain open and would likely need extensive investigations. The interested reader is directed to Chapter 19 in an edited book by David Reisner, Bionanotechnology II: Global Prospects [43] for further information.

References

[1] Prakash S, Akbelov R, Agonafer D, Armijo AD, Shannon MA. Influence of boundary conditions on sub-millimeter scale combustion. Energy Fuels 2009;23(7):3549–57.

[2] Prakash S, Armijo AD, Masel RI, Shannon MA. Flame Dynamics and Structure within Sub-Millimeter Combustors. AIChE Journal 2007;53(6):1568–77.

[3] Yang WM, Chou SK, Shu C, Li ZW, Xue H. A prototype microthermophotovoltaic power generator. Applied Phys Letters 2004;84(19). Available from: http://dx.doi.org/10.1063/1.1751614.

[4] Mehra A, Zhang X, Ayon A, Waitz I, Schmidt M, Spadaccini C. A six-wafer combustion system for a silicon micro gas turbine engine. Journal of Microelectromechanical Systems 2000;9(4):517–27.

[5] Nguyen NT, Chan SH. Micromachined polymer electrolyte membrane and direct methanol fuel cells - a review. J Micromech Microeng 2006;16:R1–12.

[6] Marton CH, Haldeman GS, Jensen KF. Portable thermoelectric power generator based on a microfabricated silicon combustor with low resistance to flow. Ind Eng Chem Res 2011;50(14):8468–75.

[7] Morse JD. Micro-fuel cell power sources. Int J Energy Res 2007;31:576–602.

[8] Sundarrajan S, Allakhverdiev SI, Ramakrishna S. Progress and perspectives in micro direct methanol fuel cell. Int J Hydrogen Energy 2012;37:8765–86.

[9] Barton SC, Gallaway J, Atanassov P. Enzymatic biofuel cells for implantable and microscale devices. Chem Rev 2004;104:4867–86.

[10] Wang H-Y, Bernarda A, Huang C-Y, Lee D-J, Chang J-S. Micro-sized microbial fuel cell: a mini-review. Bioresour Technol 2011;102:235–43.

[11] Bieberle-Hütter A, Beckel D, Infortuna A, Muecke UP, Rupp JLM, Gauckler LJ, et al. A micro-solid oxide fuel cell system as battery replacement. J Power Sour 2008;177 (1):123–30.

[12] Evans A, Bieberle-Hütter A, Rupp JLM, Gauckler LJ. Review on microfabricated micro-solid oxide fuel cell membranes. J Power Sour 2009;194:119–29.

[13] Jayashree RS, Yoon SK, Brushetta FR, Lopez-Montesinosa PO, Natarajanc D, Markoski LJ, et al. On the performance of membraneless laminar flow-based fuel cells. J Power Sour 2010;195:3569–78.

[14] Choban ER, Markoski LJ, Wieckowski A, Kenis PJA. Microfluidic fuel cell based on laminar flow. J Power Sour 2004;128:54–60.

[15] Choban ER, Spendelow JS, Gancs L, Wieckowski A, Kenis PJA. Membraneless laminar flow-based micro fuel cells operating in alkaline, acidic, and acidic/alkaline media. Electrochim Acta 2005;50:5390–8.

[16] Choban ER, Waszczuk P, Kenis PJA. Characterization of limiting factors in laminar flow-based membraneless microfuel cells. Electrochem Solid-State Lett 2005;8: A348–52.

[17] Jayashree RS, Gancs L, Choban ER, Primak A, Natarajan D, Markoski LJ, et al. Air-breathing laminar flow-based microfluidic fuel cell. J Am Chem Soc 2005;127:16758–9.

[18] Kjeang E, Michel R, Harrington DA, Djilali N, Sinton DA. Microfluidic fuel cell with flow-through porous electrodes. J Am Chem Soc 2008;130:4000–6.

[19] Brushett FR, Jayashree RS, Zhou W-P, Kenis PJA. Investigation of fuel and media flexible laminar flow-based fuel cells. Electrochim Acta 2009;54:7099–105.

[20] Gold S, Chu K-L, Lu C, Shannon MA, Masel RI. Acid loaded porous silicon as a proton exchange membrane for micro-fuel cells. J Power Sour 2004;135:198–203.

[21] Chu KL, Gold S, Subramanian V, Lu C, Shannon MA, Masel RI. A nanoporous silicon membrane electrode assembly for on-chip micro fuel cell applications. J Microelectromech Syst 2006;15(3):671–7.

[22] Chu K-L, Shannon MA, Masel RI. An improved miniature direct formic acid fuel cell based on nanoporous silicon for portable power generation. J Electrochem Soc 2006;153:A1562–7.

[23] Moghaddam S, Pengwang E, Jiang Y-B, Garcia AR, Burnett DJ, Brinker CJ, et al. An inorganic–organic proton exchange membrane for fuel cells with a controlled nanoscale pore structure. Nat Nanotechnol 2010;5:230–6.

[24] Moghaddam S, Pengwang E, Masel RI, Shannon MA. An enhanced microfluidic control system for improving power density of a hydride-based micro fuel cell. J Power Sour 2010;195:1866–71.

[25] Yeom J, Jayashree RS, Rastogi C, Shannon MA, Kenis PJA. Passive direct formic acid microfabricated fuel cells. J Power Sour 2006;160:1058–64.

[26] Zhu L, Kim D, Kim H, Masel RI, Shannon MA. Hydrogen generation from hydrides in millimeter scale reactors for micro proton exchange membrane fuel cell applications. J Power Sour 2008;185:1334–9.

[27] Zhu L, Lin KY, Morgan RD, Swaminathan VV, Kim HS, Gurau B, et al. Integrated micro-power source based on a micro-silicon fuel cell and a micro electromechanical system hydrogen generator. J Power Sour 2008;185:1305–10.

[28] Swaminathan VV, Zhu L, Gurau B, Masel RI, Shannon MA. Integrated micro fuel cell with on-demand hydrogen production and passive control MEM. Microfluid Nanofluid 2012;12:735–49.

[29] Moghaddam S, Pengwang E, Masel RI, Shannon MA. A self-regulating hydrogen generator for micro fuel cells. J Power Sour 2008;185:445–50.

[30] Zhu L, Meng DD, Kroodsma N, Yeom J, Shannon MA, editors. Transducers. Denver, CO: IEEE; 2009.

[31] Zhu L, Kroodsma N, Yeom J, Haan JI, Shannon MA, Meng DD. An on-demand microfluidic hydrogen generator with self-regulated gas generation and self-circulated reactant exchange with a rechargeable reservoir. Microfluid Nanofluid 2011;11:569–78.

[32] Zhu L, et al. An integrated microfluidic self-regulating and self-circulating hydrogen generator for fuel cells. Solid-State Sensors, Actuators and Microsystems Conference 2009;652–5.

[33] Datta, S, Conliks, AT. Curent conduction and fluid flow in nanonchannels with overlapped double layers for surface characterization and electrokinetic power generation. 48th AIAA Aerospace Sciences Meeting (2010) Orlando, Florida (4–7Jan), Article 2010-0906.

[34] van der Heyden FHJ, Bonthius DJ, Stein D, Meyer C, Dekker C. Electrokinetic energy conversion: efficiency in nanofluidic channels. Nano Lett 2006;6(10):2232–7.

[35] Davidson C, Xuan X. Electrokinetic energy conversion in slip nanochannels. J Power Sources 2008;179:297–300.

[36] Xuan X, Li D. Thermodynamic analysis of electrokinetic energy conversion. J Power Sources 2006;156:677–84.

[37] Pennathur S, Eijkel JCT, van den Berg A. Energy conversion in microsystems: is there a role for micro/nanofluidics? Lab Chip 2007;7:1234–7.

[38] Wu Y, Misra S, Karacor MB, Prakash S, Shannon MA. Dynamic response of AFM cantilevers to dissimilar functionallized silica surfaces in aqueous electrolyte solutions. Langmuir 2010;26(22):16963–72.

[39] Shannon MA, Bohn PW, Elimelech M, Georgiadis JG, Marinas BJ, Mayes AM. Science and technology for water purification in the coming decades. Nature 2008;452:301–10.

[40] Lefevre F, Chalifour A, Yu L, Chodavarapu V, Juneau P, Izquierdo R. Algal fluorescence sensor integrated into a microfluidic chip for water pollutant detection. Lab Chip 2012;12(4):787–93.

[41] Kim SJ, Ko SH, Kang KH, Han J. Direct seawater desalination by ion concentration polarization. Nat Nanotechnol 2010;5:297–301.

[42] Cao X, Huang X, Liang P, Xiao K, Zhou Y, BEA Logan. New method for water desalination using microbial desalination cells. Environ Sci Technol 2009;43 (18):7148–52.

[43] Reisner D. Bionanotechnology II: global prospects. Boca Raton, FL: CRC Press; 2011.

Select Bibliography

Asatekin A. Anti-fouling ultrafiltration membranes containing polyacrylonitrile-*graft*-poly (ethlene oxide) comb copolymer additives. J Membr Sci 2007;298:136–46.

Asatekin A, Menniti A, Kang S, Elimelech M, Morgenroth E, Mayes AM. Antifouling nanofiltration membranes for membrane bioreactors from self-assembling graft copolymers. J Membr Sci 2006;285:81–9.

Bard AJ, Faulkner LR. Electrochemical methods: fundamentals and applications. New York, NY: John Wiley & Sons; 1980.

Bhattacharya A, Ghosh P. Nanofiltration and reverse osmosis membranes: theory and application in separation of electrolytes. Rev Chem Eng 2004;20:111–73.

Bourouni K, Chaibi MT, Tadrist L. Water desalination by humidification and dehumidification of air: state of art. Desalination 2001;137:167–76.

Buros OK. The ABCs of desalting. International Desalination Association; 2000.

Cannon Jr DM, Kuo T-C, Bohn PW, Sweedler JV. Nanocapillary array interconnects for gated analyte injections and electrophoretic separations in multilayer microfluidic architectures. Anal Chem 2003;75:2224–30.

Cannon Jr DM, Kuo T-C, Feng W, Shannon MA, Sweedler JV, Bohn PW, editors. Characterization of molecular transport within gateable nanofluidic interconnects for three-dimensional microfluidic systems. Monterey, CA: Micro Total Analysis Systems; 2001.

Cath TY, Childress AE, Elimelech M. Forward osmosis: principles, applications, and recent developments. J Memb Sci 2006;281:70–87.

Childress AE, Elimelech M. Relating nanofiltration membrane performance to membrane charge (Electrokinetic) characteristics. Environ Sci Technol 2000;34:3710–6.

Clavin P, Pelce P, He L. One-dimensional vibratory instability of planar flames propagating in tubes. J Fluid Mech 1990;216:299–322.

Coster HGL, Kim KJ, Dahlan K, Smith JR, Fell CJD. Characterisation of ultrafiltration membranes by impedance spectroscopy. I. Determination of the separate electrical parameters and porosity of the skin and sublayers. J Memb Sci 1992;66:19–26.

Darwish MA. Desalting: fuel energy cost in Kuwait in view of $75/barrel oil price. Desalination 2007;208:306–20.

Darwish MA, El-Dessouky H. The heat recovery thermal vapour-compression desalting system: a comparison with other thermal desalination processes. Appl Therm Eng 1996;16:523–7.

Dzubiella J, Hansen JP. Electric field controlled water and ion permeation of a hydrophobic nanopores. J Chem Phys 2005;122(3):243706.

Elimelech M, Zhu X, Childress AE. Role of surface morphology in colloidal fouling of polymeric membranes. J Memb Sci 1997;129:101–9.

Enezi GL, Hisham E, Fawzy N. Low temperature humidication dehumidication desalination process. Energy Convers Manage 2006;47:470–84.

Epstein AH, Senturia SD, Anathasuresh G, Ayon A, Breuer K, Chen KS, et al., editors. Power MEMS and microengines.Chicago, IL: Transducers; 1997.

Ettouney H. Design and analysis of humidification dehumidification desalination processes. Desalination 2005;183:341–52.

Fermandez-Pello C. Microscale power generation using combustion: issues and approaches. Proc Combust Inst 2002;29:883–99.

Forgacs C. Polarization at ion-exchange membranes in electrodialysis. Desalination 1972;10:181–214.

Fritzmann C, Lowenberg J, Wintgens T, Melin T. State-of-the-art reverse osmosis desalination. Desalination 2007;216.

Ganley JC, Riechmann KL, Seebauer EG, Masel RI. Porous anodic alumina optimized as a catalyst support for microreactors. J Catal 2004;227(1):26–32 PubMed PMID: ISI:000223753600003.

Ganley JC, Seebauer EG, Masel RI. Porous anodic alumina microreactors for production of hydrogen from ammonia. AIChE J 2004;50(4):829–34 PubMed PMID: ISI:000221023200011.

Gardeniers H, van den Berg A. Micro- and nanofluidic devices for environmental and biomedical applications. Int J Environ Anal Chem 2004;84:809–19.

Glueckauf E. Seawater desalination - in perspective. Nature 1966;211:1227–30.

Gong X, Li J, Lu H, Wan R, Li J, Hu J, et al. A charge-driven molecular water pump. Nat Nanotechnol 2007;2:709–12.

Gray GT, McCutcheon RL, Elimelech M. Internal concentration polarization in forward osmosis: role of membrane orientation. Desalination 2006;197:1–8.

Ha S, Dunbar Z, Masel RI. Characterization of a high performing passive direct formic acid fuel cell. J Power Sour 2006;158(1):129–36.

Hasson D, Semiat R. Scale control in saline and wastewater desalination. Isr J Chem 2006;46:97–104.

Hille B. Ionic channels of excitable membranes. Sunderland, MA: Sinauer Associates; 1992.

Ho C, Qiao R, Heng JB, Chatterjee A, Timp RJ, Aluru NR, et al. Electrolytic transport through a synthetic nanometer-diameter pore. Proc Nat Acad Sci 2005;102 (30):10445–50.

Hoek EMV, Bhattacharjee S, Elimelech M. Effect of membrane surface roughness of colloid-membrane DLVO interactions. Langmuir 2003;19:4836–47.

Holloway RW, Childress AE, Dennett KE, Cath TY. Forward osmosis for concentration of anaerobi digester centrate. Water Res 2007;41:4005–14.

Holt JK, Park HG, Wang Y, Stadermann M, Artyukhin AB, Grigoropoulos CP, et al. Fast mass transport through sub-2nm carbon nanotubes. Science 2006;213:1034–7.

Hong JW, Quake SR. Integrated nanoliter systems. Nat Biotechnol 2003;21:1179–83.

Israelachvili JN. Intermolecular and surface forces. Orlando, FL: Academic Press; 1985.

Jackson RD, van Bavel CHM. Solar distillation of water from soil and plant materials: a simple desert survival technique. Science 1965;149:1377–8.

Janata J, Josowicz M, Vanysek P, DeVaney DM. Chemical sensors. Anal Chem 1998;70 (12):179R–208R.

Jensen CD. The dependence of flame/wall interactions on the composition of the walls for determining the material composition of a micro-combustor [Ms]. Urbana, IL: University of Illinois; 2000.

Jirage KB, Hulteen JC, Martin CR. Nanotubule-based molecular-filtration membranes. Science 1997;278:655–8.

Karlsson A, Karlsson M, Karlsson R, Sott K, Lundqvist A, Tokarz M, et al. Nanofluidic networks based on surfactant membrane technology. Anal Chem 2003;75:2529–37.

Karniadakis G, Beskok A, Aluru N. In: Antman SS, Marsden JE, Sirovich L, editors. Microflows and nanoflows: fundamentals and simulation. New York, NY: Springer; 2005.

Karnik R, Fan R, Yue M, Li D, Yang P, Majumdar A. Electrostatic control of ions and molecules in nanofluidic transistors. Nano Lett 2005;5(5):943–8.

Kohli P, Harrell CC, Cao Z, Gasparac R, Tan W, Martin CR. DNA-functionalized nanotube membranes with single-base mismatch selectivity. Science 2004;304:984–6.

Kovarik ML, Jacobson SC. Attoliter-scale dispensing in nanofluidic channels. Anal Chem 2007.

Ku C-K, Ho C-H, Lee Y-D. Synthesis of polyimides containing pyridine or triazole moeity to improve their adhesion to sputter-deposited copper. J Adhes Sci Technol 2005;19:909–25.

Kuo T-C, Cannon Jr DM, Chen Y, Tulock JJ, Shannon MA, Sweedler JV, et al. Gateable nanofluidic interconnects for multilayered separation systems. Anal Chem 2003;75:1861–7.

Kuo T-C, Cannon Jr DM, Shannon MA, Bohn PW, Sweedler JV. Hybrid three-dimensional nanofluidic/microfluidic devices using molecular gates. Sens Actuators A. 2003;102:223–33.

Kuo T-C, Kim H-K, Cannon Jr DM, Shannon MA, Sweedler JV, Bohn PW. Nanocapillary arrays effect mixing and reaction in multilayer fluidic structures. Angew Chem Int Ed 2004;43:1862–5.

Kyritsis DC, Roychoudhury SC, McEnally ES, Pfefferle LD, Gomez A. Mesoscale combustion: a first step towards liquid fueled batteries. Exp Therm Fluid Sci 2004;28:763–70.

Lakshminarayanaiah N. Transport phenomena in membranes. New York, NY: Academic Press; 1969.

Lange KE. Get the salt out. Nat Geog 2010;32–5.

Lee SB, Martin CR. Electromodulated molecular transport in gold-nanotube membranes. J Am Chem Soc 2002;124:11850–1.

Lennon E, Senez V, Yamamoto T, Fujita H, Fujii T, editors. Transducers. Seoul, South Korea: Transducer Research Foundation; 2005.

Li D. Electrokinetics in microfluidics. New York, NY: Academic Press; 2004.

Liu C, Naismith N, Economy J. Advanced mesoporous organosilica material containingmicroporous cyclodextrins for the removal of humic acid from water. J Chromatogr A 2004;1036:113–8.

Liu CY, Naismith N, Fu L, Economy J. Novel nanoporous hybrid organic-inorganic silica containing iminodiethanol chelating groups inside the channel pores. Chem Commun 2003;:1920–1.

Liu JW, Brown AK, Meng X, Cropek DM, Istok JD, Watson DB, et al. A catalytic beacon sensor for uranium with parts-per-trillion and million-fold selectivity. Proc Nat Acad Sci USA 2007.

Liu JW, Lu Y. A colorimetric lead biosensor using DNAzyme-directed assembly of gold nanoparticles. J Am Chem Soc 2003;125(22):6642–3.

McCutcheon JR, Elimelech M. Modeling water flux on forward osmosis: implications for improved membrane design. AIChE J 2007;53:1736–44.

McGinnis RL, Elimelech M. Energy requirements of ammonia-carbon dioxide forward osmosis desalination. Desalination 2007;207:370–82.

Mijatovic D, Eijkel JCT, van den Berg A. Technologies for nanofluidic systems: top-down vs. bottom-up-a review. Lab Chip 2005;5:492–500.

Miller JE. Review of water resources and desalination technologies. Sandia National Laboratories, New Mexico, 2003.

Mukhopadhyay R. What does nanofluidics have to offer? Plenty, say experts. Anal Chem 2006;78:7379–82.

Newman JS. Electrochemical systems. Englewood Cliffs, NJ: Prentice-Hall; 1973.

Ni Z, Seebauer E, Masel RI. Effects of microreactor geometry on performance: Differences between posted reactors and channel reactors. Ind Eng Chem Res 2005;44 (12):4267–71.

Nishizawa M, Menon VP, Martin CR. Metal nanotubule membranes with electrochemically switchable ion-transport selectivity. Science 1995;268:700–2.

Osada Y, Nakagawa T, editors. Membrane science and technology. New York, NY: Marcel Dekker; 1992.

Osterle, JF. A unified treatment of the thermodynamics of steady state energy conversion. Applied Science Resaearch, 1964, V12, 425–34.

Pedley TJ. Calculation of unstirred layer thickness in membrane transport experiments: a survey. Q Rev Biophys 1983;16:115–50.

Pilat B. Practice of water desalination by electrodialysis. Desalination 2001;139:385–92.

Popovtzer R, Neufeld T, Biran D, Ron EZ, Rishpon J, Shacham-Diamand Y. Novel integrated electrochemical nano-biochip for toxicity detection water. Nano Lett 2005;5(6):1023–7.

Prakash S, Glumac NG, Shankar N, Shannon MAOH. Concentration profiles over alumina, quartz, and platinum surfaces using laser-induced fluorescence spectroscopy in low-pressure hydrogen/oxygen flames. Combust Sci Technol 2005;177(4):793–817.

Prakash S, Karacor MB, Banerjee S. Surface modification in microsystems and nanosystems. Surf Sci Rep 2009;64:233–54.

Prakash S, Yeom J, Jin N, Adesida I, Shannon MA. Characterization of ionic transport at the nanoscale. Proc Inst Mech Eng, Part N: J Nanosyst Nanoeng 2007; 220(2):45–52.

Prakash S, Yeom J, Shannon MA, editors. A microfabricated impedance sensor for ionic transport in nanopores. Solid-state sensors, actuators, and microsystems workshop. Hilton Head Island, SC: Transducer Research Foundation; June 4–8, 2006.

Prakash S, Yeom J, Shannon MA, editors. ASME international mechanical engineering congress and exposition. Chicago, IL: ASME; 2006.

Pu Q, Yun J, Temkin H, Liu S. Ion-enrichment and ion-depletion effect of nanochannel structures. Nano Lett 2004;4(6):1099–103.

Qiao R, Aluru NR. Charge inversion and flow reversal in a nanochannel electroosmotic flow. Phys Rev Lett 2004;92:19.

Qiao R, Aluru NR. Atomistic simulation of KCl transport in charged silicon nanochannels: interfacial effects. Colloids Surf A 2005;267:103–9.

Qiao R, Aluru NR. Surface charge induced assymetric electrokinetic transport in confined silicon nanochannels. Appl Phys Lett 2005;86:143105–7.

Qiao R, Georgiadis JG, Aluru NR. Differential ion transport induced electroosmosis and internal recirculation in heterogeneous osmosis membranes. Nano Lett 2006;6(5):995–9.

Raguin G, Falkenberg CV, Prakash S, FitzHenry HR, Mensing GA, Ciobanu L, editors. Materials research society spring meeting. San Francisco, CA: MRS; 2006.

Rautenbach R, Albrecht R. Membrane processes. Frankfurt: John Wiley & Sons; 1989.

Reid CE. Principles of reverse osmosis. In: Merten U, editor. Desalination by reverse osmosis. Cambridge, MA: MIT Press; 1966. p. 1–14.

Schaevitz SB, Franz AJ, Jensen KF, Schmidt M, editors. A combustion-based MEMS thermoelectric power generator. The 11th international conferences on solid-state sensors and actuators. Munich, Germany; 2001.

Service RF. Desalination freshens up. Science 2006;313:1088–90.

Shaegh SAM, Nguyen N-T, Chan SH. A review on membraneless laminar flow-based fuel cells. Int J Hydrogen Energy 2011;36:5675–94.

Shannon MA. Fresh for less. Nat Nanotechnol 2010;5:248–50.

Shannon MA, Bohn PW, Elimelech M, Georgiadis JG, Marinas BJ, Mayes AM. Science and technology for water purification in the coming decades. Nature 2008;452:301–10.

Shannon MA, Flachsbart BR, Iannacone JM, Wong K, Cannon Jr DM, Fa K, et al., editors. Nanofluidic interconnects within a multilayer microfluidic chip for attomolar biochemical analysis and molecular manipulation. Proceedings of the 3rd annual international IEEE EMBS special topic conference on microtechnologies in medicine and biology. Oahu, Hawaii; May 12–15, 2005.

Shannon MA, Kuo T-C, Kim H-K, Cannon Jr DM, Flachsbart BR, Sweedler JV, et al., editors. Injection and ultrafast mixing of attomole samples via micro-nanofluidic gates for on-chip biochemical analysis. 7th international conference on miniautrized chemical and biochemical analysis systems. Squaw Valley, CA; October 5–9, 2003.

Shannon MA, Moore GV, Ganley J, Miesse C, Rice C, Seebauer EG, et al., editors. High-temperature microcombustion-based ammonia microchemical hydrogen generator reactors for PEM fuel cells. Technical digest solid-state sensors and actuators workshop. Hilton Head Island, SC; June 3–7, 2002.

Siwy Z, Fulinski A. Fabrication of a synthetic nanopore ion-pump. Phys Rev Lett 2002;89(19):19803–14.

Spielger KS. Salt-water purification. New York, NY: John Wiley & Sons; 1962.

Steinle ED, Mitchell DT, Wirtz M, Lee SB, Young VY, Martin CR. Ion channel mimetic micropore and nanotube membrane sensors. Anal Chem 2002;74:2416–22.

Suh KY, Jon S. Control over wettability of polyethylene glycol surfaces using capillary lithography. Langmuir 2005;21:6836–41.

Swearingen CB, Wernette DP, Cropek DM, Lu Y, Sweedler JV, Bohn PW. Immobilization of a catalytic DNA molecular beacon on Au for Pb(II) detection. Anal Chem 2005;77:442–8.

Torcellini P, Long N, Judkoff R. Consumptive water use for U.S. power production. Golden, CO: National Renewable Energy Laboratory; 2003.

Torsten Vilkner DJ, Manz A. Micro total analysis systems. Recent developments. Anal Chem 2004;76(12):3373–86.

Trieb F. Concentrating solar power for water desalination. Stuttgart: Federal Ministry for the Environment, Nature Conservation and Nuclear Energy; 2007.

Tulock JJ, Shannon MA, Bohn PW, Sweedler JV. Microfluidic separation and gateable fraction collection for mass-limited samples. Anal Chem 2004;76(21):6419–25.

UNICEF, WHO. Water for life: making it happen. Geneva: United Nations; 2005.

USDOE, Energy demands on water resources. Sandia National Laboratories; 2006.

Uyama Y, Kato K, Ikada Y. Surface modification of polymers by grafting. Adv Polym Sci 1998;137:1–39.

Veerapaneni S, Long B, Freeman S, Bond R. Reducing energy consumption for seawater desalinations. J Am Water Works Assoc 2007;99:95–106.

Wang H, Branton D. Nanopores with a spark for single-molecule detection. Nat Biotechnol 2001;19:622–3.
Wernette DP, Swearingen CB, Cropek DM, Lu Y, Sweedler JV, Bohn PW. Incorporation of a DNAzyme into Au-coated nanocapillary array membranes with an internal standard for Pb(II) sensing. Analyst 2006;131:41–7.
Wu MS, Ronney PD, Colantino R, VanZandt D. Detailed numerical simulation of flame ball structure and dynamics. Combust Flame 1999;116(3):387–97.
Wu Y, Prakash S, Gupta C, Shannon MA. Solid/water interface charge density of surface functionalized materials studies by atomic force microscope. 233rd national meeting and exposition. Chicago, IL: American Chemical Society; March 25–29, 2007.
Yang W, Chou SK, Shu C, Xue H, Li Z. A prototype microthermophotovoltaic power generator. Appl Phys Lett 2004;84:3864–6.
Yang WM, Chou SK, Shu C, Xue H, Li ZW, Li DT, et al. Microscale combustion research for application to micro thermophotovoltaic systems. Energy Convers Manage 2003;44 (16):2625–34.
Yeom J, Mozsgai GZ, Flachsbart BR, Choban ER, Asthana A, Shannon MA, et al. Microfabrication and characterization of a silicon-based millimeter scale, PEM fuel cell operating with hydrogen, methanol, or formic acid. Sens Actuators, B 2005;107(2):882–91.
Yeston J, Coontz R, Smith J, Ash C. A thirsty world. Science 2006;313:1067.
Yip NY, Tiraferri A, Phillip WA, Schiffman JD, Elimelech M. High Performance Thin-Film Composite Forward Osmosis Membrane. Environ Sci Technol 2010;44:3812–8.
Zhongren Y, Economy J. Nanoparticle and nanoporous carbon adsorbents for removal of trace organic contaminants from water. J Nanopart Res 2005;7:477–87.
Zhou Y, Tol RSJ. Evaluating the costs of desalination and water transport. Water Resour Res 2005;41(W03003):1–10.
Zhu X, Elimelech M. Fouling of reverse osmosis membranes by aluminum oxide colloids. J Environ Eng 1995;121:884–92.

APPENDIX

Review of Mathematical Concepts

A

A.1 Calculus

Derivatives

$y = ax^n$

derivative of $y = ax^n$ with respect to x:
$dy/dx = a\ n\ x^{(n-1)}$

$y = ax + b$: Slope a and y-axis intercept b

b is a constant – think of it as bx^0
its derivative is: $0bx^{-1} = 0$

derivative of $y = ax^n$ with respect to x:
$dy/dx = a\ n\ x^{(n-1)}$

For a straight line, $n = 1$ so
$dy/dx = a\ 1\ x^{(0)}$, or because $x^0 = 1$,
$dy/dx = a$

derivative of $y = ax + b$ with respect to x:
$dy/dx = a$
$dy/dx = \Delta y/\Delta x$ for straight lines

Partial Derivatives

Partial derivative of *a function* g with respect to x at a y location y_0
Notation $\partial g/\partial x|_{y=y_0}$
Treat variable not being differentiated with respect to as constant

Review of Integral Calculus

Integrating a polynomial

$$\int ax^n \partial x = a\frac{x^{(n+1)}}{n+1} + C$$

Integration by parts:

$$\int f(x)g'(x)\mathrm{d}x = f(x)g(x) - \int f'(x)g(x)\mathrm{d}x$$

Bessel Functions

Bessel's equation of order n:

$$\frac{\mathrm{d}^2A}{\mathrm{d}r^2} + \frac{1}{c}\frac{\mathrm{d}A}{\mathrm{d}r} + \left(1 - \frac{n^2}{r^2}\right)A = 0$$

Solution:

$$A(r) = c_1 J_n(r) + c_2 Y_n(r)$$

Tabulated values exist for J_n and Y_n and these are called Bessel's functions. Use of Bessel's equations and solutions is often handy in cylindrical geometries.

Wronskian relationship:

$$J_n(\beta r)Y_n'(\beta r) - Y_n(\beta r)J_n'(\beta r) = \frac{2}{\pi\beta r}$$

A.2 Vectors

Gradient, Divergence, and Curl in Cartesian Coordinates

The Vector Differential Operator Del. Also, written as ∇ is defined in Cartesian coordinates by:

$$\nabla = \mathbf{i}\frac{\partial}{\partial x} + \mathbf{j}\frac{\partial}{\partial y} + \mathbf{k}\frac{\partial}{\partial z} = \mathbf{i}_i\frac{\partial}{\partial x_i}$$

where i, j, k are the unit vectors in $i=1=x$, $i=2=y$, $i=3=z$ directions, respectively.

Gradient. The gradient of a scalar function $\phi(x, y, z)$ written as, $\nabla\phi$ or *grad* ϕ, is defined by

$$\nabla\phi = \left(\mathbf{i}\frac{\partial}{\partial x} + \mathbf{j}\frac{\partial}{\partial y} + \mathbf{k}\frac{\partial}{\partial z}\right)\phi = \mathbf{i}\frac{\partial\phi}{\partial x} + \mathbf{j}\frac{\partial\phi}{\partial y} + \mathbf{k}\frac{\partial\phi}{\partial z} = i_k\frac{\partial\phi}{\partial x_k}$$

which is a vector field.

Divergence. Let $\mathbf{V} = V_x\mathbf{i} + V_y\mathbf{j} + V_z\mathbf{k} = \mathrm{V}_i\mathbf{i}_i$ represent a vector in Cartesian coordinate system and differentiable at each point (x, y, z). Then the divergence of $\mathbf{V}$, written $\nabla\cdot\mathbf{V}$ or *div*$\mathbf{V}$, is defined by:

$$\nabla\cdot\mathbf{V} = \left(\mathbf{i}\frac{\partial}{\partial x} + \mathbf{j}\frac{\partial}{\partial y} + \mathbf{k}\frac{\partial}{\partial z}\right)\cdot(V_x\mathbf{i} + V_y\mathbf{j} + V_z\mathbf{k}) = \frac{\partial V_x}{\partial x} + \frac{\partial V_y}{\partial y} + \frac{\partial V_z}{\partial z}$$

which is a scalar.

Curl or Rotation. The curl or rotation of a differentiable vector $\mathbf{V} = V_x\mathbf{i} + V_y\mathbf{j} + V_z\mathbf{k} = V_i i_i$, written $\nabla \times \mathbf{V}$, *curl* **V**, or *rot* **V** is defined by

$$\nabla \times \mathbf{V} = \left(\mathbf{i}\frac{\partial}{\partial x} + \mathbf{j}\frac{\partial}{\partial y} + \mathbf{k}\frac{\partial}{\partial z}\right) \times (V_x\mathbf{i} + V_y\mathbf{j} + V_z\mathbf{k})$$

$$= \begin{vmatrix} \mathbf{i} & \mathbf{j} & \mathbf{k} \\ \frac{\partial}{\partial x} & \frac{\partial}{\partial y} & \frac{\partial}{\partial z} \\ V_x & V_y & V_z \end{vmatrix} = \begin{vmatrix} \frac{\partial}{\partial y} & \frac{\partial}{\partial z} \\ V_y & V_z \end{vmatrix}\mathbf{i} - \begin{vmatrix} \frac{\partial}{\partial x} & \frac{\partial}{\partial z} \\ V_x & V_z \end{vmatrix}\mathbf{j} + \begin{vmatrix} \frac{\partial}{\partial x} & \frac{\partial}{\partial y} \\ V_x & V_z \end{vmatrix}\mathbf{k}$$

Note, using coordinate transformations similar results can be obtained for cylindrical, spherical, or any other orthogonal coordinate system.

A.3 Matrices

Matrix Definition. A matrix is an array of numbers or mathematical terms. The numbers or the mathematical terms that make up the matrix are called the *elements of a matrix.* The *size* of a matrix is defined by its number of rows and columns. A matrix may consist of *m* rows and *n* columns. For example:

$$\mathbf{A} = \begin{bmatrix} 5 & 8 & 2 \\ -1 & 3 & 6 \end{bmatrix}$$

$$\mathbf{K} = \begin{bmatrix} \int_0^L x\mathrm{d}x & \int_0^L x^3\mathrm{d}x \\ \int_0^L (1+x)\mathrm{d}x & \int_0^L (1+x)^2\mathrm{d}x \end{bmatrix}$$

where **A** is a 2 by 3 matrix whose elements are numbers, **K** is a 2 by 2 matrix whose elements are integrals or mathematical terms.
Symmetric Matrix. A *symmetric matrix* is a square matrix whose elements satisfy

$$A_{i,j} = A_{j,i} \quad \text{or} \quad \text{equivalently} \quad \mathbf{A} = \mathbf{A}^T$$

where the superscript T denotes the transpose of a matrix.

That is, elements located symmetrically with respect to the principal diagonal are equal. For example:

$$\mathbf{K} = \begin{bmatrix} 1 & -1 & 0 \\ -1 & 2 & -1 \\ 0 & -1 & 1 \end{bmatrix}$$

Square Matrix. A matrix whose number of rows equals the number of columns is called a *square matrix.*

Diagonal Matrix. A diagonal matrix is a square matrix with nonzero elements only along the principal diagonal. For example:

$$\mathbf{A} = \begin{bmatrix} A_{1,1} & 0 & 0 \\ 0 & A_{2,2} & 0 \\ 0 & 0 & A_{3,3} \end{bmatrix}$$

Identity Matrix. The *identity* (or *unit*) matrix is a diagonal matrix with ones along the principal diagonal. For example:

$$\mathbf{I} = \begin{bmatrix} 1 & 0 & 0 \\ 0 & 1 & 0 \\ 0 & 0 & 1 \end{bmatrix}$$

Note that if **I** is of dimension n × n and **x** is an $(n \times 1)$ vector, then $\mathbf{Ix} = \mathbf{x}$

Matrix Addition and Subtraction

If A and B are m × n matrices, then the matrix addition or subtraction gives

$$\mathbf{C} = \mathbf{A} \pm \mathbf{B}$$

where **C** is also a $m \times n$ matrix (compatibility condition). This can be accomplished by

$$C_{i,j} = A_{i,j} \pm B_{i,j} \quad \text{for} \quad i = 1, 2, \ldots, m \quad \text{and} \quad j = 1, 2, \ldots, n.$$

Multiplication of Matrices. If **A** is a $l \times m$ matrix and **B** is a $m \times n$ matrix, then the matrix multiplication gives

$$\mathbf{C} = \mathbf{AB}$$

where **C** is a $l \times n$ matrix. This can be accomplished by

$$C_{i,j} = \sum_{k=1}^{m} A_{i,k} B_{k,j} \quad \text{for} \quad i = 1, 2, \ldots, l \quad \text{and} \quad j = 1, 2, \ldots, n.$$

Number of columns in A should be equal to the number of rows in B.

Transpose of a Matrix. The transpose of a matrix **A** is obtained by interchanging its rows and columns and is denoted by

$$\mathbf{B} = \mathbf{A}^T \quad \text{or} \quad B_{i,j} = A_{j,i}$$

Determinants. A *determinant* is a scalar computed from the entries in a square matrix. Determinants have various applications in engineering, including computing inverses and solving systems of simultaneous equations. For a 2 × 2 matrix:

$$\mathbf{A} = \begin{bmatrix} a_{1,1} & a_{1,2} \\ a_{2,1} & a_{2,2} \end{bmatrix}$$

the determinant is

$$|\mathbf{A}| = a_{1,1}a_{2,2} - a_{2,1}a_{1,2}.$$

Co-factor Matrix

If A is a square matrix, then the minor of its entry a_{ij}, also known as the i,j, or (i,j), or $(i,j)^{\text{th}}$ minor of A, is denoted by M_{ij} and is defined to be the determinant of the submatrix obtained by removing from A its i-th row and j-th column.

$$C_{ij} = (-1)^{i+j} M_{ij}$$

Inverse of a Matrix. If $\mathbf{A}$ is a $m \times m$ square matrix, then the inverse of $\mathbf{A}$ denoted as $\mathbf{A}^{-1}$ which satisfies the relationship

$$\mathbf{I} = \mathbf{A}\mathbf{A}^{-1} = \mathbf{A}^{-1}\mathbf{A}$$

where $\mathbf{I}$ is an identity matrix (square).

$$\mathbf{A}^{-1} = \frac{1}{|\mathbf{A}|}(\mathbf{C}^{\mathrm{T}})_{ij} = \frac{1}{|\mathbf{A}|}(\mathbf{C}_{ji}) = \frac{1}{|\mathbf{A}|}\begin{pmatrix} C_{11} & C_{21} & \cdots & C_{n1} \\ C_{12} & C_{22} & \cdots & C_{n2} \\ \vdots & \vdots & \ddots & \vdots \\ C_{1n} & C_{2n} & \cdots & C_{nn} \end{pmatrix}$$

C: co-factor matrix of A

2 by 2

$$\mathbf{A} = \begin{bmatrix} a_{1,1} & a_{1,2} \\ a_{2,1} & a_{2,2} \end{bmatrix}$$

3 by 3

$$\mathbf{A}^{-1} = \frac{1}{|\mathbf{A}|}\begin{bmatrix} \begin{vmatrix} a_{22} & a_{23} \\ a_{32} & a_{33} \end{vmatrix} & \begin{vmatrix} a_{13} & a_{12} \\ a_{33} & a_{32} \end{vmatrix} & \begin{vmatrix} a_{12} & a_{13} \\ a_{22} & a_{23} \end{vmatrix} \\ \begin{vmatrix} a_{23} & a_{21} \\ a_{33} & a_{31} \end{vmatrix} & \begin{vmatrix} a_{11} & a_{13} \\ a_{31} & a_{33} \end{vmatrix} & \begin{vmatrix} a_{13} & a_{11} \\ a_{23} & a_{21} \end{vmatrix} \\ \begin{vmatrix} a_{21} & a_{22} \\ a_{31} & a_{32} \end{vmatrix} & \begin{vmatrix} a_{12} & a_{11} \\ a_{32} & a_{31} \end{vmatrix} & \begin{vmatrix} a_{11} & a_{12} \\ a_{21} & a_{22} \end{vmatrix} \end{bmatrix}$$

the determinant is

$$|\mathbf{A}| = a_{1,1}a_{2,2} - a_{2,1}a_{1,2}.$$

Inverse of the above (2×2) matrix is:

$$\mathbf{A}^{-1} = \frac{1}{|\mathbf{A}|}\begin{bmatrix} a_{2,2} & -a_{1,2} \\ -a_{2,1} & a_{1,1} \end{bmatrix}$$

For inverse of a matrix to exist, its determinant must be non-zero. Matrices with zero determinants are called *singular matrices*. There are a number of methods for evaluating inverse of matrices.

Note that the inverse of a diagonal matrix **D** is obtained by simply calculating the inverse of its diagonal elements. For example:

$$\mathbf{D}^{-1} = \begin{bmatrix} 1/d_{1,1} & 0 & 0 \\ 0 & 1/d_{2,2} & 0 \\ 0 & 0 & 1/d_{3,3} \end{bmatrix}$$

where

$$\mathbf{D} = \begin{bmatrix} d_{1,1} & 0 & 0 \\ 0 & d_{2,2} & 0 \\ 0 & 0 & d_{3,3} \end{bmatrix}$$

Thus

$$\mathbf{D}^{-1}\mathbf{D} = \mathbf{D}\mathbf{D}^{-1} = \mathbf{I}$$

Multiplication of Vectors

Matrices with a single row or column represents a vector. This form of matrices is useful for numerical calculations, for example, implementation in MATLAB.

Multiplication of Vectors. Let **a** and **b** are two vectors of size m (or $m \times 1$). The *Dot* or *Inner Product* is defined as:

$$c = \mathbf{a}^T\mathbf{b} = \mathbf{b}^T\mathbf{a} = \sum_{i=1}^{n}(a_i b_i) = a_1 b_1 + a_2 b_2 + \ldots + a_n b_n$$

(a, b are column vectors, a^T and b^T are row vectors.)

The *matrix* or *tensor* product **a** and **b** are defined as:

$$\mathbf{D} = \mathbf{a}\mathbf{b}^T$$

Multiplication of a Matrix with a Vector. If **A** is a $m \times n$ matrix and **x** is a $n \times 1$ vector, then

$$\mathbf{A}\mathbf{x} = \mathbf{f}$$

where **f** is a $m \times 1$ vector. This can be accomplished by

$$f_i = \sum_{j=1}^{n} A_{i,j} x_j \quad \text{for} \quad i = 1, 2, \ldots, m$$

System of Linear Equations. The above matrix equations denote a system of n linear equations with n unknowns x_j if $m = n$. In that case, $A_{i,j}$ is called the coefficient matrix of the system of equations.
Eigenvalues and Eigenvectors. The set of linear equations of the form

$$\mathbf{AX} - \lambda\mathbf{X} = 0$$

is called an *eigenvalue problem*, which occurs in analysis of buckling problems, vibration of elastic structures, heat transfer, fluid mechanics, and electrical systems. In general, this class of problems has non-unique solutions. That is, we can establish relationships among the unknowns, and many values can satisfy these relationships. Customarily, the above equation is written as:

$$(\mathbf{A} - \lambda\mathbf{I})\mathbf{X} = 0$$

where **I** is the identity matrix having the same dimensions as **A** matrix. The unknown parameter λ is called the *eigenvalue*, and vector **X** is called the *eigenvector*. The nonzero solution for **X** will occur when $\mathbf{A} - \lambda\mathbf{I}$ is a singular matrix or

$$\det(\mathbf{A} - \lambda\mathbf{I}) = 0$$

$$\mathbf{AX} - \lambda\mathbf{X} = 0$$

The above equation is called the *characteristic equation*. We can solve it for the n roots of λ or eigenvalues $\lambda_1, \lambda_2,\ldots,\lambda_n$. For each eigenvalue λ_i, the associated eigenvector $\mathbf{X}_i$ is obtained from

$$(\mathbf{A} - \lambda_i\mathbf{I})\mathbf{X}_i = 0$$

Rank of a Tensor. A tensor of rank r is a quantity having n^r components in n-dimensional space. A tensor of rank 2 is called a *dyadic* (or *matrix*) represented by two indices, e.g.,

$$a_{ij} = \begin{bmatrix} a_{11} & a_{12} & a_{13} \\ a_{21} & a_{22} & a_{23} \\ a_{31} & a_{32} & a_{33} \end{bmatrix}$$

is a tensor of rank 2 in a 3-dimensional space.

A tensor of rank 1 is a *vector* represented by a single index, e.g.,

$$u_i = \begin{bmatrix} u_1 \\ u_2 \end{bmatrix} = \boldsymbol{u}$$

is a tensor of rank 1 in a 2-dimensional space.

A tensor of rank 0 is a scalar represented without an index, e.g.,

$$T$$

is a scalar in an n-dimensional space.

A.4 Complex variables

The following presents a brief review of complex number theory.

1. A complex number, z is defined as:

$$z = x + \mathrm{i}y$$

with $\mathrm{Re}(z) = x$, $\mathrm{Im}(z) = y$, and $\mathrm{i} = \sqrt{-1}$

2. The complex number z can be represented as a vector as shown in the figure below.
If,

$$z_1 = x_1 + \mathrm{i}y_1 \text{ and } z_2 = x_2 + \mathrm{i}y_2$$

then,

$$z_R = z_1 + z_2 = (x_1 + x_2) + \mathrm{i}(y_1 + y_2)$$

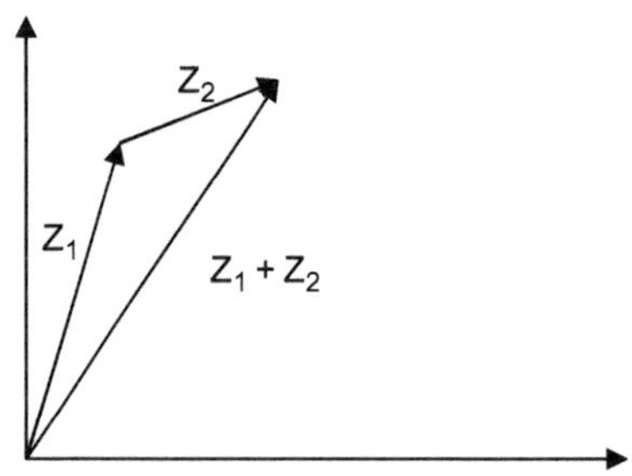

3. Multiplication is: $z_R = z_1 z_2 = (x_1 x_2 - y_1 y_2) + \mathrm{i}(x_1 y_2 + x_2 y_1)$
4. Euler's theorem: $z = \mathrm{e}^{\mathrm{i}\theta} = \cos\theta + \mathrm{i}\sin\theta$
5. Conversion to polar coordinates (where modulus is r and argument is θ):

$$x = r\cos\theta$$
$$y = r\sin\theta$$
$$z = r[\cos\theta + \mathrm{i}\sin\theta] = r\,\mathrm{e}^{\mathrm{i}\theta}$$

6. Complex conjugate: $\bar{z} = x - \mathrm{i}y = r\,\mathrm{e}^{-\mathrm{i}\theta}$
7. For $z = r\,\mathrm{e}^{\mathrm{i}\theta}$, $\ln(z) = \ln(r) + \mathrm{i}\theta = \ln(r) + \mathrm{i}(\theta + 2\pi n)$
The direct implication is that ln(z) is a multivalued function with n branches.

$$\cos(\mathrm{i}\theta) = \cosh(\theta), \quad \sin(\mathrm{i}\theta) = \mathrm{i}\sinh(\theta)$$

A.5 Coordinate transformations

Many fluid problems require working in cylindrical, spherical, or other coordinate systems. We will refer to the general system as orthogonal curvilinear coordinate systems. There are a few simple rules to follow that will allow transformation of coordinates. For the orthogonal curvilinear coordinate systems we will always have coordinate lines intersect each other at right angles.

Consider, u_1, u_2, and u_3 as the three space coordinates and $\hat{u}_1$,$\hat{u}_2$, and $\hat{u}_3$ be the unit vectors in the u_1, u_2, and u_3 directions, respectively.

$$(\mathrm{d}s)^2 = (\mathrm{d}x)^2 + (\mathrm{d}y)^2 + (\mathrm{d}z)^2 \tag{A.1}$$

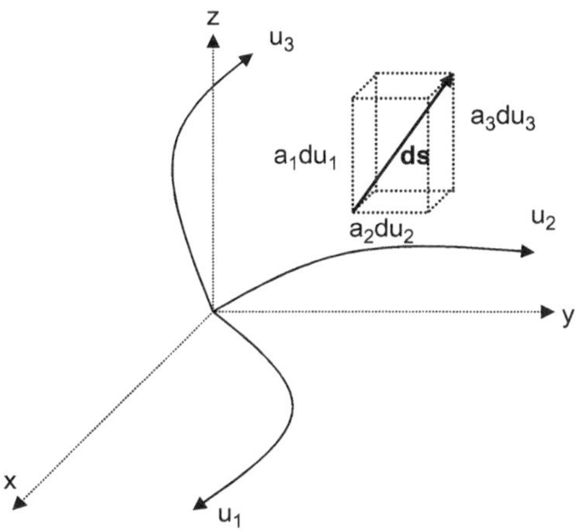

If the functional relationship between the orthogonal curvilinear coordinates (u_1, u_2, u_3) and the rectangular coordinates (x, y, z) is:

$x = \mathrm{X}(u_1, u_2, u_3)$, $y = \mathrm{Y}(u_1, u_2, u_3)$, and $z = \mathrm{Z}(u_1, u_2, u_3)$ then the differential lengths are:

$$\mathrm{d}x = \sum_{i=1}^{3}\left(\frac{\partial X}{\partial u_i}\right)\mathrm{d}u_i, \quad \mathrm{d}y = \sum_{i=1}^{3}\left(\frac{\partial Y}{\partial u_i}\right)\mathrm{d}u_i, \quad \mathrm{d}z = \sum_{i=1}^{3}\left(\frac{\partial Z}{\partial u_i}\right)\mathrm{d}u_i$$

Or plugging back in (A.1)

$$(\mathrm{d}s)^2 = a_1^2(\mathrm{d}u_1)^2 + a_2^2(\mathrm{d}u_2)^2 + a_3^2(\mathrm{d}u_3)^2 \tag{A.2}$$

Where, $a_i^2 = (\partial x/\partial u_i)^2 + (\partial y/\partial u_i)^2 + (\partial z/\partial u_i)^2$ with $i = 1, 2, 3$

Therefore, here the a_is are the scale factors allowing coordinate transformations. Consider as an example the heat flux equation,

$$\vec{q} = -k\nabla T = -k\sum_{i=1}^{3} \hat{u}_i \frac{1}{a_i}\frac{\partial T}{\partial u_i} \tag{A.3}$$

A.6 Common vector and tensor notations

Indicial, tensor, or Einstein summation: While working with *vectors*, if an index appears twice a summation is carried out over all values of the repeated index. No summation sign is explicitly written but is implied.

i.e.,

$$x'_j = x_i C_{ij} = \sum_{i=1}^{3} x_i C_{ij} = x_1 C_{1j} + x_2 C_{2j} + x_3 C_{3j} \tag{A.4}$$

The free index is j and i is the repeated or dummy index. In fact, i can be replaced with any letter except j and the meaning of the equation or term does not change. Therefore,

$$x'_j = x_i C_{ij} = x_k C_{kj} = x_p C_{pj} = \cdots \tag{A.5}$$

Also, any letter can be used for the free index as long as it is consistent and the same letter is used on both sides of the equation. Therefore, if i is the free index and k the dummy index:

$$x'_i = x_k C_{ki} \tag{A.6}$$

While working with *tensors*, the same summation rules apply but twice. Remember, a second-order tensor (e.g., stress) requires nine components for complete description. We need two indices for the two directions (hence two free indices and subsequently the summation rules operate twice)—one denotes the direction of the surface on which the stress is applied and the second denotes the direction of the force on that surface. So, the j-component of the force on a surface whose outward normal points in the i-direction would be denoted by τ_{ij}.

The tensor with nine components would be:

$$\tau_{ij} = \begin{bmatrix} \tau_{11} & \tau_{12} & \tau_{13} \\ \tau_{21} & \tau_{22} & \tau_{23} \\ \tau_{31} & \tau_{32} & \tau_{33} \end{bmatrix}$$

$$\tau'_{mn} = C_{im}C_{jn}\tau_{ij} \tag{A.7}$$

Kronecker delta defined as:

$$\delta_{ij} = \begin{cases} 1 & \text{if } i = j \\ 0 & \text{if } i \neq j \end{cases}$$

Most common operation with the Kronecker delta is:

$$\delta_{ij}u_j = \delta_{i1}u_1 + \delta_{i2}u_2 + \delta_{i3}u_3 \tag{A.8}$$

And, if $i = 1$, then RHS is u_1, for $i = 2$ it is u_2, and for $i = 3$ it is u_3. Therefore,

$$\delta_{ij}u_j = u_i \tag{A.9}$$

Permutation symbol is defined as:

$$\varepsilon_{ijk} = \begin{cases} 1 & \text{if } ijk = 123, 231, \text{ or } 312 \text{ (in cyclic order)} \\ 0 & \text{if any two indices are equal} \\ -1 & \text{if } ijk = 321, 213, \text{ or } 132 \text{ (in anti-cyclic order)} \end{cases}$$

Gradient, Divergence, and Curl operations:

$$\text{Gradient:} \quad \nabla \equiv \mathbf{a}^1 \frac{\partial}{\partial x_1} + \mathbf{a}^2 \frac{\partial}{\partial x_2} + \mathbf{a}^3 \frac{\partial}{\partial x_3} = \mathbf{a}^i \frac{\partial}{\partial x_i} \rightarrow \nabla\phi = \mathbf{a}^i \frac{\partial \phi}{\partial x_i} \tag{A.10}$$

Note: **a**s do not denote components of the vector but are themselves vectors → unit or basis vectors.

Implication of the gradient operator is that $\nabla\phi$ is perpendicular to the ϕ = constant lines and gives the magnitude and direction of the maximum spatial rate of change of ϕ. Therefore, rate of change in any other direction would be $(\partial\phi/\partial n) = (\nabla\phi) \cdot n$

$$\text{Divergence:} \quad \nabla \cdot u \equiv \frac{\partial u_i}{\partial x_i} = \frac{\partial u_1}{\partial x_1} + \frac{\partial u_2}{\partial x_2} + \frac{\partial u_3}{\partial x_3} \tag{A.11}$$

What happens if we are working with tensors?

$$(\nabla \cdot \tau)_i = \frac{\partial \tau_{ij}}{\partial x_j}$$

Note: divergence operation on a tensor decreases the order of the tensor by one and the gradient operation increases the order by one.

$$\text{Curl:} \quad (\nabla \times \mathbf{u})_i = \varepsilon_{ijk} \frac{\partial u_k}{\partial x_j} \tag{A.12}$$

A little algebra would yield the three components of the cross product as:

$$\left(\frac{\partial u_3}{\partial x_2} - \frac{\partial u_2}{\partial x_3}\right), \left(\frac{\partial u_1}{\partial x_3} - \frac{\partial u_3}{\partial x_1}\right), \left(\frac{\partial u_2}{\partial x_1} - \frac{\partial u_1}{\partial x_2}\right)$$

A.7 Potential flows summary

The following table presents a short summary for potential flows that are useful in many fluid mechanics applications to elucidate several insightful pieces of information about the flow field.

	Potential	Stream Function
Definition	$\vec{u} = \vec{\nabla}\varphi$	$\vec{u} = \vec{\nabla} \times \vec{\psi}$
Continuity	$\nabla^2\phi$	Automatically satisfied
Irrotationality	Automatically satisfied	$\vec{\nabla} \times (\vec{\nabla} \times \vec{\psi}) = \vec{\nabla}(\nabla \cdot \psi) - \nabla^2\psi$
	2-D flows ($w = 0$, $(\partial/\partial z = 0)$)	
	$\nabla^2\phi$ for continuity	$\nabla^2\psi = 0$ for irrotationality
	Cauchy–Riemann equations	
Cartesian (x, y)	$u = (\partial\phi/\partial x)$	$u = (\partial\psi/\partial x)$
	$v = (\partial\phi/\partial y)$	$v = (\partial\psi/\partial y)$
Polar (r, θ)	$u = (\partial\phi/\partial r)$	$u = (1/r)(\partial\psi/\partial\theta)$
	$v = (1/r)(\partial\phi/\partial\theta)$	$v = (\partial\psi/\partial r)$
General rules and shortcuts: *For irrorational flows it is easier to use* ϕ. *For incompressible flows it is easier to use* ψ. *For incompressible and irrotational flows use* ϕ *and* ψ.		

Select bibliography

Currie IG. Fundamental mechanics of fluids. 4th ed. Boca Raton, FL: CRC Press; 2013.
Kirby BJ. Micro- and nanoscale fluid mechanics. New York: Cambridge University Press; 2010.
Ozisik MN. Heat conduction. 2nd ed. New York: John Wiley and Sons; 1993.
Conlisk AT. Essentials of micro and nanofluidics with applications to the biological and chemical sciences. New York: Cambridge University Press; 2013.
Panton RL. Incompressible flow. 3rd ed. New York: John Wiley and Sons; 2005.
Kundu PK, Cohen IM, Dowling DR. Fluid mechanics. 5th ed. Academic Press; 2011.

APPENDIX

B

Useful Tabulated Data

Table B.1 Dielectric Properties at 25°C for Various Materials (Liquid and Solid) Used in Microfluidic and Nanofluidic Systems

Material	ε
Hydrogen fluoride (at 0°C)	84
Water	78.5
Methanol	32.6
Ethanol	24.3
Acetone	20.7
Cyclohexane	2
Polycarbonate	3
Fused silica	3.8
Soda lime glass	7
Borosilicate glass	4.5
Alumina	8.5
Air	1

ε denotes the dielectric function of the material listed. The common terminology in literature also refers to ε as the dielectric constant. It should be noted that for most materials, ε is a function of temperature and also of applied frequency in the presence of an electric field.

Table B.2 Ionic Radii of Different Ions in Dehydrated State and as Hydrated Ions[a]

Ion	Dehydrated Radius (nm)	Hydrated Radius (nm)
H_3O^+	NA	0.28
Na^+	0.09	0.36
K^+	0.13	0.33
Mg^{2+}	0.06	0.43
Ca^{2+}	0.1	0.41
OH^-	0.18	0.3
F^-	0.14	0.35
Cl^-	0.18	0.33
Br^-	0.19	0.33

[a]*Data assembled from Intermolecular and surface forces (2nd Ed.)*

Table B.3 Values for Some Essential Constants

Quantity	Quantity Name	Value	SI Units
ε_0	Permittivity of free-space	8.85×10^{-12}	C/V m
F	Faraday's constant	96450	C/mol
E	Electronic charge	1.6×10^{-19}	C
R	Universal gas constant	8.314	J/mol K
μ_0	Permeability of free space	12.56×10^{-7}	H/m
N_A	Avogadro's number	6.023×10^{23}	particles/mol
k_B	Boltzmann constant	1.38×10^{-23}	J/K
H	Planck's constant	6.63×10^{-34}	Js
C	Speed of light in vacuum	3×10^8	m/s
ρ	Density of water (25°C)	997	kg/m^3
η	Dynamic viscosity of water (25°C)	1×10^{-3}	Pa s

Table B.4 Summary of Material Properties for Several Materials Commonly Used in Fabrication of Microfluidic and Nanofluidic Systems

	Si	SiO_2	Si_3N_4	Diamond	Quartz	Al_2O_3	Polyimide	PMMA	PDMS	SU-8
Density (g/cm^3)	2.4	2.2	3.1	3.5	2.65	3.62	1.42	1.3	0.97	1.2
Melting/glass transition temperature (°C)	1415	1700	1800	3652	1610	1800	380	90	220	$50^a > 200^b$
Thermal conductivity (W/m K)	157	1.4	19	990–2000	1.4	36	0.12	0.2	0.18	0.2
Specific heat (J/g K)	0.7	1	0.7	0.6	0.787	0.8	1.09	1.3	1.3	1.6
Young's modulus (GPa)	160	73	323	1035	107	275	2.5	3	0.0001–0.003	4.4
Yield/fracture strength (GPa)	7	8.4	14	>1.2	9	15.4	0.23	0.06	<0.001	0.07
Poisson ratio	0.22	0.17	0.25	0.1	0.16	0.31	0.34	–	–	0.22
Coefficient of thermal expansion (10^{-6}/°C)	2.6	0.55	2.8	1	0.55	6.57	20	70	305	51
Relative permittivity	11.7	3.9	4–8	5.7	3.75	9	3	2.8	2.6	3.2
Breakdown voltage (kV/mm)	30	500–1000	500–1000	1000	2500–4000	1160	150–300	17	25	110

[a] *Unexposed SU-8.*
[b] *Fully crosslinked SU-8.*

Select Bibliography

Incropera FP, DeWitt DP. Fundamentals of heat and mass transfer. 4th ed. New York: John Wiley and Sons; 1996.

Israelachvili J. Intermolecular and surface forces. 2nd ed. London: Academic Press; 1991.

Index

Note: Page numbers followed by "*t*", "*b*" and "*f*" denotes table, box and figure respectively.

R

S

Kolibri 360 GmbH
Ferdinand-Jühlke-Straße 7
99095 Erfurt, Deutschland
produktsicherheit@kolibri360.de